International Conference on Vibration Problems (ICOVP 2015)

AA002925

Journal of Physics: Conference Series Volume 662

Warangal, India
18 – 20 February 2015

Editors:

Malla Reddy Perati
L. P. Raj Kumar
Paritosh Biswas

ISBN: 978-1-5108-1815-6
ISSN: 1757-6588

Printed from e-media with permission by:

Curran Associates, Inc.
57 Morehouse Lane
Red Hook, NY 12571

Some format issues inherent in the e-media version may also appear in this print version.

Printed by Curran Associates, Inc. (2016)

For permission requests, please contact the Institute of Physics at the address below.

Institute of Physics
Dirac House, Temple Back
Bristol BS1 6BE UK

Phone: 44 1 17 929 7481
Fax: 44 1 17 920 0979

techtracking@iop.org

Additional copies of this publication are available from:

Curran Associates, Inc.
57 Morehouse Lane
Red Hook, NY 12571 USA
Phone: 845-758-0400
Fax: 845-758-2634
Email: curran@proceedings.com
Web: www.proceedings.com

Table of contents

Volume 662
International Conference on Vibration Problems (ICOVP-2015)
18–20 February 2015, Warangal, India

Accepted papers received: 6 November 2015
Published online: 22 December 2015

Editorial Material

Papers

Solid Mechanics

Fluid Mechanics

Mechanical Engineering

Applied Mathematics

Physics

Journal of Physics: Conference Series **662** (2015) 011001 doi:10.1088/1742-6596/662/1/011001

Preface

Vibrations produced by operating machine cause deleterious effect including excessive stresses in mechanical components and reduce the machine performance. Hence, it is important to minimize the vibrations to improve the machine performance. Machines need the materials wherein vibration characteristics such as frequency and amplitude are lower. The vibration characteristics depend on strength and other elastic constants. Therefore, study of the relation between vibration characteristics and the elastic constants of the material is very much important. In the domain of seismology, the knowledge of vibrations associated with an earthquake is needed for the mitigation plans. With the increased use of strong and light weight structures especially in defence and aero-space engineering applications, wherein, precision i s very i mportant, problems of vibrations arise. The knowledge of quality (mechanical properties) of bones comes f rom t he s tudy of vibrations i n i t. This knowledge may, f or example, help t o answer bone t issue r emodelling problems. Unfortunately, vibrations mostly deal with destructive areas such as manufacturing industry, seismology, and bonemechanics. These days, mathematics has become a very important tool for Non-Destructive Evaluation (NDE) in the destructive areas. A very common issue in the said domains is that the pertinent problems result in non-linear coupled differential equations which are not easily solvable.

Keeping the above facts in mind, the Department of Mathematics, Kakatiya University has organized the International Conference on Vibration Problems (ICOVP-2015) from February, 18-20, 2015 in collaboration with the Department of Mechanical Engineering, Kakatiya Uni-versity, and Von-Karman Society, West Bengal. This association has already succeeded in organizing the Wave Mechanics and Vibration Conference (WMVC) in the year 2010. In the Conference, new research results were presented by the experts from eight countries. There were more than 150 participants from India and abroad. The infrastructure facilities, administrative help, and financial support provided by Kakatiya University are gratefully acknowledged. The financial support received from Telangana State Council of Higher Education, University Grant Commission (UGC), National Board for Higher Mathematics

(NBHM), Department of Atomic Energy (DAE), Ministry of Earth Sciences, and State Bank of Hyderabad are thankfully acknowledged. The thanks are due to all the speakers and participants who have come to this place by performing long journies and facing many inconveniences. Thanks are also due to International Scientific Committee members, reviewers, local organizing committee members, the members of organizing sub committees, supporting staff, research scholars, and the students for their help.

This Conference Series is the outcome of the contributions from the conference and is broadly classified into five sections containing the papers from Solid Mechanics (SM), Fluid Mechan-ics (FM), Mechanical Engineering (ME), Applied Mathematics(AM), and Physics(PH). All efforts have been made to maintain the consiouness and standards that are the Hallmarks of Institute of Physics (IOP). We appreciate the efforts of Dr.B.Sandhya Rani, Dr.G.Rajitha, and Ms. Hari Priya, research scholors in the Department of Mathematics, Kakatiya Uni-versity for checking the typos and pointing out the numerous errors. We also appreciate Dr.D.Rajaiah and Mr.D.Ranadheer, research scholors, in the Department for their efforts in the preparation of some LaTeX files.

Peer review statement

All papers published in this volume of *Journal of Physics: Conference Series* have been peer reviewed through processes administered by the proceedings Editors. Reviews were conducted by expert referees to the professional and scientific standards expected of a proceedings journal published by IOP Publishing.

Propagation of shear waves in viscoelastic heterogeneous layer overlying an initially stressed half space

Mita Chatterjee, Sudarshan Dhua and Amares Chattopadhyay

Department of Applied Mathematics, Indian School of Mines, Dhanbad-826004, Jharkhand, India.

E-mail: `mita.cht@gmail.com`

Abstract. The present paper is concerned with the propagation of shear waves in an isotropic, viscoelastic, heterogeneous layer lying over a homogeneous half space under initial stress. For the layer the inhomogeneity associated to rigidity, internal friction and density is assumed to be linear function of depth. The dispersion equation of shear waves has been obtained in closed form. The dimensionless phase velocity and damping velocity have been plotted against dimensionless wave number for different values of inhomogeneity parameter and initial stress. The effects of inhomogeneity and initial stress have been shown in the dispersion curves.

1. Introduction

Many problems in seismology can be solved by representing the Earth as a layered medium [1] that is, formed by layers of certain thickness and mechanical properties. Motivation behind choosing viscoelastic media for the present study lies in the fact that the Earth is composed of silicate and iron-alloy materials. These materials respond nearly elastically under the application of small-magnitude transient forces but viscously under the application of long-duration forces. Materials like coal, tar, salt, sediments, etc., which are buried beneath the Earth surface can be modeled as viscoelastic materials. Cooper [2] , Shaw and Bugl [3], Borcherdt [4], Romeo [5] and Chattopadhyay [6] have studied the propagation of SH waves in viscoelastic media with and without heterogeneity [7].

As pointed out by Bullen [8] the density inside the Earth varies at different rates with different layers within the Earth. Authors [9] have taken different forms of variation, like harmonic, linear, quadratic, etc., for simulating the variation in density and other geological parameters inside the Earth. For the present study, the heterogeneity is present in rigidity, density and internal friction.

A significant quantity of initial stress may develop in a medium due to many physical causes like slow process of creep, gravity variation, difference of temperature etc. In fact, the Earth is an initially stressed medium. Authors [10, 11, 12] have employed the theory of incremental deformation formulated by Biot [13] to study seismic waves in pre-stressed elastic solids.

Keeping these things in mind, we have discussed the propagation of shear waves in an inhomogeneous, viscoelastic medium overlying a homogeneous initially stressed half space.

2. Formulation of the problem

We have considered a viscoelastic, heterogeneous layer lying over a homogeneous half space under initial stress and in welded contact with each other. The inhomogeneity has been considered in density, rigidity and internal friction for the layer M_1. The semi-infinite medium M_2 is under the constant initial stress P along x-axis.The origin has been taken at the interface between the layer and the half space. The x-axis is taken along the interface and the z-axis vertically downwards. The layer is taken from $z = 0$ to $z = -h_1$. The semi-infinite space extends from $z = 0$ to $z = \infty$.
The inhomogeneity of the layer M_1 is specified by $\mu_1 = \mu_{10}(1+\alpha z)$, $\mu_1' = \mu_{10}'(1+\alpha z)$, $\rho_1 = \rho_0(1+\alpha z)$. The elastic parameters and the initial compressive stress of the homogeneous half space M_2 are represented by μ_2, ρ_2, P. Here μ_1, μ_2 are rigidities, ρ_1, ρ_2 are densities of M_1 and M_2 respectively and μ_1' is the internal friction of M_1. The inhomogeneity parameter α is real positive having dimension of inverse of length.

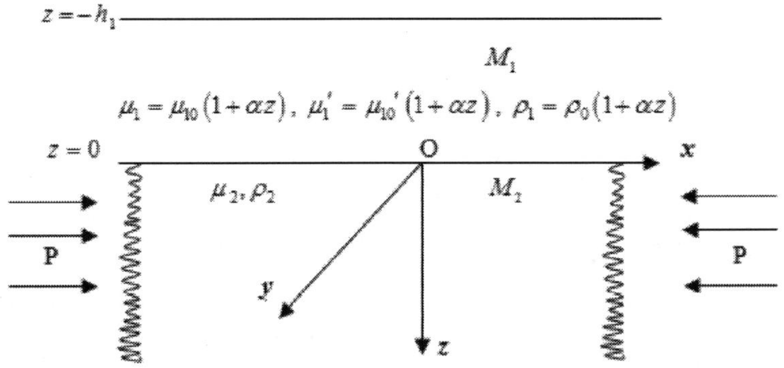

Figure 1. Geometry of the problem.

3. Solution of the problem

For SH-type wave motion $u = w = 0, v = v\,(x, z, t)$ and in the absence of body forces the equation of motion for the layer M_1 is

$$\left(\mu_1 + \mu_1' \frac{\partial}{\partial t}\right) \frac{\partial^2 v_1}{\partial x^2} + \frac{\partial}{\partial z}\left(\mu_1 + \mu_1' \frac{\partial}{\partial t}\right) \frac{\partial v_1}{\partial z} = \rho_1 \frac{\partial^2 v_1}{\partial t^2}. \tag{1}$$

If the plane waves propagate in the direction of increasing x, then we put

$$v_j\,(x, z, t) = V_j\,(z)\; e^{ik(ct-x)}, \quad j = 1,\ 2. \tag{2}$$

Using (2) in (1), we have

$$\frac{d^2 V_1}{dz^2} + \frac{\alpha}{1+\alpha z}\frac{dV_1}{dz} + k^2\left(\frac{c^2 \rho_0}{\bar{\mu}_1} - 1\right) V_1 = 0, \tag{3}$$

where $\bar{\mu}_1 = \mu_{10} + ikc\mu_{10}'$.
The solution of (3) is obtained as

$$V_1 = \left[A J_0\left(\frac{1}{\alpha}(1+\alpha z)\, k\sqrt{\frac{c^2}{\beta_1^2} - 1}\right) + B Y_0\left(\frac{1}{\alpha}(1+\alpha z)\, k\sqrt{\frac{c^2}{\beta_1^2} - 1}\right)\right], \tag{4}$$

where A and B are constants, J_0 and Y_0 are Bessel functions of the first and second kind respectively of order zero and $\beta_1^2 = \bar{\mu}_1/\rho_0$.
Hence,

$$v_1(x,z,t) = \left[A\,J_0\left(\frac{1}{\alpha}(1+\alpha z)\,k\sqrt{\frac{c^2}{\beta_1^2}-1}\right) + B\,Y_0\left(\frac{1}{\alpha}(1+\alpha z)\sqrt{\frac{c^2}{\beta_1^2}-1}\right)\right] e^{ik(ct-x)}. \quad (5)$$

In the half space M_2 the equation of motion [13] under initial stress P is

$$\left(\mu_2 - \frac{P}{2}\right)\frac{\partial^2 v_2}{\partial x^2} + \mu_2 \frac{\partial^2 v_2}{\partial z^2} = \rho_2 \frac{\partial^2 v_2}{\partial t^2}. \quad (6)$$

The above equation can be written as

$$(1-\zeta)\frac{\partial^2 v_2}{\partial x^2} + \frac{\partial^2 v_2}{\partial z^2} = \frac{1}{\beta_2^2}\frac{\partial^2 v_2}{\partial t^2}, \quad (7)$$

where $\beta_2^2 = \mu_2/\rho_2$ and $\zeta = P/2\mu_2$.
Using (2) in (7), we have

$$d^2V_2\big/dz^2 = m^2 V_2, \quad (8)$$

where $m = k\sqrt{(1-\zeta) - (c^2/\beta_2^2)}$.
The appropriate solution of (8) is

$$V_2 = C\,e^{-mz},\ z \geq 0\,. \quad (9)$$

Therefore, $v_2(x,z,t)$ of the lower half space is

$$v_2(x,z,t) = C\,e^{-mz}e^{ik(ct-x)}. \quad (10)$$

4. Boundary conditions

- The upper surface M_1 is stress free, i.e., $(\mu_1 + \mu'_1 \partial/\partial t)\,\partial v_1/\partial z = 0$ at $z = -h_1$.
- The displacement components are continuous at the interface, i.e., $v_1 = v_2$ at $z = 0$.
- The stresses are continuous at the interface, i.e., $(\mu_1 + \mu'_1\,\partial/\partial t)\,\partial v_1/\partial z = \mu_2(\partial v_2/\partial z)$ at $z = 0$.

Using (5) and (10) in all the boundary conditions and eliminating A, B and C, we have

$$\begin{aligned}
&\overline{\mu_1} k\sqrt{\tfrac{c^2}{\beta_1^2}-1}\Big[J_0'\left(\tfrac{1}{\alpha}(1-\alpha h_1)\,k\sqrt{\tfrac{c^2}{\beta_1^2}-1}\right)Y_0'\left(\tfrac{1}{\alpha}k\sqrt{\tfrac{c^2}{\beta_1^2}-1}\right)\\
&\qquad -J_0'\left(\tfrac{1}{\alpha}k\sqrt{\tfrac{c^2}{\beta_1^2}-1}\right)Y_0'\left(\tfrac{1}{\alpha}(1-\alpha h_1)\,k\sqrt{\tfrac{c^2}{\beta_1^2}-1}\right)\\
&+m\mu_2\Big[J_0'\left(\tfrac{1}{\alpha}(1-\alpha h_1)\,k\sqrt{\tfrac{c^2}{\beta_1^2}-1}\right)Y_0\left(\tfrac{1}{\alpha}k\sqrt{\tfrac{c^2}{\beta_1^2}-1}\right)\\
&\qquad -J_0\left(\tfrac{1}{\alpha}k\sqrt{\tfrac{c^2}{\beta_1^2}-1}\right)Y_0'\left(\tfrac{1}{\alpha}(1-\alpha h_1)\,k\sqrt{\tfrac{c^2}{\beta_1^2}-1}\right) = 0.
\end{aligned} \quad (11)$$

Equation (11) is the dispersion equation of SH wave propagating in viscoelastic, inhomogeneous layer overlying a homogeneous half space under initial stress.

5. Particular case

Consider $\alpha \to 0$, $\mu' = 0$, $P = 0$. When the initial stress $P = 0$, then $\zeta = 0$. Using asymptotic analysis [14], we obtain the standard dispersion equation of Love waves as

$$\tan\left(kh_1\sqrt{\frac{c^2}{\beta_{10}^2}-1}\right) = \left[\mu_2\sqrt{1-\frac{c^2}{\beta_2^2}}\right] \Big/ \left[\mu_{10}\sqrt{\frac{c^2}{\beta_{10}^2}-1}\right], \tag{12}$$

where $\beta_{10}^2 = \mu_{10}/\rho_{10}$.

6. Numerical results and discussions

Equation (11) gives the dispersion equation for shear waves propagating in a viscoelastic, heterogeneous layer lying over a homogeneous half space under initial stress. In this section, the phase velocities as well as damping velocities have been plotted from (11) for different values of the inhomogeneity parameter αh and the initial stress ζ. We have taken the following data from Gubbins [15] and Caloi [16]:

$$\begin{aligned} &\mu_{10} = 7.45 \times 10^{10}\, N/m^2,\ \ \mu_{10}/\mu'_{10} = 30\,\text{sec}^{-1},\ \ \rho_0 = 3293 Kg/m^3, \\ &\mu_2 = 7.10 \times 10^{10} N/m^2,\ \ \rho_2 = 3321 Kg/m^3. \end{aligned} \tag{13}$$

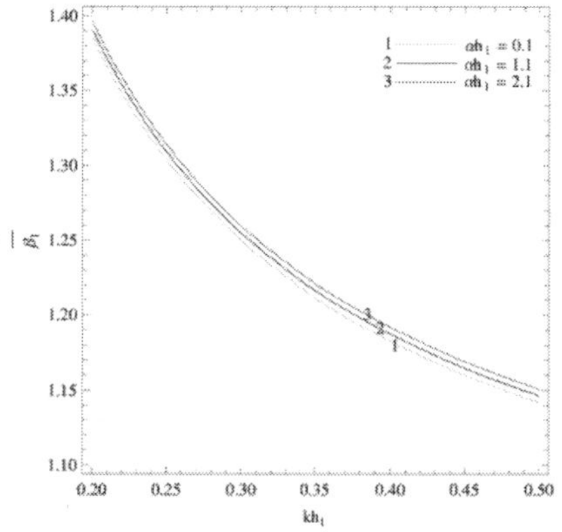

Figure 2. Dimensionless phase velocity against dimensionless wave number when $\zeta = 0.1$.

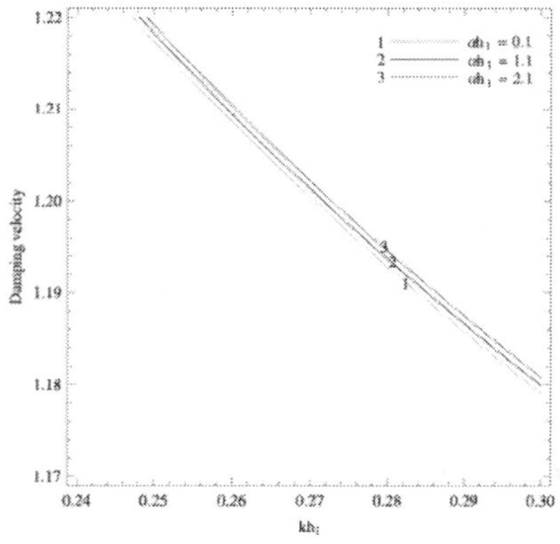

Figure 3. Dimensionless damping velocity against dimensionless wave number when $\zeta = 0.1$.

In figures 2 and 4 the dimensionless phase velocity c/β_1 has been plotted against the dimensionless wave number kh_1 whereas the damping velocity has been plotted against dimensionless wave number in figures 3 and 5.

In figures 2 and 3, we have taken different values of inhomogeneity parameter of the layer M_1 ($\alpha h_1 = 0.1,\ 1.1,\ 2.1$). From figure 2 it is observed that the phase velocity c/β_1 decreases monotonically with the increase of kh_1 but increases with the inhomogeneity. In the figure 3, the effect of inhomogeneity parameter on damping velocity have same pattern as in figure 2. It has been observed from the figure 3 that due to damping the range of kh_1 differ from without damping case (vide figure 2).

In figures 4 and 5, it has been observed that if the values of the initial stress are increased in the layer M_2 then the phase velocity as well as the damping velocity increases. Also it has been noticed that phase and damping velocities decrease monotonically with the increase of dimensionless wave number.

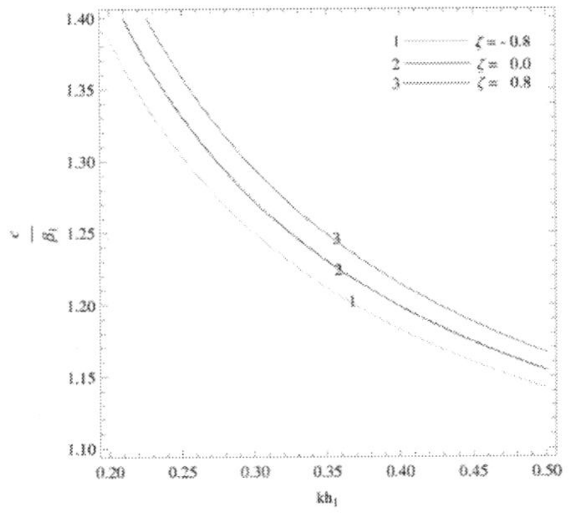

Figure 4. Dimensionless phase velocity against dimensionless wave number when $\alpha h = 0.1$.

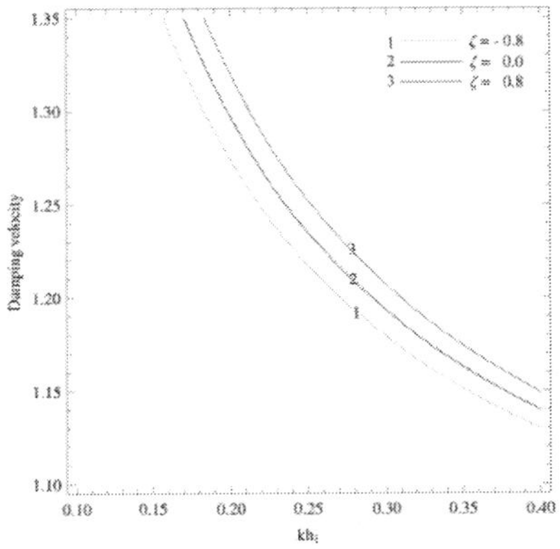

Figure 5. Dimensionless damping velocity against dimensionless wave number when $\alpha h = 0.1$.

7. Conclusion

An analytical approach is used to investigate shear wave propagation in a viscoelastic, heterogeneous layer lying over a homogeneous half space under initial stress. The dispersion equation obtained here is in agreement with the classical result of Love wave propagation when the inhomogeneity of the upper layer and the initial stress on the half space are neglected. The numerical results reflects that the phase and damping velocities decrease with wave number but increase with the inhomogeneity parameter of the upper layer and the initial stress of the half space .

References

[1] Ewing W M, Jardetzky W S and Press F 1957 *Elastic waves in layered media* (New York: McGraw-Hill)
[2] Cooper H F 1967 Reflection and transmission of oblique plane waves at a plane interface between viscoelastic media *J. Acoust. Soc. Am* **42** 1064-69
[3] Shaw R P and Bugl P 1969 Transmission of plane waves through layered linear viscoelastic media *J. Acoust. Soc. Am.* **46** 649-654
[4] Borcherdt R D 1973 Rayleigh-type surface wave on a linear viscoelastic halfspace *J. Acoust. Soc. Am.* **54** 6 1651-53
[5] Romeo M 2003 Interfacial viscoelastic SH waves *Int. J. Solids Struct.* **40** 9 2057-68
[6] Chattopadhyay A 1978 Propagation of SH waves in a viscoelastic medium due to irregularity in the crustal layer *Bull. Calcutta Math. Soc.* **70** 303-316
[7] Chatterjee M and Chattopadhyay A 2015 Propagation, reflection, and transmission of SH-waves in slightly compressible, finitely deformed elastic media *Appl. Math. Mech. -Engl. Ed.* **36** 8 1045-56
[8] Bullen K E 1940 The problem of the Earths density variation *Bull. Seismol. Soc. Am.* **30** 3 235-250
[9] Chatterjee M, Dhua S, Chattopadhyay A and Sahu S A 2015 Seismic waves in heterogeneous Crust-Mantle layers under initial stresses *J. Earthquake Eng.* DOI: 10.1080/13632469.2015.1038371
[10] Chatterjee M, Dhua S, Sahu S A and Chattopadhyay A 2014 Reflection in a highly anisotropic medium for three-dimensional plane waves under initial stresses *Int. J. Eng. Sci.* **85** 136-149
[11] Chatterjee M, Dhua S and Chattopadhyay A 2015 Quasi-P and quasi-S waves in a selfreinforced medium under initial stresses and under gravity *J. Vib. Control* DOI: 10.1177/1077546314568694
[12] Chatterjee M, Dhua S, Chattopadhyay A and Sahu S A 2015 Reflection and refraction for three-dimensional plane waves at the interface between distinct anisotropic half spaces under initial stresses *Int. J. Geomech.* DOI: 10.1061/(ASCE)GM.1943-5622.0000601
[13] Biot M A 1965 *Mechanics of incremental deformation* (New York: John Wiley and Sons, Inc.)
[14] Watson G N 1944 *A Treatise on the Theory of Bessel Functions* (London: Cambridge University Press)
[15] Gubbins D 1990 *Seismology and plate tectonics* (London: Cambridge University Press)
[16] Caloi P 1950 Comportement des ondes de Rayleigh dans un milieu firmo-lastique indfini *Publications du Bureau Central Seismologique International* Travaux Scientifiques **17** A 89-108

Journal of Physics: Conference Series **662** (2015) 012002 doi:10.1088/1742-6596/662/1/012002

Study of SH-type wave propagating in an anisotropic layer sandwiched between an orthotropic medium and an in-homogeneous half-space

Rehena Sultana and Shishir Gupta

Department of Applied Mathematics, Indian School of Mines, Dhanbad 826004, Jharkhand, India.

E-mail: rehenasultana28@gmail.com and shishir_ism@yahoo.com

Abstract. The present paper investigates the effect of initial stress and the existence of SH-type wave in an initially stressed porous stratum sandwiched between an initially stressed orthotropic medium and an in-homogeneous half-space. Exponential variations of rigidity and density have been taken in the lower half-space. The dispersion equation has been derived in closed form. It is observed that the propagation of SH type wave is influenced by the medium characteristics such as porosity, initial stress, anisotropy and in-homogeneity parameter.

1. Introduction

The study of seismic wave propagation has long been a prime interest to the seismologist as it is responsible for constituting and layering the earths interior. Earths structure is very complex in nature containing various types of layers including porous layer contribute amazing characteristics such as anisotropy, heterogeneity and stress parameter also. Generally these pores can contain oil or gas or water and rock will be saturated with one of these three. The studies of SH-type wave propagation in a fluid saturated anisotropic media have received considerable attention in the field of Earthquake engineering, Geophysical problem and soil dynamics. Scattering of SH-type wave in multi-layered media with irregular interfaces was developed by Ding and Darvinski [1]. A model for spherical SH-wave propagation in self-reinforced linearly elastic media was formulated by Chattopadhyay and Michel [2]. By the virtue of Biots dynamic theory [3, 4] on fluid saturated porous media Deresiewiez [5] was the first who studied the propagation of Love waves in a porous layer resting on an elastic homogeneous isotropic semi-infinite space. Propagation of SH-type wave in a fluid saturated elastic plate was discussed by Pradhan et al. [6]. Sharma and Gogna [7] studied the propagation of Love waves in an initially stressed medium consisting of a slow elastic layer lying over a liquid-saturated porous solid half-space. Possibility of Love wave propagation in a porous layer under the effect of linearly varying directional rigidities was introduced by Gupta et al. [8]. The present paper deals with the propagation of SH-type wave in an anisotropic layer sandwiched between orthotropic layer and an in-homogeneous half-space. SH-type waves propagate in the layer and attenuate along the thickness of the substrate so that its particle of motion and direction of propagation are in a same horizontal line. Since the composition of the Earth is a heterogeneous medium under high initial stresses so the medium in-homogeneous, orthotropic, anisotropic and initial stresses contribute a significant effect in the propagation of seismic waves.

2. Formulation of the problem

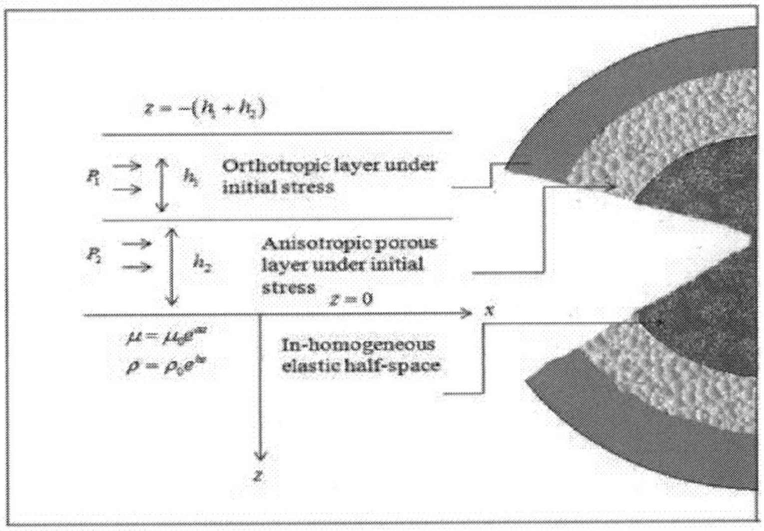

Figure 1. Schematic diagram of the physical problem.

3. Basic preliminaries

3.1. Solution for the upper layer

The upper layer is considered as the orthotropic medium with finite thickness h_2 under initial stress P_1 along x axis. Let u_1, v_1, w_1 are the displacement components, ω_x, ω_y and ω_z are the rotational component along x,y and z axis respectively. Now the dynamical equations of motion for this medium in the absence of body forces are (Biot [9]):

$$\left.\begin{aligned} \frac{\partial S_{11}}{\partial x} + \frac{\partial S_{12}}{\partial y} + \frac{\partial S_{13}}{\partial z} - P_1\left(\frac{\partial \omega_z}{\partial y} - \frac{\partial \omega_y}{\partial w_z}\right) = \rho_1 \frac{\partial^2 u_1}{\partial t^2} \\ \frac{\partial S_{21}}{\partial x} + \frac{\partial S_{22}}{\partial y} + \frac{\partial S_{23}}{\partial z} - P_1 \frac{\partial \omega_z}{\partial x} = \rho_1 \frac{\partial^2 v_1}{\partial t^2} \\ \frac{\partial S_{31}}{\partial x} + \frac{\partial S_{32}}{\partial y} + \frac{\partial S_{33}}{\partial z} - P_1 \frac{\partial \omega_y}{\partial x} = \rho_1 \frac{\partial^2 w_1}{\partial t^2} \end{aligned}\right\} \quad (1)$$

where $S_{ij}(i, j = 1, 2, 3)$ are the incremental stress components for the half-space and ρ_1 is the density of the material in this medium. Using the relations between the strain and incremental stress components with SH-type surface wave condition $u_1 = 0, w_1 = 0$ and $v_1 = v_1(x, z, t)$,the aforementioned equation of motion from (1) for the upper orthotropic medium can be written as

$$\frac{\partial}{\partial x}\left(R_3 \frac{\partial v_1}{\partial x}\right) + \frac{\partial}{\partial z}\left(R_1 \frac{\partial v_1}{\partial z}\right) - P_1 \frac{\partial}{\partial x}\left(\frac{1}{2}\frac{\partial v_1}{\partial x}\right) = \rho_1 \frac{\partial^2 v_1}{\partial t^2}. \quad (2)$$

The above equation can be restated as

$$\left(R_3 - \frac{P_1}{2}\right)\frac{\partial^2 v_1}{\partial x^2} + R_1 \frac{\partial^2 v_1}{\partial z^2} = \rho_1 \frac{\partial^2 v_1}{\partial t^2}. \quad (3)$$

The only two non-zero components are $S_{12} = 2Q_3e_{12}$ and $S_{23} = 2Q_1e_{23}$.
Here e_{ij} is the strain components defined by

$$e_{ij} = \frac{1}{2}\left(\frac{\partial u_i}{\partial x_j} + \frac{\partial u_j}{\partial x_i}\right).$$

For a wave propagating along x direction, it may be assumed that $v_1 = V_1(z)e^{ik}(x - ct)$ then the equation (3) can be converted to

$$\frac{d^2V_3}{dz^2} - m^2V = 0, \tag{4}$$

where

$$m^2 = \frac{k^2}{R_1}\left[\left(R_3 - \frac{P_1}{2}\right) - c^2\rho_1\right].$$

So, the solution for the pre-stressed upper semi-infinite orthotropic medium can be taken as

$$v_1 = \left(A_1e^{mz} + A_2e^{-mz}\right)e^{ik(x-ct)}, \tag{5}$$

where A_1, A_2 are arbitrary constants and c is the phase velocity.

3.2. Solution for the intermediate anisotropic medium

The effective equation of motion in an anisotropic porous layer under the effect of initial stress, ignoring the viscosity of fluid and body forces are given by (Biot [10])

$$\left.\begin{array}{l}
\frac{\partial\sigma_{11}}{\partial x} + \frac{\partial\sigma_{12}}{\partial y} + \frac{\partial\sigma_{13}}{\partial z} - P_2\frac{\partial\omega'_z}{\partial y} + P_2\frac{\partial\omega'_y}{\partial z} = \frac{\partial^2}{\partial t^2}\left(\rho_{11}u_2 + \rho_{12}U_x\right)\\
\frac{\partial\sigma_{21}}{\partial x} + \frac{\partial\sigma_{22}}{\partial y} + \frac{\partial\sigma_{23}}{\partial z} - P_2\frac{\partial\omega'_z}{\partial x} = \frac{\partial^2}{\partial t^2}\left(\rho_{11}v_2 + \rho_{12}V_y\right)\\
\frac{\partial\sigma_{31}}{\partial x} + \frac{\partial\sigma_{32}}{\partial y} + \frac{\partial\sigma_{33}}{\partial z} - P_2\frac{\partial\omega'_y}{\partial x} = \frac{\partial^2}{\partial t^2}\left(\rho_{11}w_2 + \rho_{12}W_z\right)\\
\frac{\partial S}{\partial x} = \frac{\partial^2}{\partial t^2}\left(\rho_{12}u_2 + \rho_{22}U_x\right),\ \frac{\partial S}{\partial y} = \frac{\partial^2}{\partial t^2}\left(\rho_{12}v_2 + \rho_{22}V_y\right),\ \frac{\partial S}{\partial z} = \frac{\partial^2}{\partial t^2}\left(\rho_{12}w_2 + \rho_{22}W_z\right)
\end{array}\right\} \tag{6}$$

where $\sigma_{ij}(i,j = 1,2,3)$ are the incremental stress components,(u_2, v_2, w_2) and (u_x, v_y, w_z) are the components of the displacement vector for solid and liquid respectively along x, y and z direction respectively,S is the stress vector due to the liquid part of porous which is related to the fluid pressure p by the relation $-S = fp$.
The angular component ω'_x, ω'_y and ω'_z are defined as

$$\omega'_x = \frac{1}{2}\left(\frac{\partial w_2}{\partial y} - \frac{\partial v_2}{\partial z}\right), \quad \omega'_y = \frac{1}{2}\left(\frac{\partial u_2}{\partial z} - \frac{\partial w_2}{\partial x}\right), \quad \omega'_z = \frac{1}{2}\left(\frac{\partial v_2}{\partial x} - \frac{\partial u_2}{\partial y}\right).$$

The mass coefficients ρ_{11}, ρ_{12} and ρ_{22} are related to the densities ρ', ρ_s and ρ_w of the layer, solid and liquid respectively by

$$\rho_{11} + \rho_{12} = (1 - f)\,\rho_s, \quad \rho_{12} + \rho_{22} = f\,\rho_w,$$

so that the mass densities of the bulk material is

$$\rho' = \rho_{11} + 2\rho_{12} + \rho_{22} = \rho_s + f\left(\rho_w - \rho_s\right).$$

These mass co-efficient also obey the following inequalities

$$\rho_{11} > 0,\ \rho_{22} > 0, \rho_{12} < 0,\ \rho_{11}\rho_{22} - \rho_{12}^2 > 0.$$

For the SH-type wave propagating along x direction, having the displacement of particle along the y direction as the displacement along the other direction vanishes, we have

$$\left.\begin{array}{lll} u_2 = 0, & w_2 = 0, & v_2 = v_2(x, z, t) \\ U_x = 0, & W_z = 0, & V_y = V(x, z, t) \end{array}\right\}. \tag{7}$$

Substituting relation(7)into dynamical system of equation (6), the equation of motion which are not identically satisfied are

$$\left.\begin{array}{l} \frac{\partial \sigma_{21}}{\partial x} + \frac{\partial \sigma_{23}}{\partial z} - P_2 \frac{\partial \omega_z{}'}{\partial x} = \frac{\partial^2}{\partial t^2} \left(\rho_{11} v_2 + \rho_{12} V_2\right) \\ \frac{\partial S}{\partial y} = 0 = \frac{\partial^2}{\partial t^2} \left(\rho_{12} v_2 + \rho_{22} V_2\right) \end{array}\right\}. \tag{8}$$

With the help of stress-strain relation $\sigma_{23} = 2Le_{yz}$ *and* $\sigma_{12} = 2Ne_{xy}$ the above equation transformed to

$$\left.\begin{array}{l} \left(N - \frac{P_2}{2}\right) \frac{\partial^2 v_2}{\partial x^2} + L \frac{\partial^2 v_2}{\partial z^2} = \frac{\partial^2}{\partial t^2} \left(\rho_{11} v_2 + \rho_{12} V\right) \\ \frac{\partial^2}{\partial t^2} \left(\rho_{12} v_2 + \rho_{22} V\right) = 0 \end{array}\right\}. \tag{9}$$

From $\frac{\partial^2}{\partial t^2} \left(\rho_{12} v_2 + \rho_{22} V\right) = 0$ and $\rho_{12} v_2 + \rho_{22} V = d''$, so that $V = \frac{d'' - \rho_{12}}{\rho_{22}}$.
Now $\frac{\partial^2}{\partial t^2} \left(\rho_{11} v_2 + \rho_{12} V\right) = \rho_2 \frac{\partial^2 v_2}{\partial t^2}$ where $\rho_2 = \rho_{11} - \frac{\rho_{12}^2}{\rho_{22}}$.
Hence equation (9) can be rewritten as

$$\left(N - \frac{P_2}{2}\right) \frac{\partial^2 v_2}{\partial x^2} + L \frac{\partial^2 v_2}{\partial z^2} = \rho_2 \frac{\partial^2 v_2}{\partial t^2}. \tag{10}$$

From equation (9), we found that the velocity of shear wave along the x direction is $\sqrt{\frac{N - \frac{P_2}{2}}{\rho_2}}$ and along the z direction is $\sqrt{\frac{L}{\rho_2}}$. The shear wave velocity in the porous medium along the x direction can be taken as

$$\beta' = \sqrt{\frac{N - \frac{P_2}{2}}{\rho_2}} = \beta_a \sqrt{\frac{1 - \xi_2}{d}},$$

where $d = \gamma_{11} - \frac{\gamma_{12}^2}{\gamma_{22}}$, $\beta_a = \sqrt{\frac{N}{\rho'}}$ is the velocity of shear wave in the corresponding initial stress free, non-porous, anisotropic, elastic medium along the x direction. $\xi_2 = \frac{P_2}{2N}$ is the non-dimensional parameter due to the initial stress P_2 and $\gamma_{11} = \frac{\rho_{11}}{\rho'}, \gamma_{12} = \frac{\rho_{12}}{\rho'}, \gamma_{22} = \frac{\rho_{22}}{\rho'}$ are the non-dimensional parameters for the material of the porous layer as obtained by Biot[11].
Thus for the SH-type wave propagating along the x direction the solution of equtaion(10)can be taken as

$$v_2\left(x, z, t\right) = V_2\left(z\right) e^{ik(x - ct)}. \tag{11}$$

Applying equation (11) in equation (10) we have

$$\frac{d^2 V_2}{dz^2} + q'^2 k^2 V_2 = 0, \tag{12}$$

where $q' = \sqrt{\frac{1}{L}\left(c^2 \rho_2 - N + \frac{P_2}{2}\right)}$.
Therefore the solution of equation (12) takes the form as

$$v_2 = \left[A_3 \cos(qz) + A_4 \sin(qz)\right] e^{ik(x - ct)}.$$

3.3. Solution for the lower half-space

The lower semi-infinite medium is considered as heterogeneous elastic half-space.The displacement components of this medium are u_3, v_3 and w_3 along x, y, z direction respectively. Using SH-type wave conditions $u_3 = w_3 = 0$ and $v_3 = v_3(x, z, t)$, the equation of motion corresponding to the displacement due to SH-type wave can be written as

$$\frac{\partial t_{21}}{\partial x} + \frac{\partial t_{23}}{\partial z} = \frac{\partial^2}{\partial t^2}(\rho_3 v_3), \tag{13}$$

where t_{ij} are the incremental stress components in the half-space and ρ_3 is the density of the material of the half-space.Using the stress-strain relation $t_{21} = 2\mu e_{xy}, t_{23} = 2\mu e_{yz}$ and the in-homogeneity conditions the equation (13) can be written as

$$\mu_0 e^{a\,z}\frac{\partial^2 v_3}{\partial x^2} + a\mu_0 e^{a\,z}\frac{\partial v_3}{\partial z} + \mu_0 e^{a\,z}\frac{\partial^2 v_3}{\partial z^2} = \rho_0 e^{b\,z}\frac{\partial^2 v_3}{\partial t^2}, \tag{14}$$

where $e_{xy} = \frac{1}{2}\left(\frac{\partial v_3}{\partial x} + \frac{\partial u_3}{\partial y}\right)$ and $e_{yz} = \frac{1}{2}\left(\frac{\partial w_3}{\partial y} + \frac{\partial v_3}{\partial z}\right)$.

Let $v_3(x, z, t) = V_3(z)\, e^{ik(x-ct)}$ be the solution of the above equation. Then equation (14) can take the form as

$$\frac{d^2 V_3(z)}{dz^2} + a\frac{dV_3(z)}{dz} - k^2\left(1 - \frac{c^2}{c_0^2}\frac{e^{bz}}{e^{az}}\right)V_3(z) = 0, \tag{15}$$

where $c_0 = \sqrt{\frac{\mu_0}{\rho_0}}$ is the shear wave velocity. Substituting $V_3(z) = \phi(z)\, e^{-\frac{a\,z}{2}}$ in equation (15) to eliminate the term $\frac{dV_3(z)}{dz}$ we obtain

$$\varphi''(z) + \left[k^2\frac{c^2}{c_0^2}\frac{e^{bz}}{e^{az}} - k^2\left(1 - \frac{a^2}{4k^2}\right)\right]\phi(z) = 0.$$

Using non-dimensional parameters $\gamma = \left(1 + \frac{a^2}{4k^2} - \frac{b}{a}\frac{c^2}{c_0^2}\right)^{\frac{1}{2}}$, $\eta = \frac{2\gamma k(1+az)}{a}$ and $\omega = kc$, the above equation reduces to

$$\phi''(\eta) + \left(\frac{R}{\eta} - \frac{1}{4}\right)\phi(\eta) = 0, R = \frac{c^2(a-b)k}{2c_0^2 a^2\gamma}. \tag{16}$$

The Solution of Whiattaker's equation (16) is given by

$$\phi(\eta) = D_1 W_{R,\frac{1}{2}}(\eta) + D_2 W_{R,\frac{1}{2}}(-\eta), \tag{17}$$

where D_1, D_2 are arbitrary constants and $W_{R,\frac{1}{2}}(\eta)$ is the Whittaker function. The solution of equation (17)may be given by Whittaker and Watson [12], satisfying the condition $\lim_{z\to\infty} V_3(z) \to 0$ i.e. $\lim_{\eta\to\infty}\phi(\eta) \to 0$ may be taken as $\phi(\eta) = D_1 W_{R,\frac{1}{2}}(\eta)$. Hence the displacement component $v_3(z)$ in the in-homogeneous half-space is given by

$$v_3(x, z.t) = D_1 W_{R,\frac{1}{2}}\left[\frac{2\gamma k}{a}(1 + az)\right] e^{-\frac{a\,z}{2}} e^{i\,k(x-ct)}.$$

4. Boundary conditions

The suitable boundary condtions according to the formulation of the problem are as follows
(i) Continuity of the displacement components $v_1 = v_2$ at $z = -h_1$ which gives

$$A_1 e^{-mh_1} + A_2 e^{mh_1} = A_3\cos(qh_1) - A_4\sin(qh_1). \tag{18}$$

(ii) Stress components are continuous at $z = -h_1$ that is

$$m\left(A_1e^{-mh_1} - A_2e^{mh_1}\right) = \frac{R_1}{L}q\left[A_3\sin(qh_1) + A_4\cos(qh_1)\right]. \tag{19}$$

(iii) At $z = 0$ the velocity components of layer and half-space are continuous that is

$$A_3 = D_1\left(\frac{2\gamma k}{a}\right)e^{-\frac{\gamma k}{a}}\left[1 + (1-R)\frac{\gamma k}{a}\right]. \tag{20}$$

(iv) At $z = 0$, the stress components of layer and half-space are continues that implies

$$A_4q = \frac{\mu_3}{\mu_2}D_1\left(\frac{2\gamma k}{a}\right)e^{-\frac{\gamma k}{a}}\left[1 + (1-R)\frac{\gamma k}{a}\right]\left[\frac{(1-R)\gamma k}{1+(1-R)\frac{\gamma k}{a}} - \left(\gamma k + \frac{a}{2}\right) + a\right]. \tag{21}$$

(v) At $z = -(h_1 + h_2)$ stress component for the upper layer vanishes i.e

$$A_1e^{-m(h_1+h_2)} + A_2e^{m(h_1+h_2)} = 0. \tag{22}$$

Eliminating A_1, A_2, A_3, A_4 and D_1 from equations (18),(19),(20),(21) and (22) we have

$$\begin{vmatrix} e^{-mh_1} & e^{mh_2} & -\cos(qh_1) & sin(qh_1) & 0 \\ me^{-mh_1} & -me^{mh_1} & -\frac{L}{R_1}q\sin(qh_1) & -\frac{L}{R_1}q\cos(qh_1) & 0 \\ 0 & 0 & 0 & q & -\frac{\mu_3}{L}\beta \\ 0 & 0 & 1 & 0 & -\alpha \\ e^{-m(h_1+h_2)} & e^{m(h_1+h_2)} & 0 & 0 & 0 \end{vmatrix} = 0,$$

where $\alpha = \left(\frac{2\gamma k}{a}\right)e^{-\frac{\gamma k}{a}}\left[1 + (1-R)\frac{\gamma k}{a}\right]$ and
$\beta = \left(\frac{2\gamma k}{a}\right)e^{-\frac{\gamma k}{a}}\left[1 + (1-R)\frac{\gamma k}{a}\right]\left[\frac{(1-R)\gamma k}{1+(1-R)\frac{\gamma k}{a}} - (\gamma k + \frac{a}{2}) + a\right]$. This gives that

$$\tan(qh_1) = \frac{mq\alpha - \frac{\mu_3}{R_1}\beta q\tanh(mh_2)}{\frac{\mu_3}{L}\beta m + \alpha\frac{L}{R_1}q^2\tanh(mh_2)}. \tag{23}$$

This is the dispersion equation of SH-type wave in an anisotropic porous layer sandwiched between an orthotropic medium and an in-homogeneous half-space.

5. Particular cases

5.1. Case 1

If we neglect the upper orthotropic layer then equation (24) can be introduced as

$$tan(qh_1) = \frac{-\mu_3\beta q}{\mu_2\alpha q^2}. \tag{24}$$

5.2. Case 2

when the lower half-space be homogeneous that is $a \to 0, b \to 0$ this implies that $\gamma = \sqrt{1 - \frac{c^2}{c_0^2}}$ and therefore equation (24) reduces to

$$tan\left[kH\left(\frac{N}{L}\right)d\left\{\frac{c^2}{\beta_p^2} - \frac{1-\xi_2}{d}\right\}\right]^{\frac{1}{2}} = \frac{\mu_3}{L}\frac{\sqrt{1-\frac{c^2}{c_0^2}}}{\sqrt{\left(\frac{N}{L}\right)d\left\{\frac{c^2}{\beta_p^2} - \frac{1-\xi_2}{d}\right\}}}. \tag{25}$$

5.3. Case 3

If the intermediate layer is initially stress free, isotropic, non-porous elastic layer ($i.e.\ \xi_2 = 0,\ N = L = \mu_2,\ d = 1$) then the above equation reduces to

$$\tan\left[kH\left\{\frac{c^2}{\beta_p^2} - 1\right\}\right]^{\frac{1}{2}} = \frac{\mu_3}{\mu_2}\frac{\sqrt{1 - \frac{c^2}{c_0^2}}}{\sqrt{\frac{c^2}{\beta_p^2} - 1}}. \tag{26}$$

This is the general equation of Love wave established by Love [13] and hence validating the solutions of the problem discussed in the current paper.

6. Numerical computation and discussion

In order to show the effect of porosity, anisotropy, in-homogeneity and initial stresses on the propagation of SH-type wave, we take the following numerical values from Gubbins [14] and Kar and Kalyani [15]):
(1)For the upper isotropic homogeneous medium
$\mu_1 = 6.54 \times 10^{10}\ N/m^2\,, \rho_1 = 1409\,kg/m^3$
(2)For the intermediate anisotropic porous layer
$N = 5.00 \times 10^{10} N/m^2,\ \rho_2 = 9890\,kg/m^3$
(3) For the lower semi-infinite orthotropic half-space
$Q_1 = 5.82 \times 10^{10} N/m^2, Q_3 = 3.99 \times 10^{10} N/m^2$.

In figures 2-5, different curves are plotted from the dispersion equation to show the effect of

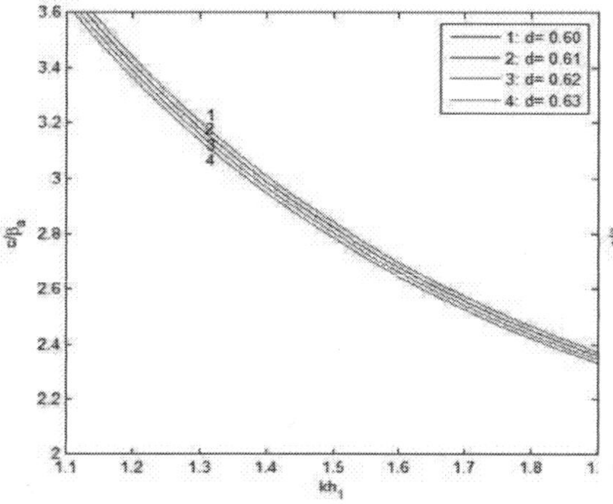

Figure 2. Dimensionless phase velocity against dimensionless wave number under the effect of porosity parameter (d).

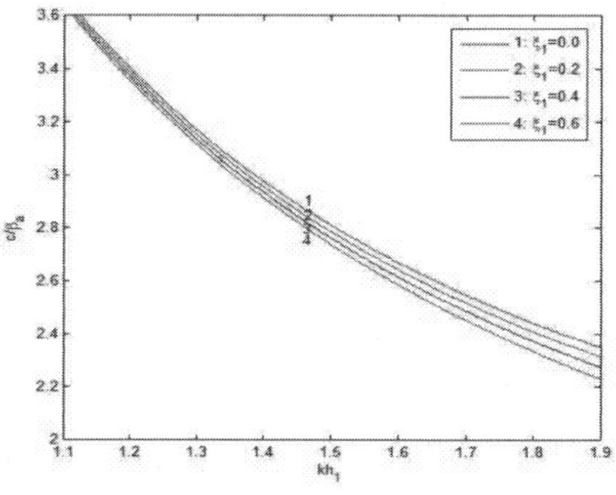

Figure 3. Effect of initial stress ξ_1 in the orthotropic layer on the propagation of SH-type waves.

various parameter.All the figures have been plotted with vertical axis as non-dimensional phase velocity $\frac{c}{\beta_a}$ against horizontal axis as dimensional wave number kh_1.Figure 2 depicts the effect of porosity on the propagation of SH-type wave. we observed that as the porosity incrases i.e as the value of d decreases, the velocity of SH-type wave increases. Figure 3 demonstrates, the effect of initial stress (ξ_1) present in the upper orthotropic layer while the intermediate porous layer (i.e. $\xi_2 = 0$)is initially stress free.It has been observed that an increases in Compressive initial

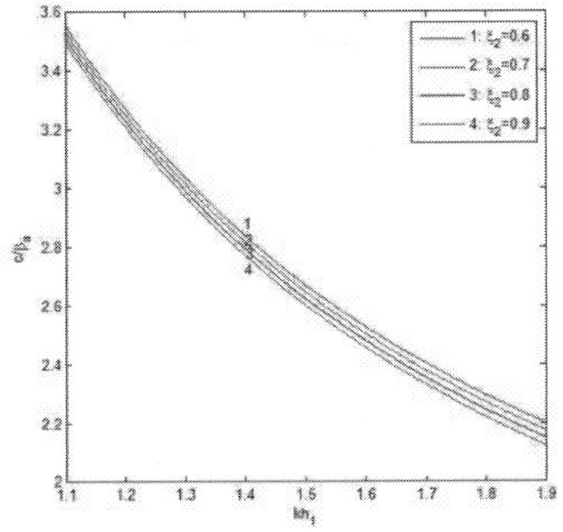

Figure 4. Effect of initial stress ξ_2 in the porous layer on the propagation of SH-type waves for d=0.6154.

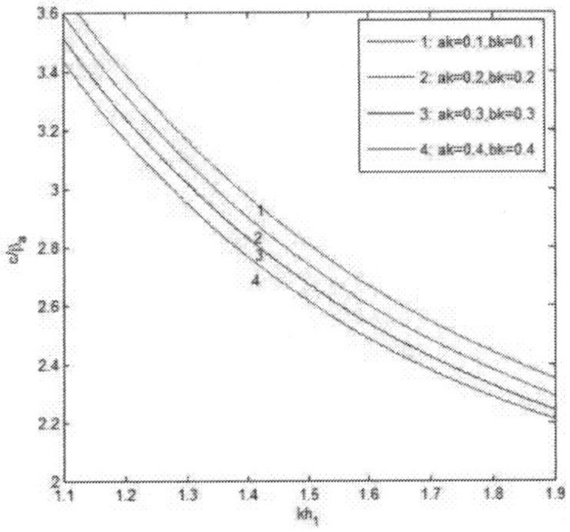

Figure 5. Effect of in-homogeneity parameter $\frac{a}{k}$, $\frac{b}{k}$, in the lower half-space on the propagation of SH-type wave for d=0.6154.

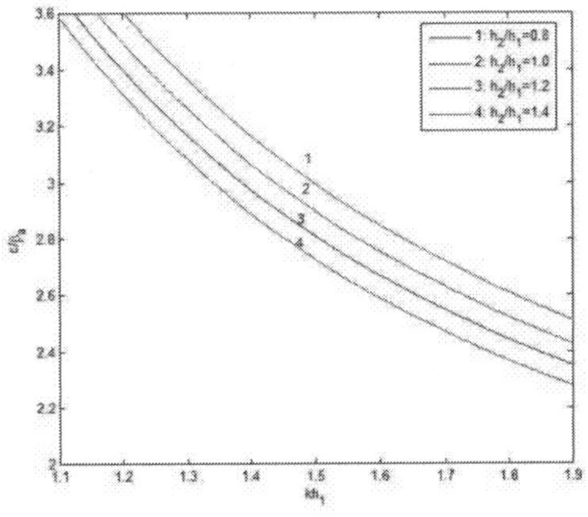

Figure 6. Dimensionless phase velocity against dimensionless wave number for different value of $\frac{h_2}{h_1}$ at a constant value of d=0.6154.

stress in the orthotropic layer decreases the velocity of SH-type wave. Figure 4 reflects the effect of initial stress in the porous layer when the stress for the upper orthotropic layer vanishes.It has been found that the phase velocity of SH-type wave decreases, as the value of initial stress parameter increases. Figure 5 shows the effect of in-homogeneity parameter $\frac{a}{k}$ and $\frac{b}{k}$ on the propagation of SH-type wave in the absence of initial stress parameters.The curves of this figure represented that the phase velocity decreases with the increase of in-homogeneity parameter. Figure 6 represents the effect of the ratio of depth parameter $\frac{h_2}{h_1}$ on the propagation of SH-type wave. It can be adduced that as the width ration increases the phase velocity decreases.

Journal of Physics: Conference Series **662** (2015) 012002 doi:10.1088/1742-6596/662/1/012002

7. Conclusion

The present paper deals with the propagation of SH-type wave in distinct vertically anisotropic double layers overlying an in-homogeneous half-space. Analytical solutions for the displacement in both the layer and half-spaces have been obtained seperately in closed form. The dispersion relation has been obtained in terms of Whittakers functions which are expanded asymptotically and the terms up to second degree are retained. It is observed from the dispersion equation that the porosity, anisotropy and initial stresses have a significant effect on the propagation of SH-type wave.

References

[1] Ding G and Darvinski M 1998 Scattering of SH-waves in multi-layered media with irregular interfaces *Earthq Eng Struct Dyn* **25**(12) 1391-1404
[2] Chattopadhyay A and Michel V 2006 A model for spherical SH-wave propagation in self-reinforced linearly elastic media *Archive of Applied Mechanics* **76** 113 -124
[3] Biot M A 1956a Theory of elastic waves in a fluid saturated porous solid, I.Low frequency range *J. Acoust. Soc. Am.* **28**(2) 168-178
[4] Biot M A 1956b Theory of elastic waves in a fluid saturated porous solid, II.High frequency range *J. Acoust. Soc. Am.* **28**(2) 179-191
[5] H. Deresiewiez 1961 The effect of boundaries on wave propagation in a liquid-filled porous solid *Bull. Seismol. Soc. Am.* **51** 51-59
[6] Pradhan A, Samal S K and Mahanti N C 2002 Shear wave in a fluid saturated elastic plate *Sadhana* **27**(6) 595-604
[7] Sharma M D and Gogna M L 1991 Propagation of Love wave in an initially stressed medium consisting of a slow elastic layer lying over a liquid saturated porous half-space *J. Acoust. Soc. Am.***89** 2584-88
[8] Gupta S, Vishwakarma, S K., Majhi D K and Kundu S 2013 *Applied Mathematical Modelling* **37** 10-11 6652-60
[9] Biot M A 1956 *Mechanics of Incremental Deformation* (New York :John Wiley and Sons)
[10] Biot M A 1965 *Mechanics of Incremental Deformation* (New York,London :John Willey and Sons)
[11] Biot M A 1956 Theory of deformation of a porous viscolelastic anisotropic solid *J.Appl. Phys* **27** 459-467
[12] Whittaker E T and Watson G N 1991 *A course of Modern Analysis* (Cambridge University Press: Cambridge)
[13] Love A E H 1911 *Some Problems of Geodynamics* (Cambridge : Cambridge University Press)
[14] Gubbins D 1990 *Seismology and Plate Tectonics* (Cambridge, UK: Cambridge University press)
[15] Kar B K and Kalyani V K 1986 Propagation of Love waves in an irregular anisotropic porous layer sandwiched between two homogeneous half-spaces *Gerlands Beitr. Geophysik Leipzig* **95** 27-35

Free torsional vibrations of carbon nanotube based on nonlocal elasticity of bi-Helmholtz type

B Yakaiah[1], **A Srihari Rao**[2]

[1] Sumathi Reddy Institute of Technology, Department of Mathematics, Warangal-506371, India.

[2] Kakatiya Institute of Technology and Science, Department of Mathematics, Warangal-506371, India.

E-mail: asriharirao@yahoo.co.in,yakaiah12@gmail.com.

Abstract. In this paper, the theory of nonlocal elasticity of bi-Helmholtz type is applied to analyze the problem of torsional vibrations of single walled carbon nanotube with higher order boundary conditions containing two nonlocal nano scale parameters.The effect of different parameters on the angular frequencies of vibrations are explained.

1. Introduction

The classical linear theory of elasticity is intrinsically size independent and it is valid only for small wave numbers with no dispersions.In this theory,the stress at any point of the body is uniquely determined by the strain at the same point.The constitutive equations of classical elasticity are simply algebraic relations between the stress- strain components.This theory cannot be be applied to study of Mechanical properties of special type bodies with carbon nanomaterials that are used in the design of MEMS and NEMS devices.The nonlocal elasticity theory used in this paper, was first developed by Eringen [1-5] and his associates.It assumes that the stress at a reference point in the elastic continuum is a function of elastic strain field at every point in that model.This theory is based on the atomic theory of lattice dynamics and on phonon dispersions.In this theory the nonlocal stress is represented by a spatial integral of weighted averages of all local stresses within the domain.The spatial weight is specified by a nonlocal modulus is known as an attenuation function which depends on a single dimensionless nonlocal parameter.This parameter is proportional to the ratio internal characteristic length to the external characteristic length.The dynamic properties of materials at nanoscale are significantly different from their behaviors at larger scales due to the influence of surfaces stress and size effects.Usually the variational principles are applied to derive for the equation of motion and corresponding higher order boundary conditions which contain a nonlocal nanoscale parameter.The constitutive equations of nonlocal elasticity for isotropic medium involve not only two elastic constants but also the nonlocal kernel with a single nonlocal parameter.In literature [6] this theory is known as nonlocal theory of Helmholtz type. Recently the theory of nonlocal elasticity of bi-Helmholtz type for isotropic materials are under discussion[7-10]. In the theory of nonlocal elasticity of bi-Helmholtz type the kernel involves two nonlocal parameters which are functions of internal characteristic length scale.The nonlocal elasticity of bi-Helmholtz type involves two limits the first limit reduces bi-Helmlholtz to Helmholtz type and the second limit from Helmholtz type to classical theory of elasticity.The nonlocal elasticity of bi-Helmholtz type is mentioned in the paper of Eringen[11] but not discussed in detail.

The main objective of this paper is to study the problem of free torsional vibration of carbon nanotubes using the theory of nonlocal elasticity of bi-Helmholtz type.The governing differential equation of fourth order is derived and a closed form solution for angular displacement involving nonlocal parameters is obtained.The frequencies of vibrations are obtained as the roots of dispersive relation which is a fourth order polynomial.The general solution is simplified by considering higher order boundary conditions at each end of the body. The effect of different nonoscale parameters on the free angular frequency are explained numerically.The corresponding problems of nonlocal elasticity of bi-Helmholtz type and of classical elasticity can be derived as special case.

2. Basic equations of nonlocal elasticity

The nonlocal elastic stress field theory first proposed by Eringen [2] concerns the state of stress at a reference point x' within domain.The nonlocal stress depends not only on the strain at that location but also on the strain at all other points within the domain in a diminishing influence away from the reference location.The nonlocal elastic field theory for homogeneous and isotropic solids is described using the following basic equations given as [13-16]

$$\sigma_{ij,i} + \rho(f_j - \ddot{u_j}) = 0, \tag{1}$$

$$\sigma_{ij}(r) = \int_v \phi(|r' - r|, \alpha) t_{ij}(r') dv(r'), \tag{2}$$

$$t_{ij}(r') = \lambda e_{kk}(r')\delta_{ij} + 2\mu e_{ij}(r'), \tag{3}$$

$$e_{ij}(r') = \frac{1}{2}\left[\frac{\partial u_j(r')}{\partial r'_i} + \frac{\partial u_i(r')}{\partial r'_j}\right], \tag{4}$$

where $\sigma_{ij}(r)$, ρ, f_j and u_j are the nonlocal stress tensor, mass density, body force density and displacement vector at time t respectively at a reference point r in the body. Here the indices i, j take the values $1(or)1, 2(or)1, 2, 3$ depending on the dimensions of the body.The linear relationship between the local stress tensor t_{ij} and classical strain tensor $e_{ij}(r')$ at any arbitrary point r' is shown in eqn(3). The Lame constants are λ, μ and δ_{ij} is the Kronnecker delta.The nonlocal kernal function denoted by $\phi(|r' - r|, \alpha)$ depends on the Euclidean distance $|r' - r|$ between the reference point r and the arbitrary point r' and on the dimensionless nanolength scale α. The dimensionless nanolength scale α is described in [3] as

$$\alpha = \frac{e_0 a}{L}, \tag{5}$$

here a is the internal characteristic length (Ex. Lattice parameter, $C-C$ bond length, granular distance), and L is an external characteristic length (Ex. Crack length, Wave length etc.)of the nanotube.Also,we take e_0 as nonlocal scaling parameter which is assumed to be constant appropriate to continuum model. The parameter e_0 is estimated such that the relations on the nonlocal elasticity model could provide satisfactory approximations to the atomic dispersion curves of the plane waves with those obtained from the atomistic lattice dynamics. The above equations is in partial integral form and generally difficult to solve analytically.Thus a differential form of nonlocal elasticity equation is often used. According to Eringen's [3] the expression of nonlocal modulus can be given as

$$\phi(|x|, \alpha) = (2\Pi l^2 \alpha^2)^{-1} k_0(\sqrt{x, x}/l\alpha), \tag{6}$$

Journal of Physics: Conference Series **662** (2015) 012003 doi:10.1088/1742-6596/662/1/012003

where k_0 is the modified Bessel function,using equations $[1-4]$ and (6), the nonlocal constitutive equations can be approximated as

$$\left[1-\epsilon^2\nabla^2+\gamma^4\nabla^4\right]\sigma_{ij}=t_{ij}, \tag{7}$$

where $\epsilon=\epsilon_0 a$ and $\gamma=\gamma_0 a$ are constants and ∇^2 is the Laplacian operator.

3. Nonlocal elasticity of bi-Helmholtz type

In the theory of nonlocal elasticity of bi-Helmholtz type the linear differential operator is in the form of

$$L=1-\epsilon^2\nabla^2+\gamma^4\nabla^4, \tag{8}$$

where ϵ and γ are nonnegative parameters of nonlocality. These two parameters have the dimension of lengths and it satisfies the equation as

$$L\phi(|x'-x|)=\delta(|x'-x|), \tag{9}$$

where ϕ is the attenuation function and δ is the usual dirac delta function. Here ∇^2 is the harmonic operator and ∇^4 is the bi-harmonic operator.In the above definition ϵ and γ are two nonlocal parameter having the dimensions of lengths. These parameters can be represented interms of characteristic length a as

$$\epsilon=\epsilon_0 a, \quad \epsilon_0\geq 0, \quad \gamma=\gamma_0 a, \quad \gamma_0\geq 0. \tag{10}$$

The nonlocal elasticity of bi-Helmholtz type has two types of limits. The first limit from $\gamma\to 0$ reduces from bi-Helmholtz type elasticity to Helmholtz type. The second limit is obtained by $\epsilon\to 0$ which is followed by first limit. The second limit reduces nonlocal elasticity of bi-Helmholtz type to classical theory of elasticity.The constitutive equations for the nonlocal elasticity of bi-Helmholtz type can be written as

$$\left[1-\epsilon^2\nabla^2+\gamma^4\nabla^4\right]\sigma_{ij}=t_{ij}. \tag{11}$$

The bi-Helmholtz operator can be factorized as

$$L=\left[1-\epsilon^2\nabla^2+\gamma^4\nabla^4\right]=\left[1-d_1\nabla^2\right]\left[1-d_2\nabla^2\right], \tag{12}$$

where $d_1+d_2=\epsilon^2, \quad d_1d_2=\gamma^4$
Solving the equation for d_1,d_2, we obtain

$$d_1=\frac{\epsilon^2}{2}\left[1+\sqrt{1-\frac{4\gamma^4}{\epsilon^4}}\right] \tag{13}$$

and

$$d_2=\frac{\epsilon^2}{2}\left[1-\sqrt{1-\frac{4\gamma^4}{\epsilon^4}}\right]. \tag{14}$$

For the validity of the roots to be real we have the condition

$$1-\frac{4\gamma^4}{\epsilon^4}\geq 0. \tag{15}$$

In order to have the roots d_1,d_2 to be distinct non-negative and satisfy the condition as $\epsilon > \sqrt{2}\gamma$. We can observe that the nonlocal elasticity of Helmholtz type due to Eringen's is obtained to the limit form $d_1 \to \epsilon^2, \quad d_2 \to 0$.

In this paper we describe a nonlocal model of lattice dynamics with next nearest neighborhood interactions. In this model we consider the interacting with the first and second neighbours in the homogeneous chain of the Brillouin zone from the dispersive relation. For such model the values of the nonlocal parameters ϵ_0, γ_0 are taken as $\epsilon_0 = 0.318, \quad \gamma_0 = 0.450$. These values satisfy the required condition $\epsilon_0 > \sqrt{2}\gamma_0$. It is described in the literature that the best physical result for the dispersive relations of nonlocal elasticity of bi-Helmholtz type with two different parameters can be obtained only from the lattice theory with next nearest neighborhood interactions.

4. Nonlocal torsional vibration of nanotube with bi-Helmholtz type operator

Consider a single walled carbon nanotube with two fixed ends undergoing torsional vibration.The influence of forces applied at one end, the nanotube is assumed to be of length L. Using by-Helmholtz type operator given by Eq.(11), the Eringen's nonlocal stress can be represented in one dimensional differential equation form as

$$\tau - (e_0 a)^2 \frac{\partial^2 \tau}{\partial x^2} + (\gamma_0 a)^4 \frac{\partial^4 \tau}{\partial x^4} \sigma_{ij} = G\gamma, \tag{16}$$

where G is the shear modulus, γ is the shear strain and τ is the shear stress of the nanotube. The resultant stress due to shear force can be written as

$$S = \int_A \tau dA, \tag{17}$$

where A is the area of the cross section of nanotube. The torque relation T due to torsional vibration can be given as

$$T = \int_A \tau z dA. \tag{18}$$

By using the equations(16)and (18) we get the constitutive relation as

$$S - (e_0 a)^2 \frac{\partial^2 S}{\partial x^2} + (\gamma_0 a)^4 \frac{\partial^4 S}{\partial x^4} = GA\gamma, \tag{19}$$

$$T - (e_0 a)^2 \frac{\partial^2 T}{\partial x^2} + (\gamma_0 a)^4 \frac{\partial^4 T}{\partial x^4} = GI_p \frac{d\Theta}{dx}, \tag{20}$$

where Θ and I_p are the angular displacement and the polar moment of inertia of the cross section respectively. By ignoring the external force, the equation of motion at an element of length dx can be written as

$$\left[T + \frac{\partial T}{\partial x} dx\right] - T = \rho I_p \frac{\partial^2 \Theta}{\partial t^2} dx, \tag{21}$$

where T is the induced torque per unit length and the term $\rho I_p \frac{\partial^2 \Theta}{\partial t^2} dx$ represents the inertia torque acting on element of the carbon nanotube. Here ρ represents the density of carbon nanotube and t indicates time respectively. From the equations(20) and (21) we have

$$T = GI_p \frac{\partial \Theta}{\partial x} + (e_0 a)^2 \rho I_p \frac{\partial^3 \Theta}{\partial t^2 \partial x} - (\gamma_0 a)^4 \rho I_p \frac{\partial^5 \Theta}{\partial t^2 \partial x^3}. \tag{22}$$

Using the equation of equilibrium (21) we obtain the governing differential equation as

$$c^2 \frac{\partial^2 \Theta}{\partial x^2} = \frac{\partial^2 \Theta}{\partial t^2} - (e_0 a)^2 \frac{\partial^4 \Theta}{\partial x^2 \partial t^2} + (\gamma_0 a)^4 \frac{\partial^6 \Theta}{\partial x^4 \partial t^2}, \tag{23}$$

where the shear speed of sound $c = \sqrt{G/\rho}$. It can be observe that small scale effects ($e_0a = 0$ and $\gamma_0 a = 0$) we get the conventional equation for torsional vibration of cylindrical tube.
We assume the solution of equation (23) as

$$\Theta(x,t) = \psi(x)e^{i\omega t}, \tag{24}$$

where ω is the angular frequency. Introducing the normalized coordinate $X = x/L$ in the above Eqn.(23) can be transformed to space variables form as

$$\frac{d^4\psi}{dX^4} + \left[\frac{\alpha^2 - \tau_1^2}{\tau_2^4}\right]\frac{d^2\psi}{dX^2} + \left[\frac{1}{\tau_2^4}\right]\psi = 0, \tag{25}$$

where $\tau_1 = \frac{e_0a}{L}$, $\tau_2 = \frac{\gamma_0 a}{L}$, and $\alpha^2 = \frac{c^2}{\omega^2 L^2}$.For the torsional vibration of a carbon with bi-Helmholtz operator the governing equation of motion can be put in the form as

$$\tau_2^4\frac{d^4\psi}{dX^4} + \left[\alpha^2 - \tau_1^2\right]\frac{d^2\psi}{dX^2} + \psi = 0. \tag{26}$$

We assume that for the single walled carbon nanotube is fixed at both ends so that a soft clamped boundary conditions [18] for the problem can be taken as

$$\Theta|_{X=0} = 0, \quad \Theta|_{X=1} = 0, \quad T|_{X=0} = 0, \quad T|_{X=1} = 0. \tag{27}$$

The general solution of the differential equation can be expressed as

$$\psi(X) = \left[c_1e^{i\beta_1 X} + c_2e^{i\beta_2 X} + c_3e^{i\beta_3 X} + c_4e^{i\beta_4 X}\right]e^{i\omega t}, \tag{28}$$

where c_i(for i=1,2,3,4) are the arbitrary constants and β_i(for i=1,2,3,4) are the roots of the dispersion relation.

$$\tau_2^4\beta^4 + \left[\alpha^2 - \tau_1^2\right]\beta^2 + 1 = 0. \tag{29}$$

The above equation is obtained by substituting $\psi(X) = Ce^{i\beta X}$ into Eq(26) where C is a nonzero constant.

5. Solution using higher order boundary conditions

The governing equations of motion consider the boundary conditions of the nanotube by Eq.(27) and the arbitrary constants c_i for most significant terms of the nonlocal effect τ by the equation (29) can be represented as

$$\begin{pmatrix} 1 & 1 & 1 & 1 \\ e^{i\beta_1} & e^{i\beta_2} & e^{i\beta_3} & e^{i\beta_4} \\ k_1 & k_2 & k_3 & k_4 \\ k_1e^{i\beta_1} & k_2e^{i\beta_2} & k_3e^{i\beta_3} & k_4e^{i\beta_4} \end{pmatrix}\begin{pmatrix} C1 \\ C2 \\ C3 \\ C4 \end{pmatrix} = 0, \tag{30}$$

where $k_i = \frac{GI_p}{L}\beta_i - \frac{(e_0a)^2\rho\rho I_p}{L}\beta_i\omega^2 - \frac{(\gamma_0 a)^4\rho I_p}{L^3}\beta_i^2\omega^2 (i = 1,2,3,4)$. For nontrivial solution of

C_i,(i=1,2,3,4)the determinant of coefficient matrix must vanish.

$$\begin{vmatrix} 1 & 1 & 1 & 1 \\ e^{i\beta_1} & e^{i\beta_2} & e^{i\beta_3} & e^{i\beta_4} \\ k_1 & k_2 & k_3 & k_4 \\ k_1e^{i\beta_1} & k_2e^{i\beta_2} & k_3e^{i\beta_3} & k_4e^{i\beta_4} \end{vmatrix} = 0. \tag{31}$$

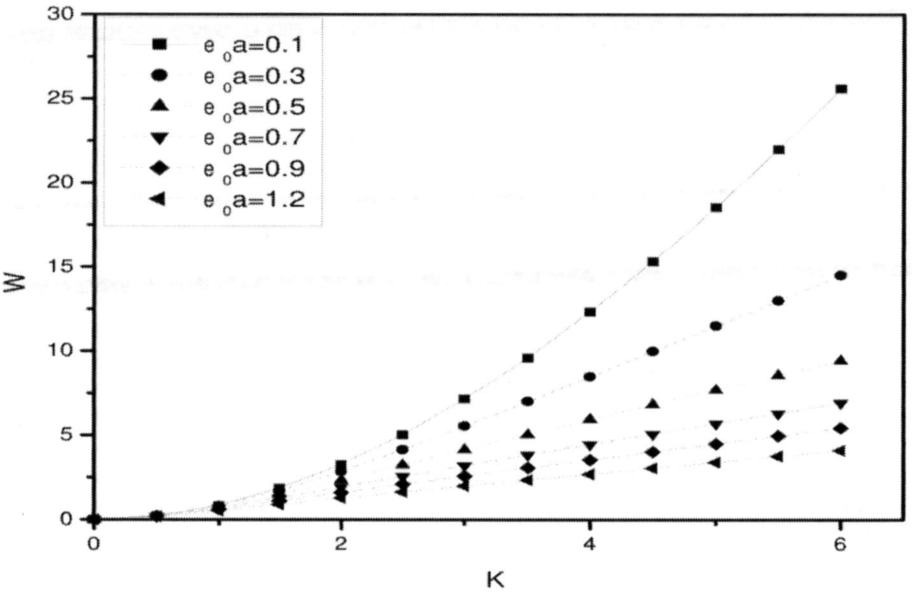

Figure 1. The effect of nonlocal parameter(e_0a)on nonlocal torsional frequency dispersion on carbon nanotube

From Eqs.(29) and (31), the relation of dimensional natural frequency and nonlocal nanoscale can be established.

6. Results and discussion

Numerical results are presented using the properties of carbon nanotubes.The following values of the parameters are used [14] $\rho = 2300kg/m^3$, $L = 10nm$, $E = 1000Gpa$, $I_p = 4.91 * 10^{-38}m^4$, $A = 7.85 * 10^{-19}m^2$.The nonlocal effects on the torsional vibration of carbon nanotubes are presented in this section.The derived expressions for torsional vibration are discussed.From Figure 1, it is found that the different values of nonlocal scale parameter(e_0a) by increasing the wave number (K) there is an increase of nonlocal frequency (ω). Also, it is observe that higher values of an nonlocal scaling parameter then the angular frequency decreases with the increase of wave number. It shows that for very small wave numbers then the influence of the nonlocal effect is also very small.

7. References

[1] Iijima S 1991 *Nature* **354** 56-58
[2] Eringen A C 1987 *Int. J. Eng. Sciences* **21** 313-342
[3] Reddy J N Pang S D 2008 *J. Appl. Phys* **103** 023511-1-16
[4] Eringen A C 1983 *Int. J. Appl.Phys.* **54** 4703-4710
[5] Eringen A C 2002 New York : Springer-Verlag
[6] Peddieson J Buchanan R G McNitt 2002 *Int. J. Engg.Sci.* **41** 305-312
[7] Lazar M Maugin A G Aifantis C E *Int. Journal of Solids Structres* **43** 1404-1421
[8] Reddy J N 2007 *Int. Journal of Engg.Sciences* **45** 288-307
[9] Lim C W and Yang Y 2009 *JoMMS* **3** 223-233
[10] Wang Q 2005 *J Appl Phys* **98** 124301
[11] Lim C W Niu J C and Yu Y M 2010 *Journal of Computational and Theoretical Nanoscience* **6** 2104-2111

[12] Wang L F Guo WL and Hu H Y 2008 *Proc.R.Soc.A-Math Phys.Eng.Sci* **464** 1423-1438
[13] Narender S and Gopalakrishnan S 2010 *Int. J.of Applied Physics* **8** 084312-1-8
[14] Pradhan S C and Phadikar J K 2009 *Journal of Sound and Vibration* **101** 206-223
[15] Murmu T Adhikari S and Wang Y C 2011 *Journal of Physica E* **43** 1276-1280
[16] Lim W C Li C and Yu J L 2012 *Journal of Sound and Vibration* **331** 2798-2808

On propagation of Love waves in an infinite transversely isotropic poroelastic layer

C Nageswara Nath[1], J Manoj Kumar[2] and S Ahmed Shah[3]

[1] Department of Mathematics, CMR Technical Campus, Hyderabad - 501 401, India
[2] Department of Mathematics, BVRIT College of Engineering for Women, Hyderabad 500 090, India
[3] Department of Mathematics, Deccan College of Engineering and Technology, Hyderabad 500 001, India

E-mail: nagesh.nath@rediffmail.com

Abstract. The present paper is devoted for the study of propagation of Loves waves in a compressible viscous liquid layer of thickness h_2 and co-efficient of viscosity η_l bounded by an infinite transversely isotropic poroelastic layer and a transversely poroelastic half-space. The equations of motion in transversely isotropic poroelastic solid have been formulated in the frame work of Biots theory. The secular equation has been obtained and is discussed in the limiting case $h_2 \to 0$, $\eta_l \to 0$ such that the ratio $h_2/\eta_l$is a constant. Further, secular equation is observed for a case of the poroelastic layer is in welded contact with half-space when the thickness of liquid layer is zero for finite co-efficient of viscosity. Several other special cases have been discussed.

1. Introduction

The study propagation of Love waves has been a topic of interest to crustal, earthquake, and engineering seismologists for several years. Recently, exploration geophysicists have become interested in Love waves as a result of current activity in shear-wave exploration. Deresiewicz [1, 2] investigated the problem of propagation of Loves waves in liquid- filled porous space. Bhangar [5] studied propagation of Love waves in a system composed of a compressible viscous liquid layer sandwiched between homogeneous, elastic layer and homogeneous isotropic half-space. Wave propagation along the interaction of a poroelastic layer and half-space is examined by Tajuddin and Ahmed [3]. Using Biot′s theory, Malla Reddy and Tajuddin [4] presented an analysis of waves propagating along the edges of a thin poroelastic flat plate of infinite length which is in a state of plane stress. Propagation of waves in a viscous liquid layer bounded by two poroelastic half-spaces is studied by Nageswara Nath et. al. [8]. Possible bonding between the poroelastic half-spaces is discussed by considering three different limiting forms of the secular equation and it is shown that these three forms respectively represent welded, smooth and loosely bonded interface of poroelastic half-spaces. It is observed that, this secular equation is same for all the three types of bonding for each permeable and impermeable surface. In the present analysis, propagation of Love waves in viscous liquid layer sandwiched between transversely isotropic poroelastic layer and half-space is studied in the frame of work Biot′s theory. The governing equations in transversely isotropic poroelastic layer have been derived. In what follows, a theory of wave propagation in a compressible viscous liquid layer of thickness

h_2 and of shear viscosity coefficient η_l bounded between a poroelastic layer of thickness h_1 and a poroelastic half-space which are homogeneous and transversely isotropic is discussed in the absence of dissipation. A secular equation for Love waves has been derived. Several limiting forms of secular equation are obtained when (i)$h_2 \to 0$, $\eta_l \to 0$ such that $\frac{h_2}{\eta_l} =$ P, a constant (ii)η_l is finite and $h_2 \to 0$ (iii)$\eta_l \to 0$ for finite thickness of h_2. It is observed that the limiting case (ii) represents Loves waves for a case of poroelastic layer in welded contact with poroelastic half-space and limiting case (iii) represents Love waves in an inviscid liquid layer of finite thickness in contact with transversely isotropic poroelastic layer.

2. Basic equations, formulation and solution of the problem

Let (x, y, z) rectangular coordinates. Consider a compressible viscous liquid layer of thickness h_2 and of shear viscosity coefficient η_l separating poroelastic layer of thickness h_1 and poroelastic half-space. Both of them are assumed to be homogeneous and transversely isotropic. The physical parameters of poroelastic layer and half-space are denoted by a superscript $j(1,2)$ enclosed in parentheses.The parameters with superscript (1), (2) refer to poroelastic layer and half-space, respectively. The parameters of liquid layer will be without any superscript. The boundaries of the viscous layer are given by $z = h_1$ and $z = h_1 + h_2$. The equations of motion of a homogeneous, transversely isotropic poroelastic solid in the presence of dissipation b are:

$$\begin{aligned}
&\left[(A+2N)\frac{\partial^2}{\partial x^2} + N\frac{\partial^2}{\partial y^2} + L\frac{\partial^2}{\partial z^2}\right]u_x + (A+2N)\frac{\partial^2 u_y}{\partial x \partial y} + (F+L)\frac{\partial^2 u_z}{\partial x \partial y} \\
&\quad + M\frac{\partial \in}{\partial x} = \frac{\partial^2}{\partial t^2}(\rho_{11}u_x + \rho_{12}U_x) + b\frac{\partial}{\partial t}(u_x - U_x), \\
&M\frac{\partial^2 u_x}{\partial x^2} + M\frac{\partial^2 u_y}{\partial x \partial y} + Q\frac{\partial^2 u_z}{\partial x \partial z} + R\frac{\partial \in}{\partial x} = \frac{\partial^2}{\partial t^2}(\rho_{12}u_x + \rho_{22}U_x) - b\frac{\partial}{\partial t}(u_x - U_x), \\
&(A+2N)\frac{\partial^2 u_x}{\partial x \partial y} + \left[N\frac{\partial^2}{\partial x^2} + (A+2N)\frac{\partial^2}{\partial y^2} + L\frac{\partial^2}{\partial z^2}\right]u_y + (F+L)\frac{\partial^2 u_z}{\partial y \partial z} \\
&\quad + M\frac{\partial \in}{\partial y} = \frac{\partial^2}{\partial t^2}(\rho_{11}u_y + \rho_{12}U_y) + b\frac{\partial}{\partial t}(u_y - U_y), \\
&M\frac{\partial^2 u_x}{\partial x \partial y} + M\frac{\partial^2 u_y}{\partial y^2} + Q\frac{\partial^2 u_z}{\partial y \partial z} + R\frac{\partial \in}{\partial y} = \frac{\partial^2}{\partial t^2}(\rho_{12}u_y + \rho_{22}U_y) - b\frac{\partial}{\partial t}(u_y - U_y), \\
&(L+F)\frac{\partial^2 u_x}{\partial x \partial y} + (L+F)\frac{\partial^2 u_y}{\partial y \partial z} + \left[L\frac{\partial^2}{\partial x^2} + L\frac{\partial^2}{\partial y^2} + C\frac{\partial^2}{\partial z^2}\right]u_z + Q\frac{\partial \in}{\partial z} = \frac{\partial^2}{\partial t^2}(\rho_{11}u_z + \rho_{12}U_z) \\
&\quad + b\frac{\partial}{\partial t}(u_z - U_z) \\
&M\frac{\partial^2 u_x}{\partial x \partial z} + M\frac{\partial^2 u_y}{\partial y \partial z} + Q\frac{\partial^2 u_z}{\partial z^2} + R\frac{\partial \in}{\partial y} = \frac{\partial^2}{\partial t^2}(\rho_{12}u_z + \rho_{22}U_z) - b\frac{\partial}{\partial t}(u_z - U_z)
\end{aligned} \tag{1}$$

where(u_x, u_y, u_z) and (U_x, U_y, U_z) are displacements of the solid and liquid media, respectively, while e and ϵ are dilatations of the solid and liquid respectively; A,N,Q,R,F,L,M and C are all poroelastic constants and are the mass coefficients [6]. The stresses σ_{ij} and the liquid pressure s of the transversely isotropic poroelastic solid[7] are

$$\begin{aligned}
\sigma_{xx} &= Pe_{xx} + Ae_{yy} + Fe_{zz} + M\in, \\
\sigma_{yy} &= Ae_{xx} + Pe_{yy} + Fe_{zz} + M\in, \\
\sigma_{zz} &= Fe_{xx} + Fe_{yy} + Ce_{zz} + Q\in, \\
\sigma_{xy} &= Ne_{xy}, \\
\sigma_{yz} &= Le_{yz}, \\
\sigma_{zx} &= Le_{zx}, \\
s &= Me_{xx} + Me_{yy} + Qe_{zz} + R\in.
\end{aligned} \tag{2}$$

For Love waves, the displacement is only along y direction thus the non-zero displacement component of the solid and liquid media are $(0,u_y,0)$ and $(0,U_y,0)$ respectively. These displacements are functions of x, y and time t. Then the equations of motion of transversely isotropic poroelastic solid [6], that is equation (1) reduces to

$$\begin{aligned} N\frac{\partial^2 u_y}{\partial x^2} + L\frac{\partial^2 u_y}{\partial z^2} &= \frac{\partial^2}{\partial t^2}\left(\rho_{11}u_y + \rho_{12}U_y\right) + b\frac{\partial}{\partial t}\left(u_y - U_y\right), \\ 0 &= \frac{\partial^2}{\partial t^2}\left(\rho_{12}u_y + \rho_{22}U_y\right) - b\frac{\partial}{\partial t}\left(u_y - U_y\right). \end{aligned} \tag{3}$$

The propagation mode shapes of solid and liquid and are taken as

$$u_y = f(z)e^{i(kx+\omega t)}, \quad U_y = g(z)e^{i(kx+\omega t)}, \tag{4}$$

where t is time,ω is circular frequency, k is wave number and i is the complex unity.
Substitution of equation (4) into equation (3) yields

$$\begin{aligned} Lf'' - k^2Nf &= -\omega^2(M_{11}f + M_{12}g), \\ 0 &= -\omega^2(M_{12}f + M_{22}g), \end{aligned} \tag{5}$$

where

$$M_{11} = \rho_{11} - \frac{ib}{\omega}, \quad M_{12} = \rho_{12} + \frac{ib}{\omega}, \quad M_{22} = \rho_{22} - \frac{ib}{\omega}. \tag{6}$$

From second equation of (5), we obtain

$$g = -\frac{M_{12}}{M_{22}}f. \tag{7}$$

Substitution of equation (7) into the first equation of (5), we obtain

$$f'' + \gamma^2 f = 0, \tag{8}$$

where

$$\gamma^2 = \left(-\frac{k^2N}{L} + \frac{\omega^2 N}{LV_3^2}\right), \quad V_3^2 = \left(\frac{NM_{22}}{M_{11}M_{22} - M_{12}^2}\right). \tag{9}$$

In equation (9), V_3 is shear wave velocity [6].
On simplification, equation (8) gives

$$f(z) = C_1e^{i\gamma z} + C_1e^{-i\gamma z}, \tag{10}$$

where C_1 and C_2 are constants. Hence from equation (7), $g(z)$ can be obtained as

$$g(z) = -\frac{M_{12}}{M_{22}}\left(C_1e^{i\gamma z} + C_1e^{-i\gamma z}\right).$$

Substituting $f(z)$ from equation (10) into first equation of (4), the displacement u_y is

$$u_y = \left(C_1e^{i\gamma z} + C_1e^{-i\gamma z}\right)e^{i(kx+\omega t)}. \tag{11}$$

Following equations (2) and (11), the only non-zero stress can be obtained as

$$\sigma_{yz} = \left[C_1\left(\frac{i\gamma L}{2}\right)e^{i\gamma z} + C_1\left(\frac{i\gamma L}{2}\right)e^{-i\gamma z}\right]e^{i(kx+\omega t)}. \tag{12}$$

From equation (4), it can be shown that the normal strains of solid and liquid are zero, hence the dilatations of solid and liquid media vanish. Since the dilatations of solid and liquid are zero, the liquid pressure s developed in the solid-liquid aggregate following equation (2) is zero.Thus, no distinction is made between a pervious and an impervious surface of the solid in case of Love waves. In the absence of body forces, the equations of motion for viscous compressible liquid are

$$\rho_l \left(\frac{\partial \mathbf{V}}{\partial t}\right) = -\nabla p + \frac{\eta_l}{3}\nabla(\nabla.\mathbf{V}) + \eta_l \nabla^2 \mathbf{V}, \tag{13}$$

where $\mathbf{V} = (u_{xl}, u_{yl}, u_{zl})$ is the velocity vector, ρ_l is density of liquid, η_l is coefficient of viscosity, p is over pressure.
For Love waves, $\mathbf{V} = (0, u_{yl}, 0)$ and $\nabla.\mathbf{V} = \mathbf{0}$. Hence, equation (13) reduces to

$$\left(\frac{\partial u_{yl}}{\partial t}\right) = \nu_l \nabla^2 u_{yl}, \tag{14}$$

where $\nu_l = \eta_l/\rho_l$ is the kinematic viscosity.
Solution of equation (14) can be obtained as

$$u_{yl} = (D_1 i\omega \cos \alpha_l z + D_2 i\omega \sin \alpha_l z)e^{i(kx+\omega t)}, \tag{15}$$

where

$$\alpha_l^2 = -k^2 - \frac{i\omega}{\nu_l}.$$

Following (15), stresses in viscous liquid layer can be shown as

$$\tau_{yz} = i\omega \eta_l \alpha_l(-D_1 \sin \alpha_l z + D_2 \cos \alpha_l z)e^{i(kx+\omega t)}. \tag{16}$$

3. Secular equation

For contact between the poroelastic half-spaces and the liquid layer, we assume that the stresses and displacement components are continuous at the interfaces z = h_1 and z = h_1+h_2 whereas, stresses vanish at free surface z = 0. Thus the boundary conditions are

$$\begin{aligned} &\sigma_{yz}^{(1)} = 0, at\ z = 0, \\ &\sigma_{yz}^{(1)} = \tau_{yz}, i\omega u_y^{(1)} = u_{yl}\ at\ z = h_1, \\ &\tau_{yz} = \sigma_{yz}^{(2)}, u_{yl} = i\omega u_y^{(2)}\ at\ z = h_1 + h_2. \end{aligned} \tag{17}$$

Eqs. (17) results in a system of five homogeneous algebraic equations in five constants. By eliminating these constants the secular equation of wave propagation in a viscous liquid layer bounded by transversely isotropic poroelastic layer and half-space can be obtained as

$$\begin{aligned} &\left[\tfrac{i\gamma^{(1)}L^{(1)}\gamma^{(2)}L^{(2)}}{2}\tan\left(\gamma^{(1)}h_1\right) - 2\eta_l^2\alpha^2\omega^2\right]\tan\left(\gamma^{(1)}h_1\right) \\ &\qquad + i\omega\eta_l\alpha_l\left[\gamma^{(1)}L^{(1)}\tan\left(\gamma^{(1)}h_1\right) - \gamma^{(2)}L^{(2)}\right] = 0. \end{aligned} \tag{18}$$

We can discuss the secular equation (18) in the following limiting cases:
(i) The thickness of liquid layer and coefficient of viscosity tends to zero such that their ratio is a constant i.e. $\eta_l \to 0$ and $h_2 \to 0$ such that $h_2/\eta_l = P$ a constant, then the secular equation (18) reduces to

$$\left[1 + \frac{1}{2\omega}\gamma^{(2)}L^{(2)}P\right]\gamma^{(1)}L^{(1)}\tan\left(\gamma^{(1)}h_1\right) = i\gamma^{(2)}L^{(2)}. \tag{19}$$

(ii) The thickness of liquid layer is zero for a finite coefficient of viscosity, the secular equation (18) reduces to

$$\gamma^{(1)}L^{(1)}\tan\left(\gamma^{(1)}h_1\right) = i\gamma^{(2)}L^{(2)}. \tag{20}$$

Here, equation (20) represents secular equation of Loves waves for a case of poroelastic layer in welded contact with poroelastic half-space, where each of them is homogeneous and transversely isotropic.

(iii) When co-efficient of viscosity tends to zero for finite thickness of liquid layer, the secular equation(19) reduces to

$$\tan\left(\gamma^{(1)}h_1\right) = 0. \tag{21}$$

From equation (21), it is clear that Love wave energy is uncoupled with poroelastic half-space hence it represents secular equation of Love waves in an inviscid liquid layer of finite thickness in contact with transversely isotropic poroelastic layer.

4. Results and discussions

Secular equations (19)-(21) are investigated numerically by considering two distinct poroelastic materials with parameters N=6.4 Gpa, L= 8 Gpa (poroelastic plate) and N=3 Gpa, L= 5 Gpa (poroelastic half-space). Poroelastic medium is dissipative in nature and thus the wave number k is complex. The waves generated obey diffusion type process and therefore get attenuated. Let $k = k_r + ik_i$, where k_r is real and k_i is the imaginary part of the wave number k. The real and imaginary part of the wave number corresponds to propagation and attenuation of waves. Hence the phase velocity C_p and attenuation δ are

$$C_p = \frac{\omega}{\mathrm{Re}(k)}, \quad \delta = \frac{\mathrm{Im}(k)}{\mathrm{Re}(k)}. \tag{22}$$

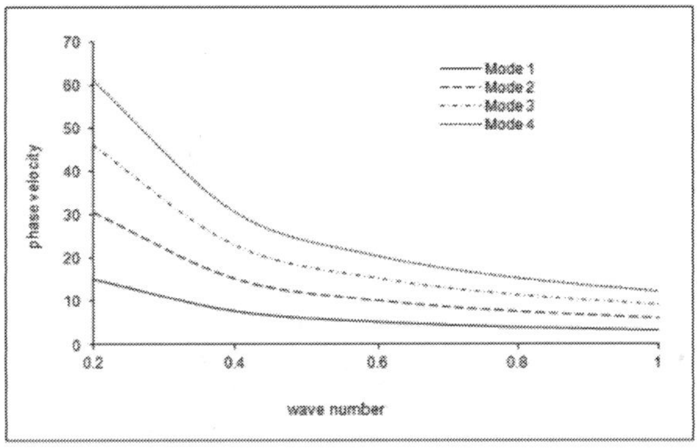

Figure 1. Phase velocity as a function of non-dimensional wave number - Viscous fluid bounded between poroelastic layer and half-space when P is a constant.

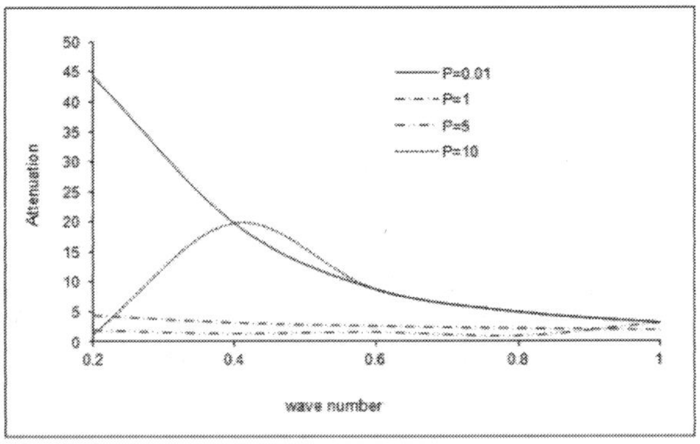

Figure 2. Attenuation as a function of non-dimensional wave number - Viscous fluid bounded between poroelastic layer and half-space when P is a constant.

The variations in phase velocity against wave number in viscous liquid layer bounded between poroelastic layer and poroelastic half-space both are transversely isotropic is depicted in figure 1. First four modes are plotted for fixed constant P=1. The phase velocity decreases gradually as wave number increases over the range 0.2 to 1. Figure 2 presents attenuation for different values of for fixed ratio P i.e. for P= 0.1, 1, 5 and 10. There is gradual decrease in attenuation for P=0.1 and it is steady for P=1 and 5. When P=10, there is an increase in attenuation in the range of wave number 0.2 to 0.4 and then a steady decrease till 1. Phase velocity at the interface of poroelastic layer and half-space as a function of wave number is presented in figure 3. From figure 1 it is clear that phase velocity in this case is same as that of viscous liquid layer bounded between poroelastic layer and half-space. Hence, it can be observed that there is no effect of viscous liquid layer on propagation of Love waves. Attenuation at the interface of poroelastic layer and half-space is presented in figure 4. It is observed that attenuation is high when wave number is 0.2 and it decreases gradually as the wave number increases from 0.2 to 1. Phase velocity in an inviscid liquid layer of finite thickness in contact with transversely isotropic

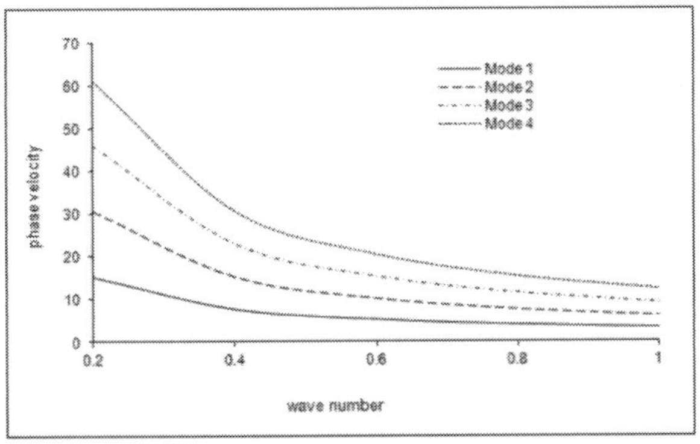

Figure 3. Phase velocity as a function of non-dimensional wave number - Interface of poroelastic layer and half-space.

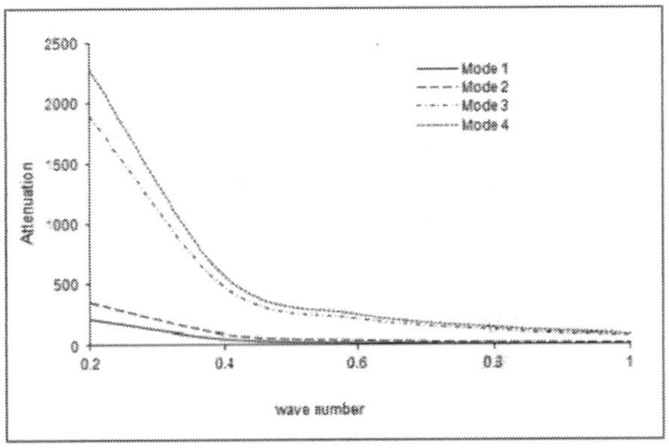

Figure 4. Attenuation as a function of non-dimensional wave number - Interface of poroelastic layer and half-space.

poroelastic layer is depicted in figure 5. A similar behavior in phase velocity is observed as in the earlier cases i.e. a gradual decrease in the range 0.2 to 1 of wave number.

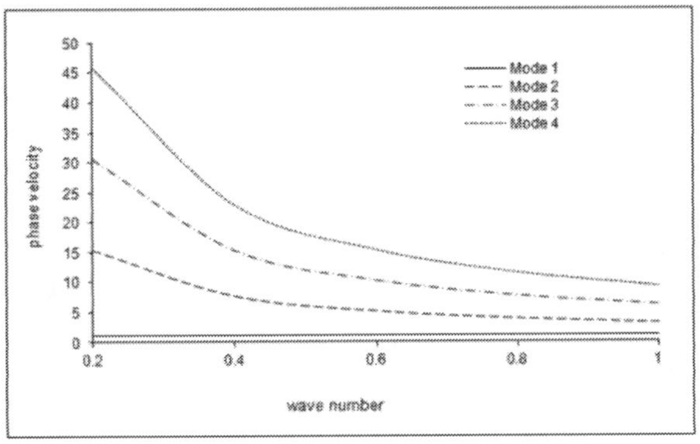

Figure 5. Phase velocity as a function of non-dimensional wave number - Inviscid liquid layer in contact with transversely isotropic poroelastic layer.

5. Conclusion

Propagation of Love waves in viscous liquid layer bounded between transversely isotropic poroelastic layer and half-space is discussed. Several limiting cases of secular equation have been discussed. For the limiting case of co-efficient of viscosity tends to zero for finite thickness of liquid layer, it is observed that Love wave energy is uncoupled with poroelastic half-space. It is found that there is no influence of viscous liquid layer on phase velocity of Love waves when the liquid layer is bounded between poroelastic layer and half-space.

6. References

[1] Deresiewicz H 1960 The effect of boundaries on wave propagation in a liquid-filled porous solid-I *Bull. Seis. Soc. Am.* **50** 599-607
[2] Deresiewicz H and Rice J T 1962 The effect of boundaries on wave propagation in a liquid-filled porous solid-II *Bull. Seis. Soc. Am.* **52** 595-626
[3] Tajuddin M and Ahmed S I 1991 Dynamic interaction of a poroelastic layer and a half-space *J. Acoust. Soc. Am.* **89** 1169-75
[4] Malla Reddy P and Tajuddin M. 2003 Edge waves in poroelastic plate under plane-stress conditions *J. Acoust. Soc. Am.* **114** 185-93
[5] Banghar A R 1978 On propagation and attenuation of Love waves *Proc. Indian Acad. Sci.* **88** 133-46
[6] Biot M A 1956 The theory of propagation of elastic waves in fluid-saturated porous solid *J. Acous. Soc. Am.* **28** 168-78
[7] Biot M A and Willis D G 1957 The elastic co-efficients of the theory of consolidation *Journal of Applied Mechanics* **24** 594-601
[8] NageswaraNath C, Manoj Kumar J and Tajuddin M 2011 On the parametric model of loose bonding between two poroelastic half spaces *Journal of Vibration and Control* **18** 1261-74

Investigation of flexural vibrations in poroelastic solid sphere in the presence of static stresses

Rajitha Gurijala

Department of Mathematics, Kakatiya University, Warangal 506009, Telangana, India.

E-mail: `rajitha.akshu@gmail.com`

Abstract. In this paper, the three dimensional wave propagation in a homogeneous isotropic poroelastic solid sphere is investigated in the presence of static stresses. The frequency equation is obtained in the framework of Biot's theory. Non-dimensional phase velocity is computed as a function of uniaxial static stress for fixed wavenumber. For the numerical results, three types of poroelastic spherical solids are employed, and the results are presented graphically.

1. Introduction

The analysis for transient problems of spherical structures is important and interesting research fields for engineers and scientists. The applications for homogeneous solid sphere have continuously increased in some engineering areas, including aerospace, frequency control filters, chemical vessels, information storage devices, and signal processing devices. Even in the human body most of the bones are in spherical shape. Mott investigated elastic motion of an isotropic medium in the presence of body forces and static stresses [1]. In the paper [1], Mott derived equations of elastic motion which include the effect of static surface forces and body forces. When the body forces imposed on a solid which is at rest gives static stresses and strains. Effect of static axial stress upon the velocity of the lowest-order flexural mode is investigated in [2]. Employing Biot's theory of poroelasticity [3], detailed investigation of wave propagation is given in [4]. From the mathematics perspective of poroelastic materials, several authors are investigated wave propagation phenomena which characterize Biot's theory [5-9]. Flexural vibrations in poroelastic elliptic cone against the angle made by the major axis of the cone in spheroconal coordinate system are given in [10]. Shah and Tajuddin [11] discussed torsional vibrations of poroelastic spheroidal shells. In this paper, authors derived frequency equations for poroelastic thin spherical shell, thick spherical shell, poroelastic solid sphere, and concluded that the frequency is same for all the three cases. Axially symmetric vibrations of fluid filled poroelastic spherical shell are studied by Shah and Tajuddin [12]. In the said paper, frequency equations are derived for radial and rotatory vibrations of fluid-filled and empty poroelastic spherical shells. A comparative study is made between the modes of composite spherical shell and its ring modes [13]. Shanker et. al. [14] investigated vibration analysis of a poroelastic composite hollow sphere. In this paper, authors derived frequency equations for radial and rotatory vibrations of fluid filled and empty poroelastic shells with rigid core. Torsional vibrations of thick-walled hollow poroelastic spheres are studied by Ahmed Shah and Tajuddin [15]. In this paper, authors derived complex frequencies of torsional vibrations of thick-walled hollow poroelastic spheres for different dissipations and it is concluded that as the dissipation increases, the propagation increases while the attenuation remains almost same. Flexural vibrations of poroelastic solids in presence of static stress are studied by Rajitha et al. [16]. In the paper, governing equations are derived in the presence of static stress, which were not available in the earlier literature. However, in all the above papers, static stress is not employed in the case of flexural vibrations of poroelastic solid sphere. Therefore, in this paper, the same is investigated in the framework of Biot's theory.

This paper is organized as follows. In section 2, governing equations, and solution of the problem are given. In section 3, the case of static uniaxial stress is considered. Numerical results are described in section 4. Finally, conclusion is given in section 5.

2. Governing equations and solution of the problem

Consider an isotropic poroelastic solid sphere in spherical coordinate system (r, θ, ϕ). Let $\vec{u}(u, v, w)$ and $\vec{U}(U, V, W)$ be the solid and fluid displacements. When the body forces are large, displacements will be large. Consequently second order coupling between stresses and strains cannot be ignored. Effective stresses must be inserted in the place of usual ones. The effective stress-strain relations are as follows [1]:

$$\begin{aligned}
\sigma'_{rr} &= (\sigma_{rr} + s) - \sigma_{r\theta}\frac{\partial u}{\partial \theta} - \sigma_{r\phi}\frac{\partial u}{\partial \phi},\\
\sigma'_{r\theta} &= \sigma_{r\theta} - (\sigma_{rr} + s)\frac{\partial v}{\partial r} - \sigma_{r\phi}\frac{\partial v}{\partial \phi},\\
\sigma'_{r\phi} &= \sigma_{r\phi} - (\sigma_{rr} + s)\frac{\partial w}{\partial r} - \sigma_{r\theta}\frac{\partial w}{\partial \theta},\\
\sigma'_{\theta\theta} &= (\sigma_{\theta\theta} + s) - \sigma_{\theta r}\frac{\partial v}{\partial r} - \sigma_{\theta\phi}\frac{\partial v}{\partial \phi},\\
\sigma'_{\theta r} &= \sigma_{\theta r} - (\sigma_{\theta\theta} + s)\frac{\partial u}{\partial \theta} - \sigma_{\theta\phi}\frac{\partial u}{\partial \phi},\\
\sigma'_{\theta\phi} &= \sigma_{\theta\phi} - \sigma_{\theta r}\frac{\partial w}{\partial r} - (\sigma_{\theta\theta} + s)\frac{\partial w}{\partial \theta},\\
\sigma'_{\phi r} &= \sigma_{\phi r} - \sigma_{\phi\theta}\frac{\partial u}{\partial \theta} - (\sigma_{\phi\phi} + s)\frac{\partial u}{\partial \phi},\\
\sigma'_{\phi\theta} &= \sigma_{\phi\theta} - \sigma_{\phi r}\frac{\partial v}{\partial r} - (\sigma_{\phi\phi} + s)\frac{\partial v}{\partial \phi},\\
\sigma'_{\phi\phi} &= (\sigma_{\phi\phi} + s) - \sigma_{\phi r}\frac{\partial w}{\partial r} - \sigma_{\phi\theta}\frac{\partial w}{\partial \theta}.
\end{aligned} \tag{1}$$

The expressions of effective stresses involve only shear components. As there are no shear components for the case of fluid pressure, the expression for the fluid pressure remains the same. In equation (1), usual stresses σ_{ij} and fluid pressure s are given under [3].

$$\begin{aligned}
\sigma_{ij} &= 2Ne_{ij} + (Ae + Q\varepsilon)\delta_{ij}, \qquad (i, j = r, \theta, \phi),\\
s &= Qe + R\varepsilon.
\end{aligned} \tag{2}$$

In equation (2), e_{ij}'s strain displacements, A, N, Q, R are poroelastic constants, e and ε are the dilatations of solid and fluid, respectively, and δ_{ij} is the well-known Kronecker delta function. Substitution of effective stresses for usual stresses, the equations of motion obtained in the following manner.

$$\frac{\partial}{\partial r}(\sigma'_{rr}(1+\frac{\partial v}{\partial\theta}+\frac{\partial w}{\partial\phi}))+\frac{1}{r}\frac{\partial}{\partial\theta}(\sigma'_{r\theta}(1+\frac{\partial u}{\partial r}+\frac{\partial w}{\partial\phi}))+\frac{1}{rsin\theta}\frac{\partial}{\partial\phi}(\sigma'_{r\phi}(1+\frac{\partial u}{\partial r}+\frac{\partial v}{\partial\theta}))+\frac{2}{r}\sigma'_{rr}(1+\frac{\partial v}{\partial\theta} +\frac{\partial w}{\partial\phi})-\frac{1}{r}\sigma'_{\theta\theta}(1+\frac{\partial u}{\partial r}+\frac{\partial w}{\partial\phi})-\frac{1}{r}\sigma'_{\phi\phi}(1+\frac{\partial u}{\partial r}+\frac{\partial v}{\partial\theta})+\frac{1}{r}\sigma'_{r\theta}(1+\frac{\partial u}{\partial r}+\frac{\partial w}{\partial\phi})cot\theta+F_r=\frac{\partial^2}{\partial t^2}(\rho_{11}u +\rho_{12}U),$$

$$\frac{\partial}{\partial r}(\sigma'_{r\theta}(1+\frac{\partial u}{\partial r}+\frac{\partial w}{\partial\phi}))+\frac{1}{r}\frac{\partial}{\partial\theta}(\sigma'_{\theta\theta}(1+\frac{\partial u}{\partial r}+\frac{\partial w}{\partial\phi}))+\frac{1}{rsin\theta}\frac{\partial}{\partial\phi}(\sigma'_{\theta\phi}(1+\frac{\partial u}{\partial r}+\frac{\partial v}{\partial\theta}))+\frac{1}{r}\sigma'_{\theta\theta}(1+\frac{\partial u}{\partial r} +\frac{\partial w}{\partial\phi})-\frac{cot\theta}{r}\sigma'_{\phi\phi}(1+\frac{\partial u}{\partial r}+\frac{\partial v}{\partial\theta})+3\sigma'_{r\theta}(1+\frac{\partial u}{\partial r}+\frac{\partial w}{\partial\phi})+F_\theta=\frac{\partial^2}{\partial t^2}(\rho_{11}v+\rho_{12}V),$$

$$\frac{\partial}{\partial r}(\sigma'_{r\phi}(1+\frac{\partial u}{\partial r}+\frac{\partial v}{\partial\theta}))+\frac{1}{r}\frac{\partial}{\partial\theta}(\sigma'_{\theta\phi}(1+\frac{\partial u}{\partial r}+\frac{\partial v}{\partial\theta}))+\frac{1}{rsin\theta}\frac{\partial}{\partial\phi}(\sigma'_{\phi\phi}(1+\frac{\partial u}{\partial r}+\frac{\partial v}{\partial\theta}))+\frac{2}{rtan\theta}(\sigma'_{\theta\phi}(1 +\frac{\partial u}{\partial r}+\frac{\partial v}{\partial\theta}))+\frac{3}{r}\sigma'_{r\phi}(1+\frac{\partial u}{\partial r}+\frac{\partial v}{\partial\theta})+F_\phi=\frac{\partial^2}{\partial t^2}(\rho_{11}w+\rho_{12}W),$$

$$Q\frac{\partial e}{\partial r}+R\frac{\partial\varepsilon}{\partial r}=\frac{\partial^2}{\partial t^2}(\rho_{12}u+\rho_{22}U),$$

$$Q\frac{\partial e}{\partial\theta}+R\frac{\partial\varepsilon}{\partial\theta}=\frac{\partial^2}{\partial t^2}(\rho_{12}v+\rho_{22}V),$$

$$Q\frac{\partial e}{\partial\phi}+R\frac{\partial\varepsilon}{\partial\phi}=\frac{\partial^2}{\partial t^2}(\rho_{12}w+\rho_{22}W). \tag{3}$$

In equation(3), ρ_{ij} are mass coefficients, F_r, F_θ, and F_ϕ are the components of the body force vector $\overrightarrow{F}$. Equations of motion and the boundary conditions together form the basis for the solution of wave propagation problems in poroelastic solids. The equations (2) and (3) have to be satisfied at every point of the poroelastic solid, and on surface of that solid. Thus the solution of the problem can be completely determined. To originate connection between the time variant and the static poroelastic quantities, the displacements and the stresses are written in the following form [1, 16]:

$$\begin{aligned}
u &= \sum u_n(r,\theta,\phi,\omega_n t) = u_0+u_1+u_2+...,\\
v &= \sum v_n(r,\theta,\phi,\omega_n t) = v_0+v_1+v_2+...,\\
w &= \sum w_n(r,\theta,\phi,\omega_n t) = w_0+w_1+w_2+...,\\
U &= \sum U_n(r,\theta,\phi,\omega_n t) = U_0+U_1+U_2+...,\\
V &= \sum V_n(r,\theta,\phi,\omega_n t) = V_0+V_1+V_2+...,\\
W &= \sum W_n(r,\theta,\phi,\omega_n t) = W_0+W_1+W_2+...,\\
\sigma_{ij} &= \sum \sigma_{ij_n}(r,\theta,\phi,\omega_n t) = \sigma_{ij_0}+\sigma_{ij_1}+\sigma_{ij_2}+...,
\end{aligned} \tag{4}$$

where ω_n is the nth angular frequency. Using equations(1), (2) and (4) in the equation(3), we get the following static equations, $(n=0)$:

$$\frac{\partial}{\partial r}(((\sigma_{rr_0}+s)-\sigma_{r\theta_0}\frac{\partial u_0}{\partial\theta}-\sigma_{r\phi_0}\frac{\partial u_0}{\partial\phi})(1+\frac{\partial v_0}{\partial\theta}+\frac{\partial w_0}{\partial\phi}))+\frac{1}{r}\frac{\partial}{\partial\theta}((\sigma_{r\theta_0}-(\sigma_{rr_0}+s)\frac{\partial v_0}{\partial\theta}-\sigma_{\theta\phi_0}\frac{\partial v_0}{\partial\phi}) (1+\frac{\partial u_0}{\partial r}+\frac{\partial w_0}{\partial\phi}))+\frac{1}{rsin\theta}\frac{\partial}{\partial\phi}((\sigma_{r\phi_0}-(\sigma_{rr_0}+s)\frac{\partial w_0}{\partial r}-\sigma_{r\theta_0}\frac{\partial w_0}{\partial\theta})(1+\frac{\partial u_0}{\partial r}+\frac{\partial v_0}{\partial\theta}))+\frac{2}{r}(((\sigma_{rr_0}+s) -\sigma_{r\theta_0}\frac{\partial u_0}{\partial\theta}-\sigma_{r\phi_0}\frac{\partial u_0}{\partial\phi})(1+\frac{\partial v_0}{\partial r}+\frac{\partial w_0}{\partial\phi}))-\frac{1}{r}(((\sigma_{\theta\theta_0}+s)-\sigma_{\theta r_0}\frac{\partial v_0}{\partial r}-\sigma_{\theta\phi_0}\frac{\partial v_0}{\partial\phi})(1+\frac{\partial u_0}{\partial r}+\frac{\partial w_0}{\partial\phi}))- \frac{1}{r}(((\sigma_{\phi\phi_0}+s)-\sigma_{\phi r_0}\frac{\partial w_0}{\partial r}-\sigma_{\phi\theta_0}\frac{\partial w_0}{\partial\theta})(1+\frac{\partial u_0}{\partial r}+\frac{\partial v_0}{\partial\theta}))+\frac{cot\theta}{r}((\sigma_{r\theta_0}-(\sigma_{\theta\theta_0}+s)\frac{\partial v_0}{\partial\theta}-\sigma_{\theta\phi_0}\frac{\partial v_0}{\partial\phi}) (1+\frac{\partial u_0}{\partial r}+\frac{\partial w_0}{\partial\phi}))+F_{r_0}=0,$$

Journal of Physics: Conference Series **662** (2015) 012005 doi:10.1088/1742-6596/662/1/012005

$$\frac{\partial}{\partial r}((\sigma_{r\theta_0}-(\sigma_{rr_0}+s)\frac{\partial v_0}{\partial r}-\sigma_{r\phi_0}\frac{\partial v_0}{\partial \phi})(1+\frac{\partial u_0}{\partial r}+\frac{\partial w_0}{\partial \phi}))+\frac{1}{r}\frac{\partial}{\partial \theta}(((\sigma_{\theta\theta_0}+s)-\sigma_{\theta r_0}-\frac{\partial v_0}{\partial r}-\sigma_{\theta\phi_0}\frac{\partial v_0}{\partial \phi})$$
$$(1+\frac{\partial u_0}{\partial r}+\frac{\partial w_0}{\partial \phi}))+\frac{1}{rsin\theta}\frac{\partial}{\partial \phi}((\sigma_{\theta\phi_0}-(\sigma_{\theta\theta_0}+s)\frac{\partial w_0}{\partial \theta}-\sigma_{\theta r_0}\frac{\partial w_0}{\partial r})(1+\frac{\partial u_0}{\partial r}+\frac{\partial v_0}{\partial \theta}))+\frac{cot\theta}{r}(((\sigma_{\theta\theta_0}+s)$$
$$-\sigma_{\theta r_0}\frac{\partial v_0}{\partial r}-\sigma_{\theta\phi_0}\frac{\partial v_0}{\partial \phi})(1+\frac{\partial u_0}{\partial r}+\frac{\partial w_0}{\partial \phi}))-\frac{cot\theta}{r}(((\sigma_{\phi\phi_0}+s)-\sigma_{\phi r_0}\frac{\partial w_0}{\partial r}-\sigma_{\phi\theta_0}\frac{\partial w_0}{\partial \theta})(1+\frac{\partial u_0}{\partial r}+\frac{\partial v_0}{\partial \theta}))$$
$$+\frac{3}{r}((\sigma_{r\theta_0}-(\sigma_{rr_0}+s)\frac{\partial v_0}{\partial r}-\sigma_{r\phi_0}\frac{\partial v_0}{\partial \phi})(1+\frac{\partial u_0}{\partial r}+\frac{\partial w_0}{\partial \phi}))+F_{\theta_0}=0,$$

$$\frac{\partial}{\partial r}((\sigma_{r\phi_0}-(\sigma_{rr_0}+s)\frac{\partial w_0}{\partial r}-\sigma_{r\theta_0}\frac{\partial w_0}{\partial \theta})(1+\frac{\partial u_0}{\partial r}+\frac{\partial v_0}{\partial \theta}))+\frac{1}{r}\frac{\partial}{\partial \theta}((\sigma_{\theta\phi_0}-(\sigma_{\theta\theta_0}+s)\frac{\partial w_0}{\partial \theta}-\sigma_{\theta r_0}\frac{\partial w_0}{\partial r})$$
$$(1+\frac{\partial u_0}{\partial r}+\frac{\partial v_0}{\partial \theta}))+\frac{1}{rsin\theta}\frac{\partial}{\partial \phi}(((\sigma_{\phi\phi_0}+s)-\sigma_{\phi r_0}\frac{\partial w_0}{\partial r}-\sigma_{\phi\theta_0}\frac{\partial w_0}{\partial \theta})(1+\frac{\partial u_0}{\partial r}+\frac{\partial v_0}{\partial \theta}))+\frac{2cot\theta}{r}((\sigma_{\theta\phi_0}-$$
$$\sigma_{\theta r_0}\frac{\partial w_0}{\partial r}-(\sigma_{\theta\theta_0}+s)\frac{\partial w_0}{\partial \theta})(1+\frac{\partial u_0}{\partial r}+\frac{\partial v_0}{\partial \theta}))+\frac{3}{r}((\sigma_{r\phi_0}-(\sigma_{rr_0}+s)\frac{\partial w_0}{\partial r}-\sigma_{r\theta_0}\frac{\partial w_0}{\partial \theta})(1+\frac{\partial u_0}{\partial r}+\frac{\partial v_0}{\partial \theta}))$$
$$+F_{\phi_0}=0,$$

$$Q\frac{\partial e}{\partial r}+R\frac{\partial \varepsilon}{\partial r}=0,$$

$$Q\frac{\partial e}{\partial \theta}+R\frac{\partial \varepsilon}{\partial \theta}=0,$$

$$Q\frac{\partial e}{\partial \phi}+R\frac{\partial \varepsilon}{\partial \phi}=0. \qquad (5)$$

In the above,

$$e=\frac{\partial^2 u_0}{\partial r^2}+\frac{2}{r}\frac{\partial u_0}{\partial r}+\frac{cot\theta}{r^2}\frac{\partial v_0}{\partial \theta}+\frac{1}{r^2}\frac{\partial^2 v_0}{\partial \theta^2}+\frac{1}{r^2sin^2\theta}\frac{\partial^2 w_0}{\partial \phi^2},$$

and

$$\varepsilon=\frac{\partial^2 U_0}{\partial r^2}+\frac{2}{r}\frac{\partial U_0}{\partial r}+\frac{cot\theta}{r^2}\frac{\partial V_0}{\partial \theta}+\frac{1}{r^2}\frac{\partial^2 V_0}{\partial \theta^2}+\frac{1}{r^2sin^2\theta}\frac{\partial^2 W_0}{\partial \phi^2}.$$

Similarly, the harmonic equations $n=1$ are obtained, which are:

$$\frac{\partial}{\partial r}(((\sigma_{rr_1}+s)-\sigma_{r\theta_1}\frac{\partial u_0}{\partial \theta}-\sigma_{r\theta_0}\frac{\partial u_1}{\partial \theta}-\sigma_{r\phi_1}\frac{\partial u_0}{\partial \phi}-\sigma_{r\phi_0}\frac{\partial u_1}{\partial \phi})(1+\frac{\partial v_0}{\partial \theta}+\frac{\partial w_0}{\partial \phi})+((\sigma_{rr_0}+s)-\sigma_{r\theta_0}$$
$$\frac{\partial u_0}{\partial \theta}-\sigma_{r\phi_0}\frac{\partial u_0}{\partial \phi})(\frac{\partial v_1}{\partial \theta}+\frac{\partial w_1}{\partial \phi}))+\frac{1}{r}\frac{\partial}{\partial \theta}((\sigma_{r\theta_1}-(\sigma_{rr_1}+s)\frac{\partial v_0}{\partial \theta}-(\sigma_{rr_0}+s)\frac{\partial v_1}{\partial \theta}-\sigma_{\theta\phi_1}\frac{\partial v_0}{\partial \phi}-\sigma_{\theta\phi_0}\frac{\partial v_1}{\partial \phi})$$
$$(1+\frac{\partial u_0}{\partial r}+\frac{\partial w_0}{\partial \phi})+(\sigma_{r\theta_0}-(\sigma_{rr_0}+s)\frac{\partial v_0}{\partial \theta}-\sigma_{\theta\phi_0}\frac{\partial v_0}{\partial \phi})(\frac{\partial u_1}{\partial r}+\frac{\partial w_1}{\partial \phi}))+\frac{1}{rsin\theta}\frac{\partial}{\partial \phi}((\sigma_{r\phi_1}-(\sigma_{rr_1}+s)$$
$$\frac{\partial w_0}{\partial r}-(\sigma_{rr_0}+s)\frac{\partial w_1}{\partial r}-\sigma_{r\theta_1}\frac{\partial w_0}{\partial \theta}-\sigma_{r\theta_0}\frac{\partial w_1}{\partial \theta})(1+\frac{\partial u_0}{\partial r}+\frac{\partial v_0}{\partial \theta})+(\sigma_{r\phi_0}-(\sigma_{rr_0}+s)\frac{\partial w_0}{\partial r}-\sigma_{r\theta_0}\frac{\partial w_0}{\partial \theta})$$
$$(\frac{\partial u_1}{\partial r}+\frac{\partial v_1}{\partial \theta}))+\frac{2}{r}(((\sigma_{rr_1}+s)-\sigma_{r\theta_1}\frac{\partial u_0}{\partial \theta}-\sigma_{r\theta_0}\frac{\partial u_1}{\partial \theta}-\sigma_{r\phi_1}\frac{\partial u_0}{\partial \phi}-\sigma_{r\phi_0}\frac{\partial u_1}{\partial \phi})(1+\frac{\partial v_0}{\partial r}+\frac{\partial w_0}{\partial \phi})+$$
$$((\sigma_{rr_0}+s)-\sigma_{r\theta_0}\frac{\partial u_0}{\partial \theta}-\sigma_{r\phi_0}\frac{\partial u_0}{\partial \phi})(\frac{\partial v_1}{\partial r}+\frac{\partial w_1}{\partial \phi}))-\frac{1}{r}(((\sigma_{\theta\theta_1}+s)-\sigma_{\theta r_1}\frac{\partial v_0}{\partial r}-\sigma_{\theta r_0}\frac{\partial v_1}{\partial r}-\sigma_{\theta\phi_1}\frac{\partial v_0}{\partial \phi}$$
$$-\sigma_{\theta\phi_0}\frac{\partial v_1}{\partial \phi})(1+\frac{\partial u_0}{\partial r}+\frac{\partial w_0}{\partial \phi})+((\sigma_{\theta\theta_0}+s)-\sigma_{\theta r_0}\frac{\partial v_0}{\partial r}-\sigma_{\theta\phi_0}\frac{\partial v_0}{\partial \phi})(\frac{\partial u_1}{\partial r}+\frac{\partial w_1}{\partial \phi}))-\frac{1}{r}(((\sigma_{\phi\phi_1}+s)$$
$$-\sigma_{\phi r_1}\frac{\partial w_0}{\partial r}-\sigma_{\phi r_0}\frac{\partial w_1}{\partial r}-\sigma_{\phi\theta_1}\frac{\partial w_0}{\partial \theta}-\sigma_{\phi\theta_0}\frac{\partial w_1}{\partial \theta})(1+\frac{\partial u_0}{\partial r}+\frac{\partial v_0}{\partial \theta})+((\sigma_{\phi\phi_0}+s)-\sigma_{\phi r_0}\frac{\partial w_0}{\partial r}-$$
$$\sigma_{\phi\theta_0}\frac{\partial w_0}{\partial \theta})(\frac{\partial u_1}{\partial r}+\frac{\partial v_1}{\partial \theta}))+\frac{cot\theta}{r}((\sigma_{r\theta_1}-(\sigma_{\theta\theta_1}+s)\frac{\partial v_0}{\partial \theta}-(\sigma_{\theta\theta_0}+s)\frac{\partial v_1}{\partial \theta}-\sigma_{\theta\phi_1}\frac{\partial v_0}{\partial \phi}-\sigma_{\theta\phi_0}\frac{\partial v_1}{\partial \phi})(1+$$
$$\frac{\partial u_0}{\partial r}+\frac{\partial w_0}{\partial \phi})+(\sigma_{r\theta_0}-(\sigma_{\theta\theta_0}+s)\frac{\partial v_0}{\partial \theta}-\sigma_{\theta\phi_0}\frac{\partial v_0}{\partial \phi})(\frac{\partial u_1}{\partial r}+\frac{\partial w_1}{\partial \phi}))+F_{r_1}=\frac{\partial^2}{\partial t^2}(\rho_{11}u_1+\rho_{12}U_1),$$

$$\frac{\partial}{\partial r}((\sigma_{r\theta_1}-(\sigma_{rr_1}+s)\frac{\partial v_0}{\partial r}-(\sigma_{rr_0}+s)\frac{\partial v_1}{\partial r}-\sigma_{r\phi_1}\frac{\partial v_0}{\partial \phi}-\sigma_{r\phi_0}\frac{\partial v_1}{\partial \phi})(1+\frac{\partial u_0}{\partial r}+\frac{\partial w_0}{\partial \phi})+(\sigma_{r\theta_0}-(\sigma_{rr_0}$$
$$+s)\frac{\partial v_0}{\partial r}-\sigma_{r\phi_0}\frac{\partial v_0}{\partial \phi})(\frac{\partial u_1}{\partial r}+\frac{\partial w_1}{\partial \phi}))+\frac{1}{r}\frac{\partial}{\partial \theta}(((\sigma_{\theta\theta_0}+s)-\sigma_{\theta r_1}\frac{\partial v_0}{\partial r}-\sigma_{\theta r_0}\frac{\partial v_1}{\partial r}-\sigma_{\theta\phi_0}\frac{\partial v_1}{\partial \phi}-\sigma_{\theta\phi_1}\frac{\partial v_0}{\partial \phi})(1$$
$$+\frac{\partial u_0}{\partial r}+\frac{\partial w_0}{\partial \phi})+((\sigma_{\theta\theta_0}+s)-\sigma_{\theta r_0}\frac{\partial v_0}{\partial r}-\sigma_{\theta\phi_0}\frac{\partial v_0}{\partial \phi})(\frac{\partial u_1}{\partial r}+\frac{\partial w_1}{\partial \phi}))+\frac{1}{rsin\theta}\frac{\partial}{\partial \phi}((\sigma_{\theta\phi_1}-(\sigma_{\theta\theta_1}+s)\frac{\partial w_0}{\partial \theta}$$
$$-(\sigma_{\theta\theta_0}+s)\frac{\partial w_1}{\partial \theta}-\sigma_{\theta r_1}\frac{\partial w_0}{\partial r}-\sigma_{\theta r_0}\frac{\partial w_1}{\partial r})(1+\frac{\partial u_0}{\partial r}+\frac{\partial v_0}{\partial \theta})+(\sigma_{\theta\phi_0}-(\sigma_{\theta\theta_0}+s)\frac{\partial w_0}{\partial \theta}-\sigma_{\theta r_0}\frac{\partial w_0}{\partial r})(\frac{\partial u_1}{\partial r}$$
$$+\frac{\partial v_1}{\partial \theta}))+\frac{cot\theta}{r}((\sigma_{\theta\theta_1}+s)-\sigma_{\theta r_1}\frac{\partial v_0}{\partial r}-\sigma_{\theta r_0}\frac{\partial v_1}{\partial r}-\sigma_{\theta\phi_1}\frac{\partial v_0}{\partial \phi}-\sigma_{\theta\phi_0}\frac{\partial v_1}{\partial \phi})(1+\frac{\partial u_0}{\partial r}+\frac{\partial w_0}{\partial \phi})+((\sigma_{\theta\theta_0}$$
$$+s)-\sigma_{\theta r_0}\frac{\partial v_0}{\partial r}-\sigma_{\theta\phi_0}\frac{\partial v_0}{\partial \phi})(\frac{\partial u_1}{\partial r}+\frac{\partial w_1}{\partial \phi}))-\frac{cot\theta}{r}(((\sigma_{\phi\phi_1}+s)-\sigma_{\phi r_1}\frac{\partial w_0}{\partial r}-\sigma_{\phi r_0}\frac{\partial w_1}{\partial r}-\sigma_{\phi\theta_1}\frac{\partial w_0}{\partial \theta}-$$
$$\sigma_{\phi\theta_0}\frac{\partial w_1}{\partial \theta})(1+\frac{\partial u_0}{\partial r}+\frac{\partial v_0}{\partial \theta})+((\sigma_{\phi\phi_0}+s)-\sigma_{\phi r_0}\frac{\partial w_0}{\partial r}-\sigma_{\phi\theta_0}\frac{\partial w_0}{\partial \theta})(\frac{\partial u_1}{\partial r}+\frac{\partial v_1}{\partial \theta}))+\frac{3}{r}((\sigma_{r\theta_1}-(\sigma_{rr_1}$$
$$+s)\frac{\partial v_0}{\partial r}-(\sigma_{rr_0}+s)\frac{\partial v_1}{\partial r}-\sigma_{r\phi_1}\frac{\partial v_0}{\partial \phi}-\sigma_{r\phi_0}\frac{\partial v_1}{\partial \phi})(1+\frac{\partial u_0}{\partial r}+\frac{\partial w_0}{\partial \phi})+(\sigma_{r\theta_0}-(\sigma_{rr_0}+s)\frac{\partial v_0}{\partial r}-\sigma_{r\phi_0}$$
$$\frac{\partial v_0}{\partial \phi})(\frac{\partial u_1}{\partial r}+\frac{\partial w_1}{\partial \phi}))+F_{\theta_1}=\frac{\partial^2}{\partial t^2}(\rho_{11}v_1+\rho_{12}V_1),$$

$$\frac{\partial}{\partial r}((\sigma_{r\phi_1} - (\sigma_{rr_1} + s)\frac{\partial w_0}{\partial r} - (\sigma_{rr_0} + s)\frac{\partial w_1}{\partial r} - \sigma_{r\theta_1}\frac{\partial w_0}{\partial \theta} - \sigma_{r\theta_0}\frac{\partial w_1}{\partial \theta})(1 + \frac{\partial u_0}{\partial r} + \frac{\partial v_0}{\partial \theta}) + (\sigma_{r\phi_0} - (\sigma_{rr_0} + s)\frac{\partial w_0}{\partial r} - \sigma_{r\theta_0}\frac{\partial w_0}{\partial \theta})(\frac{\partial u_1}{\partial r} + \frac{\partial v_1}{\partial \theta})) + \frac{1}{r}\frac{\partial}{\partial \theta}((\sigma_{\theta\phi_1} - (\sigma_{\theta\theta_1} + s)\frac{\partial w_0}{\partial \theta} - (\sigma_{\theta\theta_0} + s)\frac{\partial w_1}{\partial \theta} - \sigma_{\theta r_1}\frac{\partial w_0}{\partial r} - \sigma_{\theta r_0}\frac{\partial w_1}{\partial r})(1 + \frac{\partial u_0}{\partial r} + \frac{\partial v_0}{\partial \theta}) + (\sigma_{\theta\phi_0} - (\sigma_{\theta\theta_0} + s)\frac{\partial w_0}{\partial \theta} - \sigma_{\theta r_0}\frac{\partial w_0}{\partial r})(\frac{\partial u_1}{\partial r} + \frac{\partial v_1}{\partial \theta})) + \frac{1}{rsin\theta}\frac{\partial}{\partial \phi}(((\sigma_{\phi\phi_1} + s) - \sigma_{\phi r_1}\frac{\partial w_0}{\partial r} - \sigma_{\phi r_0}\frac{\partial w_1}{\partial r} - \sigma_{\phi\theta_1}\frac{\partial w_0}{\partial \theta} - \sigma_{\phi\theta_0}\frac{\partial w_1}{\partial \theta})(1 + \frac{\partial u_0}{\partial r} + \frac{\partial v_0}{\partial \theta}) + ((\sigma_{\phi\phi_0} + s) - \sigma_{\phi r_0}\frac{\partial w_0}{\partial r} - \sigma_{\phi\theta_0}\frac{\partial w_0}{\partial \theta})(\frac{\partial u_1}{\partial r} + \frac{\partial v_1}{\partial \theta})) + \frac{2cot\theta}{r}((\sigma_{\theta\phi_1} - \sigma_{\theta r_1}\frac{\partial w_0}{\partial r} - \sigma_{\theta r_0}\frac{\partial w_1}{\partial r} - (\sigma_{\theta\theta_1} + s)\frac{\partial w_0}{\partial \theta} - (\sigma_{\theta\theta_0} + s)\frac{\partial w_1}{\partial \theta})(1 + \frac{\partial u_0}{\partial r} + \frac{\partial v_0}{\partial \theta}) + (\sigma_{\theta\phi_0} - \sigma_{\theta r_0}\frac{\partial w_0}{\partial r} - (\sigma_{\theta\theta_0} + s)\frac{\partial w_0}{\partial \theta})(\frac{\partial u_1}{\partial r} + \frac{\partial v_1}{\partial \theta})) + \frac{3}{r}((\sigma_{r\phi_1} - (\sigma_{rr_1} + s)\frac{\partial w_0}{\partial r} - (\sigma_{rr_0} + s)\frac{\partial w_1}{\partial r} - \sigma_{r\theta_1}\frac{\partial w_0}{\partial \theta} - \sigma_{r\theta_0}\frac{\partial w_1}{\partial \theta})(1 + \frac{\partial u_0}{\partial r} + \frac{\partial v_0}{\partial \theta}) + (\sigma_{r\phi_0} - (\sigma_{rr_0} + s)\frac{\partial w_0}{\partial r} - \sigma_{r\theta_0}\frac{\partial w_0}{\partial \theta})(\frac{\partial u_1}{\partial r} + \frac{\partial v_1}{\partial \theta})) + F_{\phi_1} = \frac{\partial^2}{\partial t^2}(\rho_{11}w_1 + \rho_{12}W_1),$$

$$Q\frac{\partial e}{\partial r} + R\frac{\partial \varepsilon}{\partial r} = \frac{\partial^2}{\partial t^2}(\rho_{12}u_1 + \rho_{22}U_1),$$

$$Q\frac{\partial e}{\partial \theta} + R\frac{\partial \varepsilon}{\partial \theta} = \frac{\partial^2}{\partial t^2}(\rho_{12}v_1 + \rho_{22}V_1),$$

$$Q\frac{\partial e}{\partial \phi} + R\frac{\partial \varepsilon}{\partial \phi} = \frac{\partial^2}{\partial t^2}(\rho_{12}w_1 + \rho_{22}W_1). \tag{6}$$

In all the above, $F_{r_1}, F_{\theta_1}, F_{\phi_1}$ are the body forces which have the frequency ω. The equations (5) and (6) are static equations, and first harmonic equations respectively.

3. The case of uniaxial static stress

When the large static uniaxial stress is applied on a solid medium one can ignore the effect of body forces [1]. If we assume that applied static uniaxial stress is acting in the direction of $\phi-$axis, then we have

$$\sigma_{ijo} = e_{ijo} = 0, (i \neq j),$$

$$\sigma_{rr_0} = \sigma_{\theta\theta_0} = 0, \tag{7}$$

$$F_{r_0} = F_{\theta_0} = F_{\phi_0} = 0,$$

where $\sigma_{\phi\phi_0}$ is the applied uniform static uniaxial stress. Substituting the equation(7) in the equation(5), it can be seen that equation(5) are automatically satisfied and the static strains are obtained as follows:

$$e_{rr_0} = e_{\theta\theta_0} = -\vartheta e_{\phi\phi_0}.$$

Substituting these equations in the first equation of equation(2), the following equation is obtained:

$$e_{\phi\phi_0} = \frac{\sigma_{\phi\phi_0} - Q\varepsilon}{Y}. \tag{8}$$

In the above, $\vartheta = \frac{A}{2(A+N)}$ is the Poisson ratio, and $Y = \frac{N(P+2A)}{A+N}$ is the Young's modulus. Substituting the equations(7) and (8) in equation(6), and due to the assumption of infinitesimal deformation in the linear theory of elasticity, we can neglect the product terms consequently equations of motion in this case are obtained as follows:

$$N((\nabla^2 - \frac{2}{r}\frac{\partial}{\partial r})u + (1 + \frac{4}{r})\frac{\partial^2 u}{\partial r^2} - \frac{4}{r^2}u + (\frac{\partial v}{\partial r} - \frac{3}{r})(\frac{1}{r}\frac{\partial v}{\partial \theta} + \frac{cot\theta}{r}) + \frac{1}{rsin\theta}(\frac{\partial^2 w}{\partial r\partial \phi} - \frac{3}{r}\frac{\partial w}{\partial \phi})) + ((2 + \frac{4}{r})N \frac{\partial^2 u}{\partial r^2} + \frac{N}{r^2}\frac{\partial^2 u}{\partial \theta^2} + \frac{Ncot\theta}{r^2}\frac{\partial u}{\partial \theta} - \frac{2N}{r^2}u + ((A + Q)e + (Q + R)\varepsilon)(\frac{\partial}{\partial r} + \frac{1}{r}) + \frac{N}{r}\frac{\partial^2 v}{\partial r\partial \theta} - \frac{3N}{r^2}\frac{\partial v}{\partial \theta} + \frac{Ncot\theta}{r}(\frac{\partial v}{\partial r} - \frac{v}{r}))\frac{P}{2(A+N)}\frac{\partial w_0}{\partial \phi} + (-\frac{N}{r^2sin^2\theta}\frac{\partial^2 u}{\partial \phi^2} - \frac{2N}{r^2}u - ((A + Q)e + (Q + R)\varepsilon)(\frac{\partial}{\partial r} - \frac{2Ncot\theta}{r^2}v + \frac{N}{rsin\theta}(\frac{\partial^2 w}{\partial r\partial \phi} - \frac{3}{r}\frac{\partial w}{\partial \phi})))\frac{A}{(A+N)}\frac{\partial w_0}{\partial \phi} = \frac{\partial^2}{\partial t^2}(\rho_{11}u + \rho_{12}U),$$

$$(N(\nabla^2 - \frac{2}{r}\frac{\partial}{\partial r})v + \frac{N}{r}\frac{\partial^2 u}{\partial r\partial\theta} + \frac{N}{r^2}\frac{\partial u}{\partial\theta} + (3+cot\theta)\frac{2N}{r^2}u - \frac{N}{r^2}v + \frac{N}{r}\frac{\partial v}{\partial r} + \frac{N}{r^2}\frac{\partial^2 v}{\partial\theta^2} + \frac{Ncot\theta}{r^2}\frac{\partial v}{\partial\theta} + \frac{N}{r^2 sin\theta}\frac{\partial w}{\partial\phi}$$
$$(\frac{\partial w}{\partial\theta} + cot\theta) + \frac{1}{r}((A+Q)e + (Q+R)\varepsilon)(\frac{\partial}{\partial\theta} - cot\theta) - \frac{2Ncot\theta}{r}(\frac{\partial w}{\partial\phi} + usin\theta + vcos\theta) + (\frac{N}{r}\frac{\partial^2 u}{\partial r\partial\theta}$$
$$-\frac{N}{r^2}\frac{\partial u}{\partial\theta} + N\frac{\partial^2 v}{\partial r^2} - \frac{N}{r^2}v + \frac{N}{r}\frac{\partial v}{\partial r} + \frac{2N}{r^2}(\frac{\partial^2 v}{\partial\theta^2} + \frac{\partial u}{\partial\theta}) + \frac{1}{r}\frac{\partial}{\partial\theta}((A+Q)e + (Q+R)\varepsilon) + (3+cot\theta)\frac{2N}{r^2}$$
$$(\frac{\partial v}{\partial\theta} + u))\frac{P}{2(A+N)}\frac{\partial w_0}{\partial\phi} + (\frac{N}{r^2 sin\theta}(\frac{1}{sin\theta}\frac{\partial^2 v}{\partial\phi^2} + \frac{\partial^2 w}{\partial\theta\partial\phi} + cot\theta\frac{\partial w}{\partial\phi}) - \frac{2Ncot\theta}{r}(\frac{\partial w}{\partial\phi} + usin\theta + vcos\theta)$$
$$-\frac{cot\theta}{r}((A+Q)e + (Q+R)\varepsilon)))(\frac{-A}{A+N}\frac{\partial w_0}{\partial\phi}) = \frac{\partial^2}{\partial t^2}(\rho_{11}v + \rho_{12}V),$$

$$N\nabla^2 w + \frac{1}{rsin\theta}((A+Q)\frac{\partial e}{\partial\phi} + (Q+R)\frac{\partial\varepsilon}{\partial\phi}) + N(\frac{1}{rsin\theta}\frac{\partial^2 u}{\partial r\partial\phi} - \frac{2}{r^2 sin\theta}\frac{\partial u}{\partial\phi} - \frac{1}{r}\frac{\partial w}{\partial r} - \frac{4w}{r^2} + \frac{1}{r^2 sin\theta}$$
$$\frac{\partial^2 v}{\partial\theta\partial\phi} + \frac{1}{r^2}\csc\theta\cot\theta\frac{\partial v}{\partial\phi} + \frac{w}{r^2}\csc^2\theta) + \frac{N}{r^2 sin^2\theta}\frac{\partial^2 w}{\partial\phi^2} + \frac{2N}{r^2 sin^2\theta}(\frac{\partial u}{\partial\phi}sin\theta + \frac{\partial v}{\partial\phi}\cos\theta))(\frac{-A}{A+N}\frac{\partial w_0}{\partial\phi})$$
$$= \frac{\partial^2}{\partial t^2}(\rho_{11}w + \rho_{12}W),$$

$$Q\frac{\partial e}{\partial r} + R\frac{\partial\varepsilon}{\partial r} = \frac{\partial^2}{\partial t^2}(\rho_{12}u + \rho_{22}U), \tag{9}$$

$$Q\frac{\partial e}{\partial\theta} + R\frac{\partial\varepsilon}{\partial\theta} = \frac{\partial^2}{\partial t^2}(\rho_{12}v + \rho_{22}V),$$

$$Q\frac{\partial e}{\partial\phi} + R\frac{\partial\varepsilon}{\partial\phi} = \frac{\partial^2}{\partial t^2}(\rho_{12}w + \rho_{22}W),$$

where $\nabla^2 = \frac{\partial^2}{\partial r^2} + \frac{1}{r^2}\frac{\partial^2}{\partial\theta^2} + \frac{1}{r^2\sin^2\theta}\frac{\partial^2}{\partial\phi^2} + \frac{2}{r}\frac{\partial}{\partial r} + \frac{\cot\theta}{r^2}\frac{\partial}{\partial\theta}$, the subscript one on the displacement components and the stress components have been dropped, and $F_{r_1} = F_{\theta_1} = F_{\phi_1} = 0$. Now, we can assume the solution to the equation(9) in the frequency domain are as follows:

$$u(r,\theta,\phi) = C_1 e^{j\omega t - j(k_1 r + k_2\theta + k_3\phi)},$$

$$v(r,\theta,\phi) = C_2 e^{j\omega t - j(k_1 r + k_2\theta + k_3\phi)},$$

$$w(r,\theta,\phi) = C_3 e^{j\omega t - j(k_1 r + k_2\theta + k_3\phi)},$$

$$U(r,\theta,\phi) = C_4 e^{j\omega t - j(k_1 r + k_2\theta + k_3\phi)}, \tag{10}$$

$$V(r,\theta,\phi) = C_5 e^{j\omega t - j(k_1 r + k_2\theta + k_3\phi)},$$

$$W(r,\theta,\phi) = C_6 e^{j\omega t - j(k_1 r + k_2\theta + k_3\phi)}.$$

In all the above $C_1, C_2, C_3, C_4, C_5, C_6$ are arbitrary constants, j is the complex unity, and $k_i (i = 1,2,3)$ is the wave number in the ith direction such that the wavenumber $k = \sqrt{k_1^2 + k_2^2 + k_3^2}$. Substituting equation(10) in equation(9), the equations of motion in terms of displacements are as follows:

$$(j^2k_1^2(N + 1 + \frac{4}{r} + (2N + \frac{4N}{r} + \frac{2}{r} + \frac{A}{r})T_1 + \frac{A}{r}T_2) + (j^2k_2^2(\frac{N}{r^2}(1+T_1))) + (j^2k_3^2(\frac{N}{r^2\sin\theta}(1-T_2))$$
$$\frac{2N}{r}k_1 - \frac{Ncot\theta}{r^2}k_2 - \frac{4N}{r^2}) + (\frac{Ncot\theta}{r^2}jk_2 - \frac{2N}{r^2})T_1 + \frac{2}{r^2}(N - Ajk_1)T_2 + A(jk_1)^3T_1 - j^2\rho_{11}\omega^2)C_1 +$$
$$(j^2k_1k_2(\frac{N}{r}(1 + T_1(1 + \frac{A\cot\theta}{r^2}))) + jk_1(-\frac{Ncot\theta}{r}(1+T_1)) + jk_2(\frac{3N}{r^2}(1+T_1) - \frac{Acot\theta}{r^3}(T_2 - T_1))+$$
$$j^2k_2^2(\frac{A}{r^2}(T_1 - \frac{T_2}{r})) + \frac{2N\cot\theta}{r^2}(T_2 - 2T_1))C_2 + (j^2k_1k_3(\frac{N}{r\sin\theta}(1 - T_2) + jk_3(\frac{9NT_2}{r^3\sin\theta}) + j^2k_3^2$$
$$(-\frac{A+Q}{r^3\sin^2\theta}(T_1 + T_2)) - \frac{Aj^3}{r^2\sin^2\theta}k_1k_3^2T_1))C_3 + ((\frac{Q}{r}(jk_1^3) - \frac{2Q}{r^2}jk_1)T_2 + ((1 + \frac{3}{r})Qj^2k_1^2)T_1 - \rho_{12}j^2$$
$$\omega^2)C_4 + (jk_2(\frac{Q\cot\theta}{r^3}(T_1 - 1)) + j^2k_2^2\frac{Q}{r^3}(T_2 - T_1) - \frac{Qj^3k_1k_2^2}{r^2}T_1)C_5 + ((j^2k_3^2(\frac{Q+R}{r^3\sin^2\theta}) - j^3k_1k_3^2$$
$$\frac{Q}{r^2\sin^2\theta})T_1)C_6 = 0,$$

$$((\tfrac{N}{r}j^2k_1k_2 + \tfrac{N}{r^2}jk_1 + (3+\cot\theta)\tfrac{2N}{r^2})(1+T_1) - \tfrac{4N\cot\theta}{r^2} + \tfrac{A}{r}(-j^3k_1^2k_2 + \tfrac{2}{r}j^2k_1k_2)(1+T_1-\cot\theta\, T_2))C_1 + (j^2k_1^2(N(1+T_1)) + j^2k_2^2(\tfrac{2N}{r^2} + \tfrac{cot\theta}{r^3}(A+1) + (\tfrac{2N}{r^2} + \tfrac{A\cot\theta}{r^3})T_1 - \tfrac{\cot\theta}{r^3}T_2) + j^2k_3^2(\tfrac{N}{r^2\sin^2\theta}(1+T_2)) + jk_2(\tfrac{2N\cot\theta}{r^2} + \tfrac{3N}{r^2} - \tfrac{\cot^2\theta}{r^2}(1+T_2) - \tfrac{A\csc^2\theta}{r^3} + (\cot\theta+3)\tfrac{2NT_1}{r^2}) + \tfrac{N}{r}jk_1 - \tfrac{N}{r^2}(1+v) + \tfrac{A}{r^4}j^3k_1^3 + \rho_{12}j^2\omega^2)C_2 + (j^2k_3^2(\tfrac{-Q\csc\theta\cot\theta(1+T_1)}{r^3}) - \tfrac{\cot\theta}{r^3\sin^2\theta}(Q+AT_2) + jk_3(\tfrac{-N\cot\theta}{r^2\sin\theta}(\tfrac{1}{r} - 2T_2)) + j^2k_2k_3\tfrac{N}{r^2\sin\theta}(1+T_1) - j^3k_2k_3^2\tfrac{QT_1}{r^3\sin\theta})C_3 + (\tfrac{R}{r}(-j^3k_1^2k_2 + \tfrac{2}{r}j^2k_1k_2)(1+T_1) - \tfrac{A\cot\theta}{r}(j^2k_1^2 + \tfrac{2}{r}jk_1)(1+T_2))C_4 + (\tfrac{R}{r}(\tfrac{\cot\theta}{r^2}j^2k_2^2 + \tfrac{\csc^2\theta}{r^2}jk_2)(1+T_1) - \tfrac{Q\cot\theta}{r}(\tfrac{\cot\theta}{r^2}jk_2 + \tfrac{1}{r^2}j^2k_2^2)(1+T_2) - \rho_{22}\omega^2)C_5 + (\tfrac{R}{r}(\tfrac{-\csc\theta\cot\theta}{r^2}j^2k_3^2 + \tfrac{1}{r^2\sin^2\theta}j^3k_2k_3^2)(1+T_1) - ((\tfrac{R\cot\theta}{r^2\sin^2\theta}j^2k_3^2(1+T_2)))C_6 = 0,$$

$$((\tfrac{2}{r}j^2k_1k_3 - j^3k_3k_1^2)\tfrac{A}{r\sin\theta} + \tfrac{N}{r\sin\theta}(j^2k_1k_3 - (\tfrac{1}{r} - \tfrac{2N}{r^2} - \tfrac{3N}{r^2\sin\theta})jk_3))C_1 + (\tfrac{\cot\theta}{r^2}j^2k_2k_3 - \tfrac{1}{r^2}j^3k_2^2k_3 + \tfrac{1}{r^2\sin\theta}j^2k_2k_3 - \tfrac{\cot\theta}{r^2\sin\theta}jk_2 + \cos\theta jk_3 + \tfrac{2N\cos\theta}{r^2\sin^2\theta}jk_3)C_2 + (N(j^2k_1^2 + \tfrac{1}{r^2}j^2k_2^2 + \tfrac{1}{r^2\sin^2\theta}j^2k_3^2 + \tfrac{1}{r}jk_1 + \tfrac{2\cot\theta}{r^2})jk_2) - \tfrac{j^2k_3^2}{r^2\sin^2\theta} - \tfrac{2N}{r^2} - \tfrac{N\cot\theta}{r^2}jk_2 + \tfrac{N\csc^2\theta}{r^2}(1+j^2k_3^2) + \tfrac{2N}{r}jk_1 - j^2\omega^2\rho_{11})C_3 + (\tfrac{Q}{r\sin\theta}(\tfrac{2}{r}j^2k_1k_3 - j^3k_1^2k_3))C_4 + (\tfrac{Q}{r^2\sin\theta}(\cot\theta j^2k_2k_3 - j^3k_2^2k_3))C_5 - (\tfrac{Q}{r^3\sin^3\theta}j^3k_3^3 + j^2\omega^2\rho_{12})C_6 = 0,$$

$$Q((-j^3k_1^3 + \tfrac{2}{r^2}jk_1 + \tfrac{2}{r^2}j^2k_1^2 - \rho_{12}j^2\omega^2)C_1 + (\tfrac{\cot\theta}{r^2}j^2k_1k_2 + \tfrac{2\cot\theta}{r^3}jk_2 - \tfrac{2}{r^3}j^2k_2^2 - \tfrac{1}{r^2}j^3k_1k_2^2)C_2 - (\tfrac{2}{r^3\sin^2\theta}j^2k_3^2 + \tfrac{1}{r^3\sin^2\theta}j^3k_1k_3^2)C_3) + R((-j^3k_1^3 + \tfrac{2}{r^2}jk_1 + \tfrac{2}{r^2}j^2k_1^2 - \rho_{22}j^2\omega^2)C_4 + (\tfrac{\cot\theta}{r^2}j^2k_1k_2 + \tfrac{2\cot\theta}{r^3}jk_2 - \tfrac{2}{r^3}j^2k_2^2 - \tfrac{1}{r^2}j^3k_1k_2^2)C_5 - (\tfrac{2}{r^3\sin^2\theta}j^2k_3^2 + \tfrac{1}{r^3\sin^2\theta}j^3k_1k_3^2)C_6) = 0,$$

$$Q((-j^3k_1^2k_2 + \tfrac{2}{r^2}j^2k_1k_2)C_1 + (\tfrac{\cot\theta}{r^2}j^2k_2^2 + \tfrac{\csc^2\theta}{r^2}jk_2 - \tfrac{1}{r^2}j^3k_2^3 - \rho_{12}j^2\omega^2)C_2 - (\tfrac{\csc\theta\cot\theta}{r^2}j^2k_3^2)C_3) + R((-j^3k_1^2k_2 + \tfrac{2}{r^2}j^2k_1k_2)C_4 + (\tfrac{\cot\theta}{r^2}j^2k_2^2 + \tfrac{\csc^2\theta}{r^2}jk_2 - \tfrac{1}{r^2}j^3k_2^3 - \rho_{22}j^2\omega^2)C_5 - (\tfrac{\csc\theta\cot\theta}{r^2}j^2k_3^2)C_6) = 0,$$

$$Q((-j^3k_1^2k_3 + \tfrac{2}{r}j^2k_1k_3)C_1 + (\tfrac{\cot\theta}{r^2}j^2k_2k_3 - \tfrac{1}{r^2}j^3k_2^2k_3)C_2 - (\tfrac{j^3k_3^3}{r^2\sin^2\theta} - \rho_{12}j^2\omega^2)C_3) + R((-j^3k_1^2k_3 + \tfrac{2}{r}j^2k_1k_3)C_4 + (\tfrac{\cot\theta}{r^2}j^2k_2k_3 - \tfrac{1}{r^2}j^3k_2^2k_3)C_5 - (\tfrac{j^3k_3^3}{r^2\sin^2\theta}\rho_{22}j^2\omega^2)C_6)) = 0. \tag{11}$$

In the above, $\frac{P}{2(A+N)}\frac{\sigma_{\phi\phi_0}}{Y} = T_1$ and $\frac{A}{(A+N)}\frac{\sigma_{\phi\phi_0}}{Y} = T_2$.

4. Numerical results

For the numerical results, we consider the wave propagation in the the $\phi-$ direction and at a point $r = a$. In this case, $k_1 = k_2 = 0$, and equation (11) reduces to the following matrix form:

$$[A_{lm}][C_l] = 0, \qquad (l, m = 1, 2, 3, 4, 5). \tag{12}$$

Equation(12) results in a system of five homogeneous equations in five arbitrary constants C_1, C_2, C_3, C_4, C_5. For a non trivial solution, determinant of coefficients is zero. Accordingly we obtain the following frequency equation.

$$\mid A_{lm} \mid = 0, \qquad (l, m = 1, 2, 3, 4, 5). \tag{13}$$

Equation (13) degenerated as follows:

$$A_{33} = 0, \tag{14}$$

or

$$\begin{vmatrix} A_{11} & A_{12} & A_{14} & 0 \\ A_{21} & A_{22} & 0 & A_{25} \\ A_{41} & 0 & A_{44} & 0 \\ A_{51} & 0 & 0 & A_{55} \end{vmatrix} = 0. \tag{15}$$

here

$$
\begin{aligned}
A_{33} &= \frac{-2k_3^2}{sin^2\theta}(N + A\sigma_{\phi\phi_0} + (\csc^2\theta - 4)(\frac{N}{a^2} + \frac{A}{a^2(P+2A)}\sigma_{\phi\phi_0}) + \omega^2 N V_s^{-2}, \\
A_{11} &= \frac{-Nk_3^2}{a^2 sin^2\theta} + \frac{Ak_3^2}{a^2 sin^2\theta}\sigma_{\phi\phi_0} + \rho_{11}\omega^2, \\
A_{12} &= \frac{-3Ncot\theta}{a^2} - \frac{Pcot\theta}{(P+2A)2a^2}\sigma_{\phi\phi_0} + \frac{2A}{(P+2A)}cot\theta\sigma_{\phi\phi_0}, \\
A_{21} &= \frac{6N}{a^2} + \frac{2A}{(P+2A)}\frac{cos\theta}{a^2}\sigma_{\phi\phi_0} + \frac{P(cot\theta+3)}{a^2(P+2A)}\sigma_{\phi\phi_0}, \\
A_{22} &= \frac{k_3^2}{\sin^2\theta}(-N + \frac{A}{(P+2A)}\sigma_{\phi\phi_0}) + \frac{2A}{(P+2A)}\frac{cos^2\theta}{a^2 sin\theta}\sigma_{\phi\phi_0} + \rho_{11}\omega^2, \\
A_{14} &= A_{25} = A_{41} = A_{51} = \rho_{12}\omega^2, \\
A_{44} &= A_{55} = \rho_{22}\omega^2,
\end{aligned} \tag{16}
$$

where V_s is the shear wave velocity [3]. Equations (14) and (15) are investigated by introducing the non-dimensionalised quantities given below:

$$
\begin{aligned}
& a_1 = \frac{P}{H}, \quad a_2 = \frac{Q}{H}, \quad a_3 = \frac{R}{H}, \quad a_4 = \frac{N}{H}, \\
& d_1 = \frac{\rho_{11}}{\rho}, \quad d_2 = \frac{\rho_{12}}{\rho}, \quad d_3 = \frac{\rho_{22}}{\rho}, \quad \bar{z} = (\frac{V_0}{V_s})^2, \\
& \rho = \rho_{11} + 2\,\rho_{12} + \rho_{22}, \quad H = P + 2\,Q + R, \quad V_0^2 = \frac{H}{\rho}, \\
& m = \frac{c}{c_0}, \quad c = \frac{\omega}{k}, \quad c_0^2 = \frac{N}{\rho}.
\end{aligned} \tag{17}
$$

In equation(17), c is the phase velocity, c_0 and V_0 are the reference velocities, k is wavenumber and m is non-dimensional phase velocity. Using the equation (17) in equation (14), we obtain implicit relation between non-dimensional phase velocity (m), non-dimensional static uniaxial stress and non-dimensional wavenumber k_3a as follows.

$$
\begin{aligned}
m = [&\frac{1}{(\csc^2\theta-4)a_{42}(k_3a)^2\bar{z}}[2\csc^2\theta(a_4(k_3a)^2 + \frac{(a_1-2a_4)\sigma_{\phi\phi_0}}{3a_1-2a_4H}) - (\csc^2\theta - 4) \\
&(a_4 + \frac{(a_1-2a_4)\sigma_{\phi\phi_0}}{3a_1-2a_4H})]]^{\frac{1}{2}}.
\end{aligned} \tag{18}
$$

Employing the non-dimensional quantities in the frequency equation (15), we obtain implicit relation between non-dimensional phase velocity m, non-dimensional static uniaxial stress and non-dimensional wavenumber k_3a. For numerical process, three poroelastic solids are considered and then discussed. Of three poroelastic solids, two are sandstone saturated with water and kerosene, respectively [17,18], and the third one is bony element. The physical parameter values of first two materials pertaining to the equation(17) are given in the Table 1. Further, the values of bone poroelastic parameters A, N, Q, R and its mass coefficients are computed following the paper [19]. The values of Young's modulus and Poisson ratio are taken to be 3×10^6 lb/inch^2, and 0.28, respectively as suggested in the paper [19]. The value of θ is taken to be $\frac{\pi}{3}$ respectively. Phase velocity is computed using the bisection method implemented in MATLAB, and the results are depicted in the figure 1. Figure 1 shows the plots of non-dimensional phase velocity against the static uniaxial stress for fixed wavenumber. From this figure it is clear that material-1 values are much greater than that of material-2. This discrepancy is due to fluid present in the materials. The values of bone are much greater than that of material-1 and material-2.

Table 1. Material parameters

Material parameter	Material-1	Material-2
a_1	0.843	0.96
a_2	0.65	0.006
a_3	0.28	0.028
a_4	0.234	0.412
d_1	0.901	0.887
d_2	-0.001	0
d_3	0.101	0.123
$\tilde{z}$	3.851	2.129

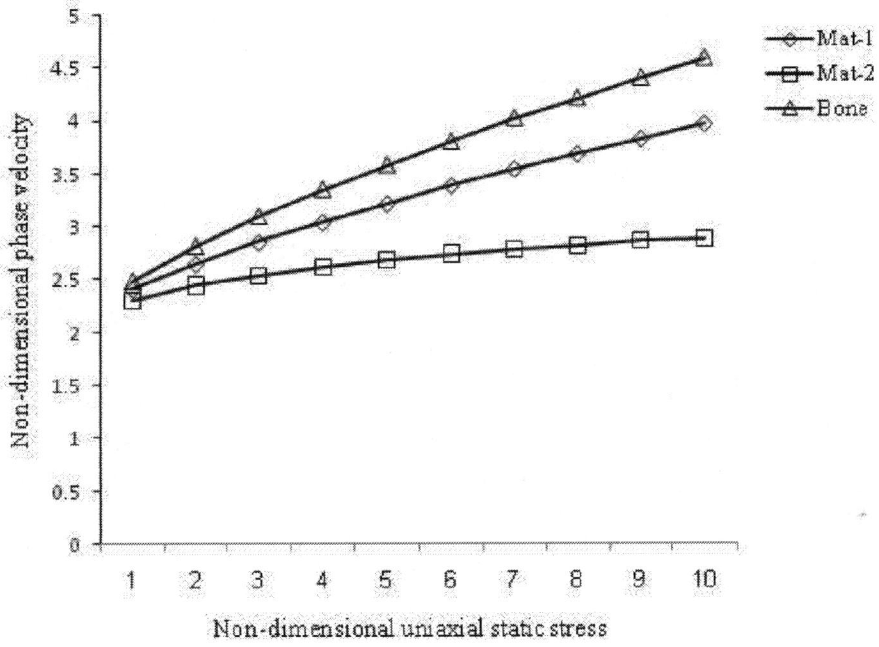

Figure 1. Variation of non-dimensional phase velocity with the non-dimensional static uniaxial stress.

5. Conclusion

Employing Biot's theory, flexural vibrations of poroelastic solid sphere in presence of static stress are investigated. Non-dimensional phase velocity against the non-dimensional static uniaxial stress for fixed wavenumber is computed for three types of poroelastic solids. Material-1 value is much greater than that of material-2 for the same value of applied uniaxial static stress. Both the materials are sandstone related and differ in only the fluid part. Hence, it can be inferred that the fluid part is causing the above discrepancy. This kind of analysis can be made for any poroelastic solid sphere if the values of parameters are available.

References

[1] Mott G 1971 Equations of elastic motion of an isotropic medium in the presence of body forces and static stresses *Journal of Acoustical Society of America* **50(3)** 856-868

[2] Mott G 1973 Elastic wave guide propagation in an infinite isotropic solid cylinder that is subjected to static axial stress and strain *Journal of Acoustical Society of America* **53(4)** 1129-33

[3] Biot MA 1956 The theory of propagation of elastic wave in fluid-saturated porous solid *Journal of Acoustical Society of America* **28** 168-178

[4] Tolstoy I 1973 Wave Propagation (International Series in the Earth and Planetary Sciences)(New York : McGraw-Hill Inc)

[5] Tajuddin M and Syed Iqbal Ahmed 1991 Dynamic interactions of a poroelastic layer and a half-and a half-space *Journal of Acoustical Society of America* 1 **89(3)** 1169-75

[6] Malla Reddy P and Tajuddin M 2000 Exact analysis of the plane strain vibrations of thick walled hollow poroelastic cylinders *International Journal of Solids and Structures* **37** 3439-56

[7] Malla Reddy P and Tajuddin M 2010 Axially symmetric vibrations of a poroelastic composite cylinder in the context of fretting fatigue *Special Topics and Reviews in Porous Media* **1(4)** 311-320

[8] Shah SA and Tajuddin M 2009 Axially symmetric vibrations of finite composite poroelastic cylinders *Internatinal Journal of Applied Mechanics and Engineering* **14(3)** 865-877

[9] Tajuddin M and Ahmed Shah 2006 circumferential waves of infinite hollow poroelastic cylinders *Journal of Applied Mechanics (ASME)* **73(4)** 705-708

[10] Rajitha G and Malla Reddy P 2014 Investigation of Flexural Vibrations in Poroelastic Elliptic Cone *Proceedings of International Conference on Mathematical Sciences, ICMS-2014* Sathyabhama University Chennai (Elseveir) pp 340-344

[11] Shah SA and Tajuddin M 2011 Torsional vibrations of poroelastic prolate spheroids *International Journal of Appllied Mechanics and Engineering* **16** 521-529

[12] Shah SA and Tajuddin M 2011 On axially symmetric vibrations of fluid filled poroelastic spherical shells *Open Journal of Acoustica* **1** 15-26

[13] Rajitha G and Malla Reddy P 2014 Axially symmetric vibrations of composite poroelastic spherical shell *International Journal of Engineering Mathematics* **2014**

[14] Shanker, Nageswara Nath, Ahmed Shah and Manoj Kumar 2013 Vibration analysis of a poroelastic composite hollow sphere *Acta Mechanica* **224(2)** 327-341

[15] Shah SA and Tajuddin M 2011 Torsional vibrations of thick-walled hollow poroelastic spheres *Bulletin of Calcutta Mathematical Society* **103** 161-170

[16] Rajitha, Srisailam and Malla Reddy 2015 Flexural vibrations of poroelastic solids in the presence of static stress *Journal of Vibration and Control* **21(11)** 2266-72

[17] Fatt I 1957 The Biot-Willis elastic coefficient for a sand stone *Journal of Appllied Mechanics* 296-297

[18] Yew CH and Jogi PN 1976 Study of wave motions in fluid-saturated porous rocks *Journal of Acoustical Society of America* **60** 2-8

[19] Nowinski JL and Davis CF 1971 Propagation of longitudinal waves in circularly cylindrical bone elements *Transactions of ASME Journal of Applied Mechanics* 578-584

Study of torsional vibrations in an initially stressed composite poroelastic cylinders

Sandhya Rani B, Balu Ch and Malla Reddy P

Department of Mathematics, Kakatiya University, Warangal 506009, Telangana, India.

E-mail: sandhyarani.bandari@gmail.com

Abstract. This paper investigates torsional vibrations in an initially stressed composite poroelastic cylinder in the framework of Biot's theory of wave propagation in poroelastic solids. Poroelastic composite cylinder consists of two concentric cylindrical layers made of different poroelastic materials. The governing equations are formulated from the Biot's incremental deformation theory. The non-dimensional frequency is computed as a function of ratio of thickness to wavelength. The limiting cases of a poroelastic solid cylinder and poroelastic hollow cylinder are discussed. The results are presented graphically for two poroelastic composite cylinders and then compared with the published results.

1. Introduction

The studies of wave propagation in poroelastic solids have many applications in various fields such as Seismology, Soil-mechanics, Bio-mechanics, Civil engineering, and Mechanical engineering. From the real time experiences, one may find that buildings, bridges and some manmade structures consist two or more material that could be combined to take advantage of the good characteristics of each of the materials. In the frame work of Biot's theory [1], the effect of boundaries on torsional vibrations in a poroelastic composite cylinder is reported in several papers [2-3]. Axially symmetric vibrations of composite poroelastic cylinder are investigated by Malla Reddy and Tajuddin [4]. The study of torsional vibrations of an elastic solid is important in several applications such as transmission of power through shafts with flanges at the ends as integral parts of the shafts. The other use of torsional vibrations is measurement of the shear modulus of elastic solids. The basic literature on the propagation of elastic waves is given by Ewing et.al. [5]. Torsion waves of an elastic composite infinite circular solid rod of two different materials are studied by Armenkas [6]. Investigation of axially symmetric wave propagation in a two layered elastic cylinder is made by Whitter and Jones [7]. Torsional vibrations in a poroelastic cylinder are reported in several papers [8-10]. The problem related to pre-stressed elastic solids has been a subject of continued interest due to its importance in the said areas. A detailed discussion about the theory of elastic medium under initial stress is given in Biot's incremental theory [11]. Torsional wave propagation in an initially stressed elastic cylinder is studied by Dey and Dutta [12]. In the paper [12], governing equations are formulated for the Biot's incremental deformation theory. In the said paper, the velocities of torsional wave propagation due to the presence of initial stress are calculated for different extension ratios. Selim [13] investigated torsional wave propagation in dissipative elastic solid cylinder subjected to initial stress. In the paper [13], the effect of damping on the propagation of torsional waves in incompressible cylinder of infinite length is discussed and it is proved that the damping of the medium has strong effect on the propagation of torsional waves. Propagation of Rayleigh waves in an initially stressed incompressible half space under a rigid layer are studied by Dey et.al. [14]. It has been shown that Rayleigh waves cannot propagate in an isotropic medium without tensile initial stress [14]. Plane strain deformation of an initially stressed orthotropic elastic medium is studied using eigen value approach [15]. Surface wave propagation in an initially stressed transversely isotropic thermoelastic solid is studied by Baljeet Singh and Renu [16]. Torsional surface waves in an initially stressed anisotropic poroelastic layer over a semi-infinite heterogeneous half space with linearly varying rigidity and density due to irregularity at the interface is studied in [17]. The propagation of plane waves is investigated in a general anisotropic elastic medium in the presence of initial stress [18]. Love wave propagation in a porous rigid layer lying over an initially stressed half space is studied in [19]. In the paper [19], the propagation of Love waves in fluid saturated, anisotropic, porous rigid layer over a pre stressed, non-homogeneous, elastic half space is investigated. In all the said papers, initial stress is not considered in composite poroelastic solid cylinder. The presence of initial stresses in solid materials can have a substantial effect on their subsequent response to applied loads that is very different from the corresponding response in the absence of initial stresses. In the present paper, torsional vibrations in the composite poroelastic solid cylinder of infinite extent consists of an inner solid circular cylinder of one material bounded by and bounded to a circular core made of another poroelastic material under initial stress are investigated. The non-dimensional frequency is computed as a function of ratio of thickness to wavelength for thin and thick coating.

The rest of the paper is organized as follows. In section 2, governing equations and solution of the problem are discussed. Boundary conditions and frequency equation are presented in section 3. Particular cases are given in section 4. Numerical results are discussed in section 5. Finally, conclusion is given in section 6.

2. Governing equations and solution of the problem

Let (r, θ, z) be the cylindrical polar coordinates. Consider a composite concentric isotropic infinite poroelastic solid cylinder with inner and outer radii r_1 and r_2, respectively, subjected to initial stress. The $z-$axis coincides with the axis of the cylinder. The substrate is a circular solid cylinder with radius r_1 and the coating is a thick-walled hollow cylinder having thickness $h(= r_2 - r_1)$. The equations of motion under initial compression stress $\sigma_{zz} = -p$ along the axis of the cylinder are [11]:

$$\begin{aligned}
&\frac{\partial\sigma_{rr}}{\partial r} + \frac{1}{r}\frac{\partial\sigma_{r\theta}}{\partial\theta} + \frac{\partial\sigma_{rz}}{\partial z} + \frac{\sigma_{rr}-\sigma_{\theta\theta}}{r} - p\frac{\partial\omega_\theta}{\partial z} = \frac{\partial^2}{\partial t^2}(\rho_{11}u_1 + \rho_{12}U_1) + b\frac{\partial}{\partial t}(u_1 - U_1),\\
&\frac{\partial\sigma_{r\theta}}{\partial r} + \frac{1}{r}\frac{\partial\sigma_{\theta\theta}}{\partial\theta} + \frac{\partial\sigma_{\theta z}}{\partial z} + \frac{2\sigma_{r\theta}}{r} + p\frac{\partial\omega_r}{\partial z} = \frac{\partial^2}{\partial t^2}(\rho_{11}u_2 + \rho_{12}U_2) + b\frac{\partial}{\partial t}(u_2 - U_2),\\
&\frac{\partial\sigma_{rz}}{\partial r} + \frac{1}{r}\frac{\partial\sigma_{\theta z}}{\partial\theta} + \frac{\partial\sigma_{rz}}{\partial z} + \frac{\sigma_{rz}}{r} - p(\frac{\partial\omega_\theta}{\partial r} - \frac{\partial\omega_r}{\partial\theta}) = \frac{\partial^2}{\partial t^2}(\rho_{11}u_3 + \rho_{12}U_3) + b\frac{\partial}{\partial t}(u_3 - U_3),\\
&Q\frac{\partial e}{\partial r} + R\frac{\partial\varepsilon}{\partial r} = \frac{\partial^2}{\partial t^2}(\rho_{12}u_1 + \rho_{22}U_1) - b\frac{\partial}{\partial t}(u_1 - U_1),\\
&\frac{Q}{r}\frac{\partial e}{\partial r} + \frac{R}{r}\frac{\partial\varepsilon}{\partial r} = \frac{\partial^2}{\partial t^2}(\rho_{12}u_2 + \rho_{22}U_2) - b\frac{\partial}{\partial t}(u_2 - U_2),\\
&Q\frac{\partial e}{\partial z} + R\frac{\partial\varepsilon}{\partial z} = \frac{\partial^2}{\partial t^2}(\rho_{12}u_3 + \rho_{22}U_3) - b\frac{\partial}{\partial t}(u_3 - U_3),
\end{aligned} \tag{1}$$

where (u_1, u_2, u_3) and (U_1, U_2, U_3) are the displacement components of solid and fluid respectively, $\sigma_{rr}, \sigma_{\theta\theta}, \sigma_{zz}, \sigma_{rz}, \sigma_{r\theta}$, and $\sigma_{\theta z}$ are the stress components, ρ_{ij} are mass coefficients,

b is the dissipative coefficient and ω_r, ω_θ are rotational components given by,

$$\omega_r = \frac{1}{2}\left(\frac{\partial u_3}{\partial \theta} - \frac{\partial u_2}{\partial z}\right),$$
$$\omega_\theta = \frac{1}{2}\left(\frac{\partial u_1}{\partial z} - \frac{\partial u_3}{\partial r}\right). \tag{2}$$

The stress component σ_{ij} and fluid pressure s [1] are

$$\sigma_{ij} = 2Ne_{ij} + (Ae + Q\varepsilon)\delta_{ij}, (i,j = 1,2,3),$$
$$s = Qe + R\varepsilon. \tag{3}$$

In the above, e_{ij} are strain components given by

$$e_{ij} = \frac{1}{2}\left(\frac{\partial u_i}{\partial x_j} + \frac{\partial u_j}{\partial x_i}\right), (i,j = 1,2,3).$$

In the case of torsional vibrations, the equations of motion is reduced to the following equations:

$$\frac{\partial \sigma_{r\theta}}{\partial r} + \frac{\partial \sigma_{\theta z}}{\partial z} + \frac{2\sigma_{r\theta}}{r} - \frac{p}{2}\frac{\partial^2 u_2}{\partial z^2} = \frac{\partial^2}{\partial t^2}(\rho_{12}u_2 + \rho_{22}U_2) + b\frac{\partial}{\partial t}(u_2 - U_2),$$
$$0 = \frac{\partial^2}{\partial t^2}(\rho_{12}u_2 + \rho_{22}U_2) - b\frac{\partial}{\partial t}(u_2 - U_2). \tag{4}$$

In the case of harmonic wave solution $e^{ik(z-ct)}$, the solution for circumferential displacements in the cylinder become,

$${}_ju_2(r,z,t) = {}_jf(r)e^{ik(z-ct)}, \ {}_jU_2(r,z,t) = {}_jF(r)e^{ik(z-ct)}, \ j = 1,2. \tag{5}$$

In Eq. (5), the quantities with subscripts 1 refer to the substrate while 2 refers to the coating. c is the torsional wave velocity, k is wavenumber, i is complex unity, and t is time. Substituting Eq. (5) and Eq. (3) in Eq. (4), one obtains

$$N\frac{d^2\,{}_jf}{dr^2} - (N - \frac{p}{2})k^2\,{}_jf + \frac{N}{r}\frac{d\,{}_jf}{dr} - \frac{N}{r^2}\,{}_jf = -kc({}_jf\rho_{11} + {}_jF\rho_{12}) - bikc({}_jf - {}_jF),$$
$$0 = -k^2c^2({}_jf\rho_{12} + {}_jF\rho_{12}) + bikc({}_jf - {}_jF). \tag{6}$$

Solutions of Eq. (6) are

$${}_2f(r) = C_1J_1({}_2qr) + C_2Y_1({}_2qr),$$
$${}_1f(r) = C_3J_1({}_1qr). \tag{7}$$

In Eq. (7),

$${}_jq^2 = \frac{1}{{}_jN}(k^2c^2\,{}_j\rho_{11} + k^2c^2\,{}_j\rho_{12}(\frac{-{}_j\rho_{12}kc + ib}{-{}_j\rho_{22}kc + ib}) + bikc(1 - (\frac{-{}_j\rho_{12}kc + ib}{-{}_j\rho_{22}kc + ib})) - k^2({}_jN - \frac{p}{2})), j = 1,2.$$

The non-zero stresses both for the substrate and the coating are

$${}_2\sigma_{r\theta} = {}_2N\,{}_2q(C_1J_2({}_2qr) + C_2Y_2({}_2qr))e^{ik(z-ct)},$$
$${}_1\sigma_{r\theta} = {}_1N\,{}_1q(C_3J_2({}_1qr)e^{ik(z-ct)}. \tag{8}$$

3. Boundary conditions and frequency equation

The boundary conditions for stress free outer surface $r = r_2$ and perfect bonding between the substrate and the coating at the interface $r = r_1$ are

$$
\begin{aligned}
{}_2\sigma_{r\theta} &= 0 \quad at \quad r = r_2 \\
{}_2\sigma_{r\theta} &= {}_1\sigma_{r\theta} \quad at \quad r = r_1, \\
{}_2u_2 &= {}_1u_2 \quad at \quad r = r_1.
\end{aligned} \tag{9}
$$

Substitution of Eqs. (7), (8) and (5) in Eq. (9) gives three homogeneous equations in three unknowns C_1, C_2, and C_3. A non-trivial solution can be obtained if the determinant of the coefficient matrix vanishes. Accordingly, one obtains the following frequency equation:

$$
\mid A_{ij} \mid = 0, \quad (i, j = 1, 2, 3) \tag{10}
$$

where,

$$
\begin{aligned}
&A_{11} = \frac{-2\,{}_2N}{r_2} J_1({}_2qr_2) + {}_2N\,{}_2qJ_0({}_2qr_2), \quad A_{12} = \frac{-2\,{}_2N}{r_2} Y_1({}_2qr_2) + {}_2N\,{}_2qY_0({}_2qr_2), \\
&A_{13} = 0, \quad A_{23} = \frac{-2\,{}_1N}{r_1} J_1({}_1qr_1) - {}_1N\,{}_1qJ_0({}_1qr_1), \\
&A_{31} = J_1({}_2qr_1), \quad A_{32} = Y_1({}_2qr_1), \quad A_{33} = -J_1({}_1qr_1),
\end{aligned} \tag{11}
$$

A_{21}, A_{22} are similar expressions as A_{11}, A_{12} with r_2 replaced by r_1.

4. Particular cases

The composite poroelastic cylinder will be reduced to the poroelastic solid cylinder and the poroelastic hollow cylinder under some special cases.

4.1. Poroelastic solid cylinder

When the poroelastic parameters of the substrate and coating are of same material, then the composite cylinder will be reduced to solid cylinder of one material. Setting ${}_2N = {}_1N = N$, ${}_2q = {}_1q = q$. Then the frequency equation Eq. (10) reduces to

$$
J_2(qr_2) = 0, \tag{12}
$$

which is the frequency equation of torsional vibrations in poroelastic solid cylinder similar to that of [10].

4.2. Poroelastic hollow cylinder

When the material constants of a substrate vanish, the composite poroelastic cylinder will become a hollow poroelastic cylinder. Setting ${}_1N = 0$, ${}_2N = N$, ${}_2q = q$, ${}_1q = {}_2q$. at the interface $r = r_1$, then the frequency equation Eq. (10) reduces to

$$
J_2(qr_2)Y_2(qr_1) - J_2(qr_1)Y_2(qr_2) = 0, \tag{13}
$$

which is the frequency equation of torsional vibrations in poroelastic hollow cylinder similar to that of [10].

5. Numerical results

Due to presence of dissipative nature of the solids, waves are attenuated. Attenuation presents some difficulty in the definition of phase velocity. If the dissipation coefficient is non-zero, the wavenumber, densities are complex. Consequently velocities of dilatational waves and shear waves are complex valued. Finally, frequency equations will be complex valued and implicit. Therefore, b is made to be zero so that frequency equation will be real valued and the roots will be obtained easily that explicitly give phase velocity. Even if b is zero, problem would be poroelastic in nature as the coefficients A, N, Q, R would not vanish. The frequency equation Eq. (10) is investigated by introducing the non-dimensional quantities given below:

$$\begin{aligned} &d_1 = \tfrac{2\rho_{11}}{{}_1\rho}, \quad d_2 = \tfrac{2\rho_{12}}{{}_1\rho}, \quad d_3 = \tfrac{2\rho_{22}}{{}_1\rho}, \quad a_4 = \tfrac{2N}{{}_1H}, \\ &g_1 = \tfrac{{}_1\rho_{11}}{{}_1\rho}, \quad g_2 = \tfrac{{}_1\rho_{12}}{{}_1\rho}, \quad g_3 = \tfrac{{}_1\rho_{22}}{{}_1\rho}, \quad b_4 = \tfrac{{}_1N}{{}_1H}, \\ &{}_1\rho = {}_1\rho_{11} + 2\,{}_1\rho_{12} + {}_1\rho_{22}, \quad {}_1H = {}_1P + 2\,{}_1Q + {}_1R, \\ &\delta = \tfrac{h}{L}, \quad c = \tfrac{\omega}{k}, \quad \Omega = \tfrac{\omega h}{{}_1c_0} \quad g = \tfrac{r_2}{r_1}. \end{aligned} \tag{14}$$

In the Eq.(14), h is the thickness of the cylinder and L is wavelength, ω is the frequency, ${}_1c_0$ is reference velocities (${}_1c_0^2 = {}_1N_1\rho^{-1}$), Ω is the non-dimensional frequency. For the numerical process, two types of composite poroelastic cylinders are considered, namely composite cylinder-I and composite cylinder-II. Composite cylinder-I consists of the substrate made of sandstone saturated with water [20] and the coating made of sandstone saturated with kerosene [21]; while in composite cylinder-II, the substrate is made of sandstone saturated with kerosene and the coating with sandstone saturated with water. The physical parameters of the said materials pertaining to Eq. (10) are given in Table 1.

Table 1. Material parameters

Material parameter	Composite cylinder-I	Composite cylinder-II
d_1	0.887	0.891
d_2	-0.001	0
d_3	0.099	0.125
a_4	0.123	0.780
g_1	0.887	0.901
g_2	0	-0.001
g_3	0.123	0.101
b_4	0.412	0.780

Employing the non-dimensional quantities in Eq.(10), one obtains an implicit relation between non-dimensional frequency (Ω) ratio of thickness to wavelength, ratio of radii and initial stress. The numerical values are depicted in figures 1-2. Variation of non-dimensional frequency as a function of ratio of thickness to wavelength for composite cylinder-I and composite cylinder-II is computed in the cases of thin coating and thick coating when initial stress is one. The numerical results are compared with the published results (when initial stress is zero, Ref. [3]) in the case of thin coating ($g = 1.01$) and thick coating ($g = 4$) for composite cylinder-I and composite cylinder-II. From figures 1 and 2, it is observed that the frequency values of composite cylinder II, in general, greater than that of composite cylinder-I. Also, it is observed that the frequency values lower when initial stress is present in the solid. From the numerical results, one can infer that the frequency depends on the presence of initial stress.

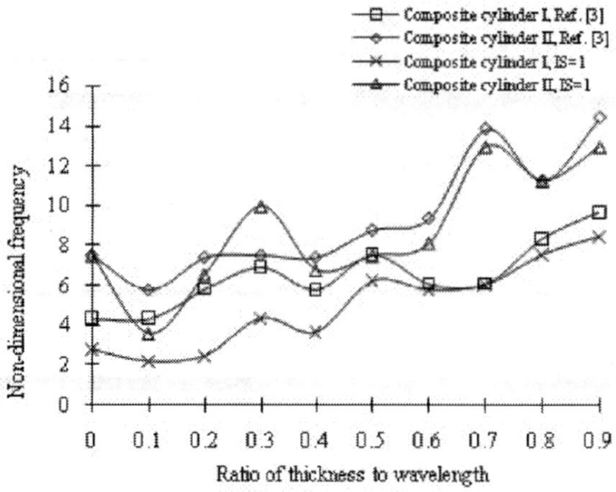

Figure 1. Variation of non-dimensional frequency with ratio of thickness to wave length in the case of thin coating ($g = 1.01$).

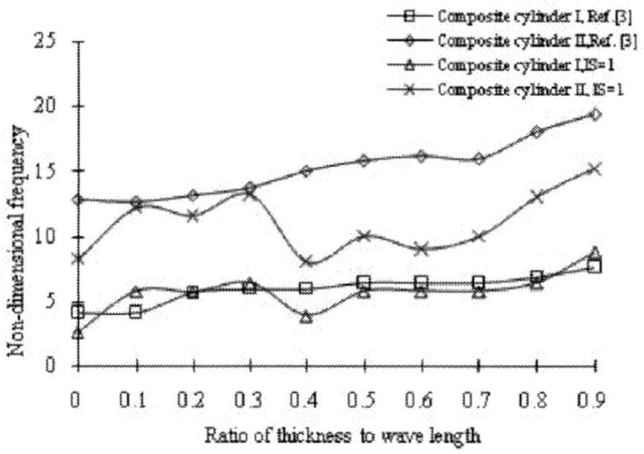

Figure 2. Variation of non-dimensional frequency with ratio of thickness to wave length in the case of thick coating ($g = 4$).

6. Conclusion

Torsional vibrations in composite poroelastic solid cylinder in the presence of initial stress are investigated. The governing equations are formulated from Biot's incremental deformation theory. Non-dimensional frequency against ratio of thickness to wavelength is computed for two composite poroelastic cylinders. From the paper, it is clear that the frequency of waves depends on the initial stress present in the solid.

References

[1] Biot M A 1956 Theory of propagation of elastic waves in a fluid-saturated porous solids-I *J. Acoust. Soc. America* **28** 168-178

[2] Tajuddin M 1982 Torsional vibration of finite composite poroelastic cylinder *Indian J. Pure Appl. Math.* **13** 3 375-381

[3] Ahmed Shah S and Tajuddin M 2010 Torsional vibrations of infinite composite poroelastic cylinders *International Journal of Engineering Science and Technology* **2** 6 150-161

[4] Malla Reddy P and Tajuddin M 2010 Axially symmetric vibrations of poroelastic composite cylinder in the context of fretting fatigue *Special Topics and Reviews in Porous Media* **1** 4 311-320
[5] Ewing W M, Jardetzky W S and Press F 1957 Elastic waves in layered media (New York: Mc Graw- Hill)
[6] Armanekas A E 1965 Torsional waves in composite rods *J. Acoust. Soc. Am.* **38** 439-446
[7] Whitter J S and Jones J P 1982 Axially symmetric wave propagation in two layered cylinder *International Journal of Solids and Structures* **3** 657-675
[8] Tajuddin M and Sarma K S 1978 Torsional vibrations of finite hollow poroelastic circular cylinder *Def. Sci. J.* **28** 2 97-102
[9] Tajuddin M and Sarma K S 1980 Torsional vibrations of poroelastic cylinder *J. Appl. Mechanics* **47** 214-216
[10] Tajuddin M and Shah S A 2007 On torsional vibrations of infinite hollow poroelastic cylinders *Journal of Mechanics of Materials and Structures* **2** 1 189-200
[11] Biot M A 1965 *Mechanics of Incremental Deformation* (New York :John Willey and Sons)
[12] Dey S and Dipasree Dutta 1992 Torsional wave propagation in an initially stressed cylinder *Proc. Indianatn. sci. Acad.* **58A** 5425-429
[13] Selim M M 2007 Torsional waves propagation in an initially stressed dissipative cylinder *Applied Mathematical Science* **1** 29 1419-1427
[14] Dey S, Nandita Roy and Sushmita Ghosh 1987 Propagation of Rayleigh waves in an initial stressed incompressible half space under a rigid layer: A Mathematical model showing the existence of non-seismic zone and explanation of two Rayleigh wave fronts *Indian J. pure appl. Math.* **18** 6 567-576
[15] Selim M M and Ahmed M K 2006 Plane strain deformation of an initially stressed orthotropic elastic medium *Applied Mathematics and Computation* **175** 221-237
[16] Baljeet Singh and Renu 2012 Surface wave propagation in an initially stressed transversely isotropic thermoelastic solid *Adv. Studies Theor. Phys.* **6** 6 263-271
[17] Gupta S, Chattopadhyay A and Majhi D K 2010 Effect of irregularity on the propagation of torsional surface waves in an initially stressed anisotropic poroelastic layer *Applied Mathematics and Mechanics* **31** 481-492
[18] Sharma M D 2007 Effect of initial stress on reflection at the free surface of anisotropic elastic medium *J. Earth. Syst. Sci.* **116** 537-551
[19] Kundu S, Gupta S, Chattopadhyay A and Majhi D K 2013 Love wave propagation in porous rigid layer lying over an initially stressed half-space *Itenrnational Journal of Applied Physics and Mathematics* **3** 140-142
[20] Yew C H and Jogi P N 1976 Study of wave motions in fluid-saturated porous rocks*Journal of Acoustical Society of America* **60** 2-8
[21] Fatt I 1959 The Biot-Willis elastic coefficients for sandstone *Journal of Applied Mechanics* **26** 296-297

Axially symmetric vibrations in poroelastic solid cylindrical panel resting on elastic foundation

Manjula Ramagiri and P Malla Reddy

Department of Mathematics, Kakatiya University, Warangal 506009, Telangana, India.

E-mail: manjularamagiri@gmail.com and mperati@yahoo.com

Abstract. The axially symmetric vibrations of an isotropic poroelastic cylindrical panel resting on elastic foundations are investigated in the framework of Biot's theory. The effects of the surrounding elastic medium are considered using the spring constant of the Wrinkler type and the shear constant of the Pasternak type. The frequency equation is obtained for both pervious and impervious surfaces. Non dimensional phase velocity is computed as a function of wavenumber. Numerical results are presented graphically.

1. Introduction

The study of elastic foundations in cylindrical structures has wide applications in the field of Engineering, Geology and Biomechanics. Cylindrical panel plays an important role in Aerospace as they are of high specific strength and high specific stiffness. Even in human body, a number of muscoskeletal models of knee joint and skin employ different forms of elastic Wrinkler foundations. Propagation of axially symmetric waves in finite elastic cylinder is studied in [1]. In the said paper, vibrations corresponding to the different propagation constants for the same frequency parameter are applied to satisfy the boundary conditions. In [2], authors studied wave propagation in a homogenous isotropic cylindrical panel embedded on an elastic medium. In the said paper, three dimensional wave propagation in cylindrical panel embedded in an elastic medium (Wrinkler model) is investigated in the context of linear theory of elasticity. Free vibrations of circular cylindrical shell on Wrinkler and Pasternak foundations is studied in [3]. In said paper, it is concluded that elastic foundation affects radial vibrations mode frequency while pertaining to torsional and longitudinal remains are unaffected. Employing Biot's theory [4], axially symmetric vibrations of a poroelastic composite solid cylinder is investigated in [5]. In the said paper, frequency equation is obtained for nonaxially and axially symmetric vibrations each for pervious and impervious surfaces, and the results are compared with that of rule of mixture (RoM). In [6] authors investigated vibration characteristics of fluid filled cylindrical shells based on elastic foundations. In the said paper, frequencies are strongly affected when a cylindrical shell is attached with elastic foundation. Axially symmetric vibrations of fluid filled poroelastic circular shells is studied in [7]. In [8], authors investigated axially symmetric vibrations of finite composite poroelastic cylinders. In the said paper, non-dimensional phase velocity for propagating modes is computed as a function of ratio of length of cylinders in the absence of dissipation. To the best of knowledge, axially symmetric vibrations in poroelastic solid cylindrical panel resting on elastic foundation are not yet investigated. Therefore, in this paper the same is investigated in the framework of Biot's theory. Frequency equations are obtained for both pervious and impervious surfaces. Non-dimensional phase velocity against the wavenumber for different elastic foundation are computed for three solids namely sandstone saturated with kerosene, water, and bone.

This paper is organized as follows. In section 2, governing equations, and solution of the problem are given. In section 3, frequency equations are derived. Numerical results are done in section 4. Finally, conclusions is given in section 5.

2. Governing equations and solution of the problem

The equations of motion of a poroelastic solid [4] in presence of dissipation (b) which in terms of displacement vectors are

$$\begin{aligned} &N\nabla^2\vec{u} + (A+N)\nabla e + Q\epsilon = \tfrac{\partial^2}{\partial t^2}(\rho_{11}\vec{u} + \rho_{12}\vec{U}) + b\tfrac{\partial}{\partial t}(\vec{u} - \vec{U}), \\ &\nabla(Qe + R\epsilon) = \tfrac{\partial^2}{\partial t^2}(\rho_{12}\vec{u} + \rho_{12}\vec{U}) - b\tfrac{\partial}{\partial t}(\vec{u} - \vec{U}), \end{aligned} \tag{1}$$

where, ∇^2 is the Laplace operator, $\vec{u}(u, v, w)$ and $\vec{U}(U, V, W)$ are solid and fluid displacements, e and ϵ are the dilatations of solid and fluid respectively; the symbols A, N, Q, R are all poroelastic constants; ρ_{ij} are mass coefficients. The constitutive relations are

$$\begin{aligned} &\sigma_{ij} = 2Ne_{ij} + (Ae + Q\epsilon)\delta_{ij}, (i, j = 1, 2, 3), \\ &s = Qe + R\epsilon. \end{aligned} \tag{2}$$

In eq. (2), e_{ij}'s are strain displacements, σ_{ij}'s are solid stresses and fluid pressure s, δ_{ij} the well known Kroneckar delta function. For axially symmetric vibrations, the displacements of solid $\vec{u}(u, 0, w)$ and fluid $\vec{U}(U, 0, W)$ which in terms of potential functions $\phi's$ and $\psi's$ are given below by

$$\begin{aligned} &u = \tfrac{\partial\phi_1}{\partial r} - \tfrac{\partial\psi_1}{\partial z}, \qquad && w = \tfrac{\partial\phi_1}{\partial z} + \tfrac{\partial\psi_1}{\partial r} + \tfrac{\psi_1}{r}, \\ &U = \tfrac{\partial\phi_2}{\partial r} - \tfrac{\partial\psi_2}{\partial z}, \qquad && W = \tfrac{\partial\phi_2}{\partial z} + \tfrac{\partial\psi_2}{\partial r} + \tfrac{\psi_2}{r}. \end{aligned} \tag{3}$$

For free harmonic waves travelling in the z- direction, we take

$$\begin{aligned} &\phi_1 = F_1(r)coskze^{i\omega t}, \qquad && \phi_2 = F_2(r)coskze^{i\omega t}, \\ &\psi_1 = G_1(r)sinkze^{i\omega t}, \qquad && \psi_2 = G_2(r)sinkze^{i\omega t}. \end{aligned} \tag{4}$$

In eq.(4), k is the wavenumber. ω is the frequency of the wave, i is the complex unity. By substituting eq. (4) in eq. (3), we obtain solid displacements as follows.

$$\begin{aligned} &u = -(C_1pK_1(pr) + C_2qK_1(qr) + A_1kK_1(dr))coskze^{i\omega t}, \\ &w = -(C_1kK_0(pr) + C_2kK_0(qr) + A_1dK_0(dr))sinkze^{i\omega t}. \end{aligned} \tag{5}$$

In eq. (5) C_1, C_2, A_1 are all arbitrary constants, $K_n(x)$ is the modified Bessel functions of second kind of order n and $p, q, d = k(1-\xi_1^2)$, $\xi_i = \frac{\omega}{kV_i}, i = 1, 2, 3$. Here $V_i(i = 1, 2, 3)$ are the dilatational wave velocities of first and second kind and shear wave velocity, respectively. Making use of eq.

(2) and eq. (5), we obtain the relevant stresses and fluid pressure that are given below

$$
\begin{aligned}
&\sigma_{rr} + s - Ku - G\Delta u = (A_{11}(r)C_1 + A_{12}(r)C_2 + A_{13}(r)A_1) coskze^{i\omega t},\\
&\sigma_{rz} = (A_{21}(r)C_1 + A_{22}(r)C_2 + A_{23}(r)A_1) sinkze^{i\omega t},\\
&s = (A_{31}(r)C_1 + A_{32}(r)C_2) coskze^{i\omega t},\\
&\tfrac{\partial s}{\partial r} = (B_{31}(r)C_1 + B_{32}(r)C_2) coskze^{i\omega t},
\end{aligned} \tag{6}
$$

where,

$$
\begin{aligned}
&A_{11} = \tfrac{2Np}{r} K_1(pr) + (2Np^2 + (P - 2N + Q\delta_1^2)(p^2 - k^2) + (Q + R\delta_1^2)(p^2 - k^2) K_0(pr) - \tfrac{2Np}{r} K_1(pr),\\
&A_{13} = \tfrac{2Np}{r} K_1(dr) + 2NkK_0(dr) - \tfrac{2Np}{r} K_1(dr),\\
&A_{21} = 2NkpK_1(pr),\\
&A_{23} = N(d^2 - k^2) K_1(dr),\\
&A_{31} = (Q + R\delta_1^2)(p^2 - k^2) K_0(pr),\\
&A_{33} = 0,\\
&\delta_1^2 = \tfrac{(PR - Q^2)V_1^{-2} - (Rm_{11} - Qm_{12})}{Rm_{12} - Qm_{22}},
\end{aligned} \tag{7}
$$

δ_2^2=similar expression as δ_1^2 with V_1 replaced by V_2,A_{12}, A_{22}, A_{32}=similar expression as A_{11}, A_{21}, A_{31},with p and δ_1^2,replaced by q and δ_2^2,respectively, and $m_{11} = \rho_{11} - ib\omega^{-1}$, $m_{12} = \rho_{12} + ib\omega^{-1}$, $m_{22} = \rho_{22} - ib\omega^{-1}$.

3. Boundary conditions and frequency equation

The boundary condition for stress free surface at $r = a$ for the pervious surface is

$$\sigma_{rr} + s - Ku - G\Delta u = \sigma_{rz} = s = 0. \tag{8}$$

The boundary condition for stress free surface at $r = a$ for an impervious surface is

$$\sigma_{rr} + s - Ku - G\Delta u = \sigma_{rz} = \tfrac{\partial s}{\partial r} = 0. \tag{9}$$

In eq. (8)and (9), $\triangle = \frac{\partial^2}{\partial r^2} + \frac{1}{r}\frac{\partial}{\partial r} + \frac{\partial^2}{\partial z^2}$, K is the foundation modulus given by $K = \frac{E}{(1+\mu)a}$, where E is the Young's modulus, μ is the poisson's ratio [9]. If the shear modulus G goes to zero, the Pasternak foundation will reduce to Wrinkler foundation [10]. Eq. (8)and (9) gives a system of three homogenous equations in the three arbitrary constants C_1, C_2, A_1 each for a pervious surface and impervious surface. A nontrivial solution can be obtained if the determinant of the coefficient vanishes. Accordingly, the frequency equation in the case of pervious surface is

$$|a_{ij}| = 0, i, j = 1, 2, 3, \tag{10}$$

where $a_{ij'}s$ are same as $A_{ij'}s$ with r is replaced by a in the eq. (7). In the case of an impervious surface the frequency equation becomes

$$|b_{ij}| = 0, i, j = 1, 2, 3, \tag{11}$$

here,

$$b_{31} = (Q + R\delta_1^2)p(p^2 - k^2)K_1(pa),$$

$$b_{32} = (Q + R\delta_2^2)q(q^2 - k^2)K_1(qa), \quad (12)$$

$$b_{33} = 0,$$

where $b_{ij} = a_{ij} \qquad i, j = 1, 2, j = 1, 2, 3.$

4. Numerical results

For the sake of numerical work, the dissipative coefficient b is taken zero and hence we obtained only real phase velocity. To analyze the frequency equations of axially symmetric vibrations of poroelastic cylindrical panel, it is convenient to introduce the following non dimensional parameters

$$a_1 = \tfrac{P}{H}, \quad a_2 = \tfrac{Q}{H}, \quad a_3 = \tfrac{R}{H}, \quad a_4 = \tfrac{N}{H},$$

$$d_1 = \tfrac{\rho_{11}}{\rho}, \quad d_2 = \tfrac{\rho_{12}}{\rho}, \quad d_3 = \tfrac{\rho_{22}}{\rho},$$

$$\tilde{x} = (\tfrac{V_0}{V_1})^2, \quad \tilde{y} = (\tfrac{V_0}{V_2})^2, \quad \tilde{z} = (\tfrac{V_0}{V_3})^2, \quad (13)$$

$$\rho = \rho_{11} + 2\rho_{12} + \rho_{22}, \quad H = P + 2Q + R, \quad V_0^2 = \tfrac{H}{\rho},$$

$$m = \tfrac{c}{c_0}, \quad c = \tfrac{\omega}{k}, \quad c_0^2 = \tfrac{N}{\rho}.$$

In the eq.(13), c is the phase velocity, m is the non-dimensional phase velocity, ka is the non-dimensional wavenumber. Employing the non-dimensional quantities in the frequency equation, we obtain a implicit relation between non-dimensional phase velocity and non-dimensional wavenumber. For the numerical process, three types of poroelastic solids are considered and then discussed. Of three poroelastic solids, two are sandstone saturated with kerosene and water, respectively [11, 12] and the third one is bony element. The physical parameters of first two materials pertaining to eq. (13) are given in the Table 1. Further, the values of bone poroelastic parameters A, N, Q, R and its mass coefficients ρ_{ij} are computed following the paper [13].The values of Young's modulus and Poisson ratio are taken to be 3×10^6 and 0.28, respectively as suggested in [13]. Phase velocity is computed using the bisection method implemented in MATLAB and the results are depicted in the figures1-8. In all the cases, curves are periodic and coincide in nature. Figures 1-8 show the plots of non-dimensional phase velocity against the wavenumber pertaining to different elastic foundations 10, 20, 50, and 100 respectively. From figures 1,3,5,7, it is observed that phase velocity of Material-I values are greater than phase velocity of Material-II for both pervious and impervious surface. From figure 1, 3, it is clear that phase velocity of Material-I pervious values are greater than phase velocity of Material-I impervious surface. Also, it is observed that the phase velocity for both pervious and impervious surface of Material-II is almost steady and constant beyond the wavenumber 1. From figures 5 and 7 it is observed that the phase velocities of Material-II both pervious and impervious surface is almost constant beyond the wavenumber 5. Figure-2 show the plot of non-dimensional phase velocity against non-dimensional wavenumber in the case of bone. From this figure, it is clear that phase velocity is same for both pervious and impervious surface beyond the wavenumber 7. From figure-4, it is observed that phase velocity is constant for both pervious and impervious surface beyond the wavenumber 5. From figure 6 and 8, it is clear that phase velocity is same for both pervious and impervious surface beyond the wavenumber 3.

Journal of Physics: Conference Series **662** (2015) 012007 doi:10.1088/1742-6596/662/1/012007

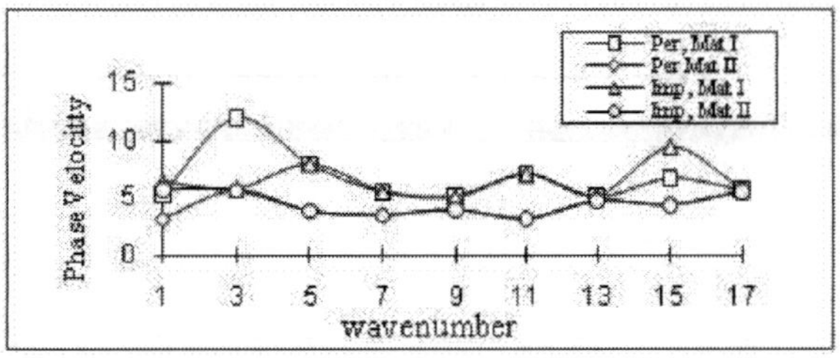

Figure 1. Variation of non-dimensional phase velocity with the wavenumber at elastic foundation ($K = 10$).

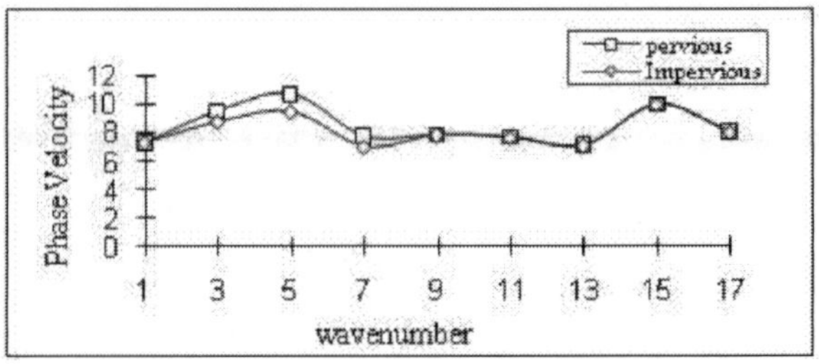

Figure 2. Variation of non-dimensional phase velocity with the wavenumber in the case of bone at elastic foundation ($K = 10$).

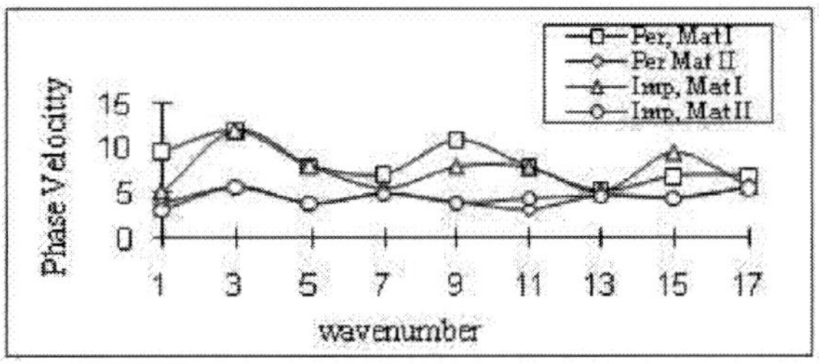

Figure 3. Variation of non-dimensional phase velocity with the wavenumber at elastic foundation ($K = 20$).

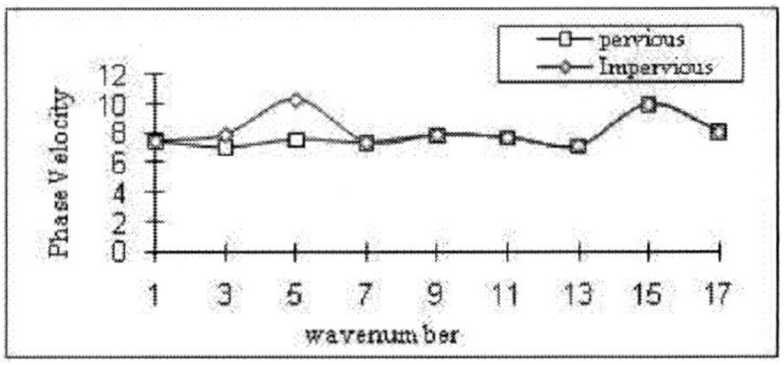

Figure 4. Variation of non-dimensional phase velocity with the wavenumber in the case of bone at elastic foundation ($K = 20$).

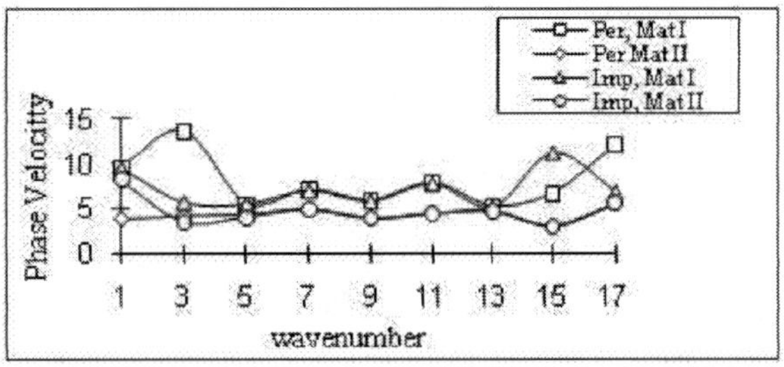

Figure 5. Variation of non-dimensional phase velocity with the wavenumber at elastic foundation ($K = 50$).

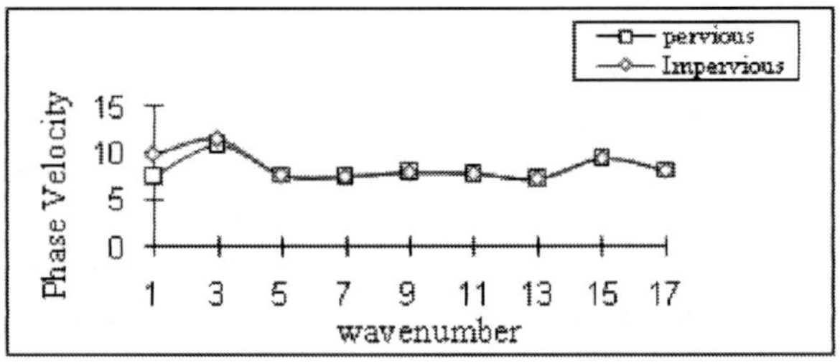

Figure 6. Variation of non-dimensional phase velocity with the wavenumber at elastic foundation ($K = 100$).

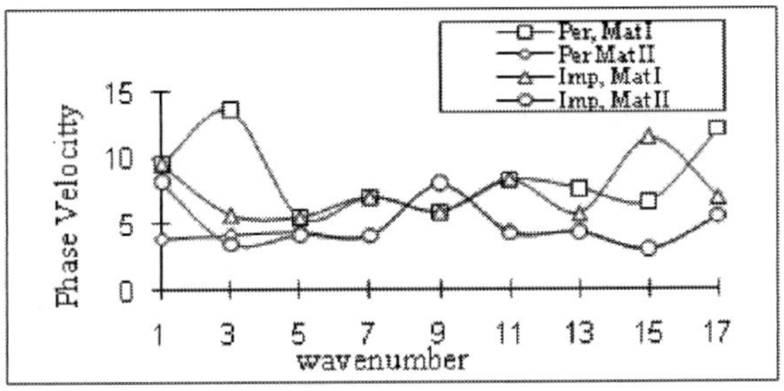

Figure 7. Variation of non-dimensional phase velocity with the wavenumber at elastic foundation ($K = 100$).

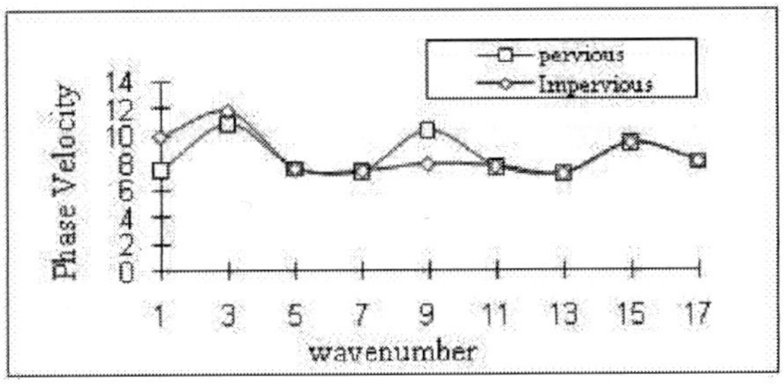

Figure 8. Variation of non-dimensional phase velocity with the wavenumber in the case of bone at elastic foundation ($K = 100$).

Table 1. Material parameters

Material parameters	Materia-I	Material-II
a_1	0.843	0.96
a_2	0.065	0.006
a_3	0.027	0.0289
a_4	0.234	0.412
d_1	0.901	0.876
d_2	-0.001	0
d_3	0.1	0.124
$\tilde{x}$	4.869	4.2977
$\tilde{y}$	0.998	0.912
$\tilde{z}$	3.85	2.126

5. Conclusion

Axially symmetric vibrations in poroelastic solid cylindrical panel resting on elastic foundation are investigated in the framework of Biot's theory. Non-dimensional phase velocity against non-dimensional wavenumber is computed for three types of poroelastic solids for different elastic foundations. Of three poroelastic solids, two are sandstone and third one is bony material. From the results, we can infer nature of surface has little influence over the values in the case of bone unlike sandstone cylinders.

References

[1] RamKumar 1965 Axially symmetric wave propagation in two layered elastic cylinder *J. Acous.Soc. of America* **38** 851-854

[2] Ponnuswamy P and Selvamani R 2011 Wave propagation in a homogenous isotropic cylindrical panel embedded on elastic medium *Int. J. of Math and Sci Computing* **1** 106-111

[3] Paliwal DN Pandey RK and Nath T 1996 Free vibrations of circular cylindrical shell on Wrinkle and Pasternak foundations *Int. J. of Pressure Vessel Pipe* **69** 79-89

[4] Biot MA 1956 Theory of propagation of elastic waves in a fluid-saturated porous solid *J. Acous Soc of America* **28** 168-191

[5] MallaReddy P and Tajuddin M 2010 Axially symmetric vibrations of poroelastic composite cylinder in the context of fretting fatigue *Spec. Top and Reviews in Porous Media* 1311-320

[6] Abdul Ghafar shah, TahirMahmood, Muhammad N.Naeem and Shahid H. Arshad 2011 Vibration characteristics of fluid filled cylindrical shells based on elastic foundations *Acta Mech* **216** 17-28

[7] Ahmed Shah S 2008 Axially symmetric vibrations of fluid filled poroelastic circular cylindrical shells*J. of Sou and Vibr* **318** 389-405

[8] Shah SA and Tajuddin M 2009 Axially symmetric vibrations of finite composite poroelastic cylinder *Int. J. of App. Mech. and Engineering* **14** 865-877
[9] Ghorbanpur Arani A, Mosallaie Barzoki, Kolahchi and Loghman A 2011 Pasternak foundation effect on the axial and torsional waves propagation in embedded DWCNTs using nonlocal elasticity cylindrical shell theory *J. of Mech Sci and Tech* **25** 2385-91
[10] Pronk C, Marion E VandenBol 1998 Wrinkler-Pasternak-Foundations, Notes on boundary conditions
[11] Fatt I 1957 The Biot-Willis elastic coefficient for a sand stone *J. App. Mech* 296-297
[12] Yew CH and Jogi PN 1976 Study of wave motions in fluid-saturated porous rocks *J. of the Acou. Soc. of America, USA* **60** 2-8
[13] JL Nowinski, CF Davis 1971 Propagation of longitudinal waves in circularly cylindrical bone elements *Transactions of ASME, J. of App.Mech* 578-584

Torsional surface waves in an inhomogeneous layer over a gravitating anisotropic porous half-space

Shishir Gupta and Abhijit Pramanik

Department of Applied Mathematics, Indian School of Mines, Dhanbad-826004, Jharkhand, India.

E-mails: shishir_ism@yahoo.com , abhijit_pramanik@yahoo.in

Abstract. The present work aims to deal with the propagation of torsional surface wave in an inhomogeneous layer over a gravitating anisotropic porous half space. The inhomogeneous layer exhibits the inhomogeneity of quadratic type. In order to show the effect of gravity the equation for the velocity of torsional wave has been obtained. It is also observed that for a layer over a homogeneous half space without gravity, the torsional surface wave does not propagate. An attempt is also made to assess the possible propagation of torsional surface waves in that medium in the absence of the upper layer. The effects of inhomogeneity factors and porosity on the phase velocity are depicted by means of graphs.

1. Introduction

The study of surface waves propagating over the surface of homogeneous and inhomogeneous half space is important to seismologists and in understanding the causes, estimation of damage due to earthquakes. A quite good amount of valuable information about the propagation of elastic waves is the monograph written by Ewing *et al.* [1]. Numerous papers on the subject have been published in various journals. A lot of information on the effect of heterogeneity in the study of surface waves due to line-load was given by Vrettos [2, 3]. Unfortunately the literature available on torsional waves in an inhomogeneous layer over an inhomogeneous half space are very much less than that of Rayleigh Waves, Love waves and Stoneley Waves. Lord Rayleigh [4], in his remarkable paper, showed that the isotropic homogenous elastic half-space does not allow a Torsional surface wave to propagate. Bhattacharya [5] has investigated the torsional wave propagation in a two layered circular cylinder with imperfect bond. Later on, Meissner [6] pointed out that in an inhomogeneous elastic half space with quadratic variation of shear modulus and density varying linearly with depth, torsional surface waves do exist. Torsional surface waves in an initially stressed cylinder have been studied by Dey and Dutta [7], and the existence and propagation of torsional surface waves in an elastic half space with void pores has been discussed by Dey *et al.* [8]. The propagation of torsional surface waves in non-homogeneous and anisotropic medium with polynomial and exponential variation in rigidity and constant density has been discussed by Dey *et al.* [9]. Whether torsional surface waves can propagate in the presence of a gravity field if the medium is elastic or dry sandy has been studied by Dey *et al.* [10]. The same authors [11] studied torsional surface waves will always propagate if the medium is taken as sandy or elastic in gravitating earth under initial stress. The presence of initial stress effects on the propagation of torsional surface waves in a non-homogeneous anisotropic medium which has been studied by Dey *et al.* [12].

The present paper attempts to study the possibility of propagation of torsional surface waves in an inhomogeneous layer over a gravitating anisotropic porous half space. It is observed that the presence of a gravity field always allow the torsional surface wave to propagate. The anisotropy also has a great effect in enhancing the velocity of torsional waves. The effect of inhomogeneity, porosity and gravity on the velocity of torsional wave was studied graphically.

2. Formulation of the problem

Let us consider a model which consists of an inhomogeneous elastic layer of thickness H, over anisotropic porous half space under gravity as shown in Figure 1. The inhomogeneity has been considered both in rigidity and density. The surface of contact is the plane $z = 0$ and the z-axis is taken to be positive downwards. The wave is assumed to propagate along radial direction.

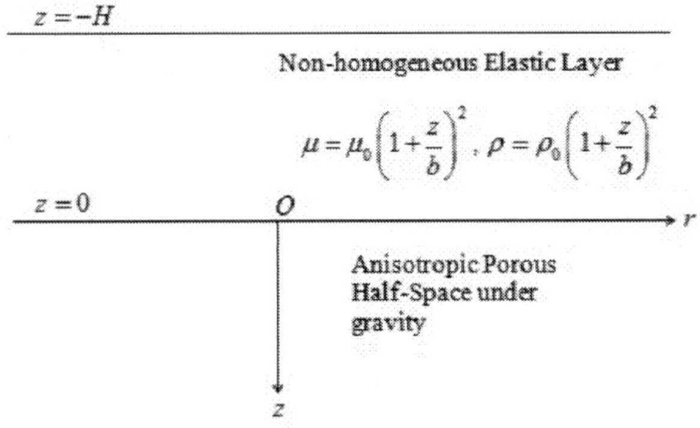

Figure 1. Geometry of the problem.

3. Solution of the problem

3.1. Solution for non-homogeneous elastic layer

The upper medium has been considered as a non-homogeneous elastic layer. The dynamical equation of motion for the torsional surface waves propagating in the radial direction can be written as

$$\frac{\partial \sigma_{r\theta}}{\partial r} + \frac{\partial \sigma_{z\theta}}{\partial z} + \frac{2\sigma_{r\theta}}{r} = \rho(z)\frac{\partial^2 v}{\partial t^2}, \tag{1}$$

where $v(r, z, t)$ is the displacement along the θ (azimuthal) direction. The stresses are related to the displacement component by

$$\sigma_{r\theta} = \mu(z)\left(\frac{\partial v}{\partial r} - \frac{v}{r}\right), \quad \sigma_{z\theta} = \mu(z)\left(\frac{\partial v}{\partial z}\right). \tag{2}$$

Using equation (2), equation (1) takes the form

$$\mu(z)\left(\frac{\partial^2}{\partial r^2} + \frac{1}{r}\frac{\partial}{\partial r} - \frac{1}{r^2}\right)v + \frac{\partial}{\partial z}\left(\mu(z)\frac{\partial v}{\partial z}\right) = \rho(z)\frac{\partial^2 v}{\partial t^2}. \tag{3}$$

where μ and ρ are the rigidity and the mass density of the media respectively, assumed to be function of depth z. Harmonic wave solution of Eq. (3) is of the form

$$v = V(z)\,J_1(kr)\exp(i\omega t), \tag{4}$$

where $V(z)$ is the solution of the following equation:

$$\frac{d^2V(z)}{dz^2}+\frac{\mu'(z)}{\mu(z)}\frac{dV(z)}{dz}-k^2\left(1-\frac{c^2}{{c_s}^2}\right)V(z)=0. \tag{5}$$

In the above equation, $c(=\omega/k)$ is the torsional wave velocity, $c_s=\sqrt{\mu(z)/\rho(z)}$, ω is the angular frequency, k is the angular wave number, and $J_1(Kr)$ is the first order Bessel function of first kind. For the layer we have considered

$$\mu=\mu_0\left(1+\frac{z}{b}\right)^2,\quad \rho=\rho_0\left(1+\frac{z}{b}\right)^2. \tag{6}$$

Assume

$$V=\frac{V_1}{\sqrt{\mu}}. \tag{7}$$

Now using (6) and (7), the equation for the upper layer, equation (5) takes the form

$$\frac{d^2V_1}{dz^2}-\lambda^2V_1=0, \tag{8}$$

where

$$\lambda^2=k^2\left(1-\frac{c^2}{c_0^2}\right),c_0=\sqrt{\frac{\mu_0}{\rho_0}}. \tag{9}$$

So, the solution of equation (8) is obtained as

$$V_1=A_1e^{\lambda z}+A_2e^{-\lambda z}. \tag{10}$$

Therefore

$$v=v_0\,(say)=\frac{A_1e^{\lambda z}+A_2e^{-\lambda z}}{\sqrt{\mu_0}\,(1+\frac{z}{b})}J_1(kr)\,e^{i\omega t}. \tag{11}$$

3.2. Solution for the porous half space

The half space is under the action of gravity field. The origin of the co-ordinate system is located at the considered point source on the free surface as shown in figure 1. Let r and be the radial and circumferential co-ordinates, respectively. It is assumed that the torsional wave travels in the radial direction and that all mechanical properties associated with it are independent of . For the torsional wave $u=w=0$ and $v=v(r,\,z,\,t)$, and the equation of motion in anisotropic porous medium under gravity may be written as Biot [13]:

$$\frac{\partial\sigma_{r\theta}}{\partial r}+\frac{\partial\sigma_{z\theta}}{\partial z}+\frac{2\sigma_{r\theta}}{r}-\frac{\partial}{\partial z}\left(d'gze_{z\theta}\right)-d'gz\frac{\partial}{\partial r}\left[\frac{1}{2}\left(\frac{\partial v}{\partial r}+\frac{v}{r}\right)\right]=d'\frac{\partial^2v}{\partial t^2}, \tag{12}$$

where $v(r,z,t)$ is displacement along the direction and g is acceleration due to gravity. For a gravitating anisotropic porous medium the stress is related to strain by

$$\sigma_{r\theta}=2Ne_{r\theta},\quad \sigma_{z\theta}=2Le_{z\theta},\quad e_{r\theta}=\frac{1}{2}\left(\frac{\partial v}{\partial r}-\frac{v}{r}\right),\quad e_{z\theta}=\frac{1}{2}\left(\frac{\partial v}{\partial z}\right), \tag{13}$$

where N and L are the shear modulus of elasticity along the r and z directions, respectively. Equation (12) using the above relation takes the form

$$\left(N-\frac{1}{2}d'gz\right)\left(\frac{\partial^2}{\partial r^2}+\frac{1}{r}\frac{\partial}{\partial r}-\frac{1}{r^2}\right)v+\left(L-\frac{1}{2}d'gz\right)\frac{\partial^2v}{\partial z^2}-\frac{d'g}{2}\frac{\partial v}{\partial z}=d'\frac{\partial^2v}{\partial t^2}, \tag{14}$$

where $d' = \rho_{11} - \frac{\rho_{12}^2}{\rho_{22}}$ and the dynamic co-efficients ρ_{11}, ρ_{12}, ρ_{22} take into account the inertia effects of the moving fluid and are related to the mass densities of the solid ρ_s and fluid ρ_w as written by Biot [14, 15] in the equations

$$\rho_{11} + \rho_{12} = (1 - f)\,\rho_s, \quad \rho_{12} + \rho_{22} = f\rho_w, \tag{15}$$

so that the mass densities of the bulk material are

$$\rho' = \rho_{11} + 2\rho_{12} + \rho_{22} = \rho_s + f\,(\rho_w - \rho_s)\,. \tag{16}$$

The dynamic co-efficients, moreover, obey the inequalities, as written by Biot [14]

$$\rho_{11} > 0,\ \rho_{22} > 0,\ \rho_{12} < 0,\ \rho_{11}\rho_{22} - \rho_{12}^2 > 0,$$

where f is the porosity of the layer. The parameters may be non-dimensionalized as

$$\gamma_{11} = \frac{\rho_{11}}{\rho'},\ \gamma_{12} = \frac{\rho_{12}}{\rho'},\ \gamma_{22} = \frac{\rho_{22}}{\rho'},\ \text{and}\, d = \frac{d'}{\rho'} = \gamma_{11} - \frac{\gamma_{12}^2}{\gamma_{22}}.$$

Thus one obtains the following:

(i) $d \to 1$, when the layer is non-porous solid;

(ii) $d \to 0$, when the layer is fluid;

(iii) $0 < d < 1$ when the layer is poro-elastic.

For the wave propagating along the r direction one may assume the solution of (14) as

$$v = V\,(z)\,J_1\,(kr)\exp\,(i\omega t)\,. \tag{17}$$

Here V is the solution of

$$V'' - \frac{Gkd}{2\left(1 - \frac{Gkdz}{2}\right)}V' - k^2\left[\frac{N\left(1 - \frac{GkdLz}{2N}\right)}{L\left(1 - \frac{Gkdz}{2}\right)} - \frac{c^2 d}{c_1^2\left(1 - \frac{Gkdz}{2}\right)}\right]V = 0, \tag{18}$$

where $c = \frac{\omega}{k}$ is the velocity of the torsional wave, $c_1 = \left(\frac{L}{\rho'}\right)^{\frac{1}{2}}$ is the velocity of the shear wave in an anisotropic elastic medium along the r direction, $G = \frac{g\rho'}{Lk}$ is Biots gravity parameter, k is the wave number, and $J_1\,(kr)$ is the Bessel function of the first kind and of order one. Substituting $V = \frac{\phi(z)}{\left(1 - \frac{Gkdz}{2}\right)^{\frac{1}{2}}}$ in equation (18), we have

$$\phi''\,(z) + \left[\frac{k^2G^2d^2}{\left(1 - \frac{Gkdz}{2}\right)^2} - k^2\left\{\frac{N\left(1 - \frac{GkdLz}{2N}\right)}{L\left(1 - \frac{Gkdz}{2}\right)} - \frac{c^2 d}{c_1^2\left(1 - \frac{Gkdz}{2}\right)}\right\}\right]\phi\,(z) = 0. \tag{19}$$

Again substituting $\delta = \frac{4}{Gd}\left(1 - \frac{Gdkz}{2}\right)$ in Equation (19), it may be reduced to

$$\phi''\,(\delta) + \left[-\frac{1}{4} + \frac{m}{\delta} + \frac{1}{4\delta^2}\right]\phi\,(\delta) = 0, \tag{20}$$

where $m = \frac{c^2}{c_1^2 G} + \frac{1}{Gd}\left(1 - \frac{N}{L}\right)$. Equation (20) is known as the Whittaker equation [16], whose solution is

$$\phi\,(\delta) = A_3 W_{-m,0}\,(-\delta) + A_4 W_{m,0}\,(\delta)\,. \tag{21}$$

As the solution should vanish at $z \to \infty$, that is, for $\delta \to -\infty$ we may take the solution as

$$\phi(\delta) = A_3 W_{-m,0}(-\delta).$$

Hence, the solution of Equation (14) may be written as

$$v = v_1\,(\text{say}) = \frac{A_3 W_{-m,0}\left[-\frac{4}{Gd}\left(1-\frac{Gdkz}{2}\right)\right]}{\left(1-\frac{Gdkz}{2}\right)^{1/2}} J_1(kr)\, e^{i\omega t}. \tag{22}$$

4. Boundary conditions

The appropriate boundary conditions are as follows:

- Upper surface is stress free i.e., at the upper boundary $z = -H$,

$$\mu_0 \frac{\partial v_0}{\partial z} = 0. \tag{23}$$

- At the interface $z = 0$
  Continuity of the displacement components,

$$v_0 = v_1. \tag{24}$$

  Continuity of stress components,

$$\mu_0\left(\frac{\partial v_0}{\partial z}\right) = L\left(\frac{\partial v_1}{\partial z}\right). \tag{25}$$

5. Results

Expanding the Whittaker function up to the quadratic term (as in the expansion, depth (z) occurs in the denominator of each term with higher powers and surface waves vanish with increase in depth), and substituting into boundary conditions (23), (24) and (25), we get

$$A_1 e^{-\lambda H}\left(\lambda\left(1-\frac{H}{b}\right)-\frac{1}{b}\right) - A_2 e^{\lambda H}\left(\lambda\left(1-\frac{H}{b}\right)+\frac{1}{b}\right) = 0, \tag{26}$$

$$\frac{A_1 + A_2}{\sqrt{\mu_0}} = A_3 e^{\frac{2}{Gd}}\left(-\frac{4}{Gd}\right)^{-m} Y_1, \tag{27}$$

$$\sqrt{\mu_0}\left[A_1\left(-\lambda+\frac{1}{b}\right) + A_2\left(\lambda+\frac{1}{b}\right)\right] = 2A_3 Lk e^{\frac{2}{Gd}}\left(-\frac{4}{Gd}\right)^{\frac{1}{2}} Y_2. \tag{28}$$

Now eliminating $A_i\,(i = 1,2,3)$ form (26), (27) and (28), we get

$$\frac{X_2\left(\lambda-\frac{1}{b}\right)e^{\lambda H} - X_1\left(\lambda+\frac{1}{b}\right)e^{-\lambda H}}{X_2 e^{\lambda H} + X_1 e^{-\lambda H}} = 2k\left(\frac{L}{\mu_0}\right)\left(\frac{Y_2}{Y_1}\right), \tag{29}$$

where X_1, X_2, Y_1, Y_2 are defined in Appendix.

6. Particular cases

6.1. Case I

When $N = L = \mu_1\,(say)$, $b \to \infty$ and $G \to 0$, that is, when the both of layer and half-space become homogeneous and isotropic and no gravity be present, then equation (29) takes the form

$$\tan\left[kH\sqrt{\frac{c^2}{c_0^2}-1}\right] = \left(\frac{\mu_1}{\mu_0}\right)\frac{\sqrt{1-\frac{c^2}{c_1^2}}}{\sqrt{\frac{c^2}{c_0^2}-1}},$$

which is well-known dispersion equation of Love type wave in a homogeneous crustal layer over a homogeneous half-space.

6.2. Case II

In the absence of the upper layer ($H \to 0$), then the velocity equation of (29) takes the form

$$\frac{c}{c_1} = \left[\frac{G}{(2+Gd)} + \frac{1}{d}\left(\frac{N}{L}-1\right)\right]^{\frac{1}{2}}.$$

This result coincides with Gupta *et al.* [17]. The above equation gives the following informations:

(i) In the absence of G, the wave is still available whose velocity is $\frac{c}{c_1} = \left[\frac{1}{d}\left(\frac{N}{L}-1\right)\right]^{\frac{1}{2}}$.
(ii) In case of isotropic medium, in the absence of layer and gravity (G) wave does not propagate.
(iii) For poro-elastic medium $0 < d < 1$, it is expected that the velocity will be more than that of elastic medium (i.e., $d = 1$).
(iv) When the medium is perfectly liquid ($d \to 0$), the velocity of torsional wave tends to zero, because we have, $d = d'/\rho'$ as $d \to 0$, $\rho' \to \infty$.

So $c_1 = \sqrt{\frac{L}{\rho'}} \to 0$, i.e., velocity of the shear wave in anisotropic elastic medium is zero.

7. Numerical computation

Based on the dispersion equation (29), numerical results of phase velocity are provided to show the effect of various inhomogeneity on the propagation of torsional surface waves for the following values of elastic constants.
For the upper layer [18]:

$$\mu_0 = 93\times10^9 N/m^2,\ \rho_0 = 7450\,Kg/m^3.$$

For the half-space [19]:

$$L = 7.5\times10^9\,N/m^2,\ \ N = 6\times10^9\,N/m^2,\ \rho' = 3364\,Kg/m^3.$$

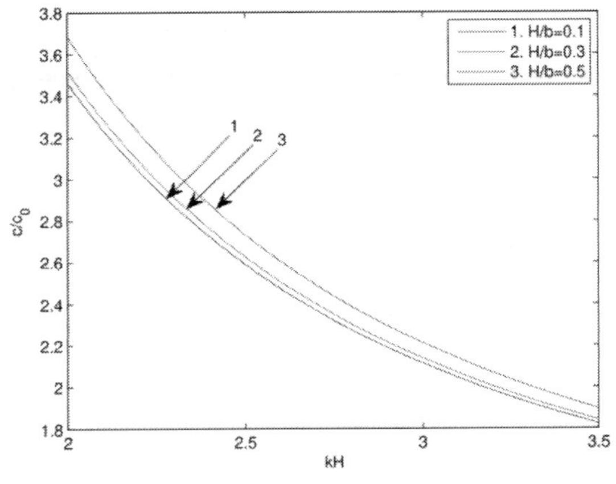

Figure 2. Effect of inhomogeneity (H/b) in the propagation of torsional wave.

Figure 2 is representing the different phase velocity of torsional waves for the different values of the inhomogeneity parameter (H/b) where the Biots Gravity parameter and the porosity of the half-space is taken to be fixed as 0.5 and 0.1, respectively, and it has been found that as the value of the inhomogeneity parameter (H/b) increases, the phase velocity of the torsional wave increases for a particular dimensionless wave number (kH).

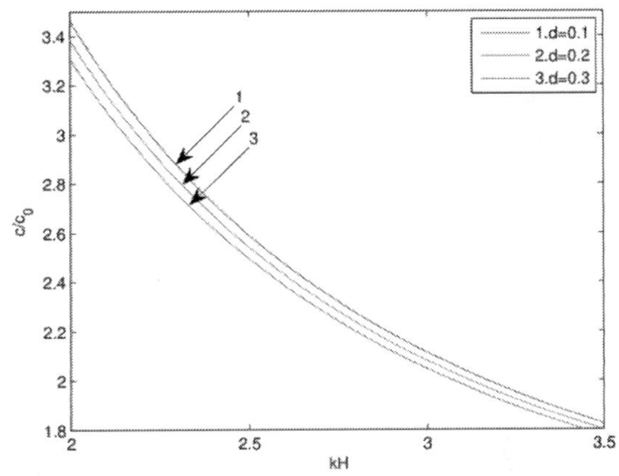

Figure 3. Effect of porosity (d) in the propagation of torsional wave.

In Figure 3, attempt has been made to show the effect of porosity (d) on the propagation of torsional waves. The graph has been plotted for different values of porosity where the inhomogeneity parameter and Biots Gravity parameter is kept fixed at 0.1 and 0.5, respectively. This shows that as the value of the porosity increases, the phase velocity of the torsional wave decreases for a particular dimensionless wave number (kH).

The graph in Figure 4 has been drawn to show the effect of Biots Gravity parameter on torsional wave propagation. The value of the inhomogeneity parameter and the porosity are 0.1 and 0.3, respectively. Here we see that as the value of the Biots Gravity parameter increases, the phase velocity of the torsional wave decreases for a particular dimensionless wave number (kH).

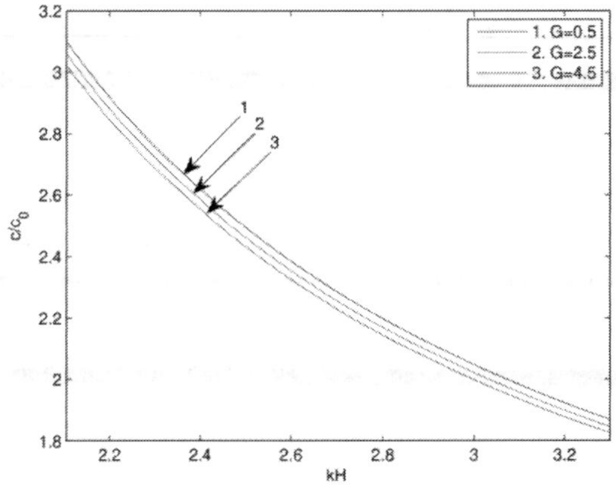

Figure 4. Effect of Biots Gravity parameter (G) in the propagation of torsional wave.

8. Conclusion

In this problem the propagation of torsional surface waves in an inhomogeneous layer of finite thickness over a gravitating porous half-space has been studied. The displacements in the layer and in the half space have been derived separately. Then using the asymptotic expansion of the Whittakers function the dispersion relation has been computed. Effects of inhomogeneity parameter, porosity and gravity factor on the phase velocity for fixed wave number are depicted by means of Graphs. And these show that the dimensionless phase velocity decreases as the non-dimensional wave number increases. As the gravity and porosity increases the phase velocity decreases and as the anisotropy increases the phase velocity increases.

Appendix

$$X_1 = \lambda\left(1 - \frac{H}{b}\right) - \frac{1}{b},$$
$$X_2 = \lambda\left(1 - \frac{H}{b}\right) + \frac{1}{b},$$
$$Y_1 = 1 + \frac{(m+0.5)^2}{\frac{4}{Gd}} + \frac{(m+0.5)^2(m+1.5)^2}{2\left(\frac{4}{Gd}\right)^2},$$
$$Y_2 = \left(\frac{1}{2} - \frac{m}{\frac{4}{Gd}}\right)\left\{1 + \frac{(m+0.5)^2}{\frac{4}{Gd}} + \frac{(m+0.5)^2(m+1.5)^2}{2\left(\frac{4}{Gd}\right)^2}\right\} - \left\{\frac{(m+0.5)^2}{\left(\frac{4}{Gd}\right)^2} + \frac{(m+0.5)^2(m+1.5)^2}{\left(\frac{4}{Gd}\right)^3}\right\}.$$

References

[1] Ewing W M, Jardetzky W S and Press F 1957 *Elastic waves in layered media* (New York: McGraw-Hill)
[2] Vrettos Ch 1990 In plane vibrations of soil deposits with variable shear modulus I. Surface waves *Int J Numer Anal Methods Geomech* **14** 209-222
[3] Vrettos Ch 1990 In plane vibrations of soil deposits with variable shear modulus II. Line load *Int J Numer Anal Methods Geomech* **14** 649-662
[4] Rayleigh Lord 1945 *Theory of sound* (New York: Dover)
[5] Bhattacharya R C 1975 On the torsional wave propagation in a two-layered circular cylinder with imperfect bonding *Proc. Indian Natn. Sci. Acad.* **41**(6) 613-619
[6] Meissner E 1921 Elastic oberflachenwellen mit dispersion in einem inhomogenen medium *Viertelgahrsschriftder Naturforschenden Gesellschaft, Zurich* **66** 181-195
[7] Dey S and Dutta D 1992 Torsional wave propagation in an initially stressed cylinder *Proc Indian Nat Sci Acad A* **58** 425-429
[8] Dey S, Gupta S and Gupta A K 1993 Torsional surface wave in an elastic half space with void pores *Int J Numer Anal Methods Geomech* **17** 197-204

[9] Dey S, Gupta A K and Gupta S 1996 Torsional surface waves in non-homogeneous and anisotropic medium *J Acoust Soc Am* **99** 2737-41
[10] Dey S, Gupta A K and Gupta S 1998 Propagation of torsional surface waves in dry sandy medium under gravity *J Math Mech Solids* **3** 229-235
[11] Dey S, Gupta A K, Gupta S and Prasad A M 2000 Torsional surface waves in non-homogeneous anisotropic medium under initial stress *J Eng Mech* **125** 1120-23
[12] Dey S, Gupta A K and Gupta S 2002 Effect of gravity and initial stress on torsional surface waves in dry sandy medium *J Eng Mech* **128** 1115-18
[13] Biot M A 1956 *Mechanics of incremental deformation* (New York: John Willey & Sons)
[14] Biot M A 1956 Theory of propagation of elastic waves in a fluid-saturated porous solid *J Acoust Soc Am* **28** 168-191
[15] Biot M A 1962 Mechanics of deformation and acoustic propagation in porous medium *J Appl Phys* **33** 1482-98
[16] Whittaker E T and Watson G N 1991 *A Course of Modern Analysis* (Cambridge University Press, Cambridge)
[17] Gupta A K and Gupta S 2011 Torsional surface waves in gravitating anisotropic porous half space *Mathematics and Mechanics of Solids* **16**(4) 445-450
[18] Gubbins D 1990 *Seismology and Plate Tectonics* (Cambridge University Press, Cambridge)
[19] Chattopadhyay A, Gupta S, Sahu S A and Dhua S 2013a Torsional surface waves in heterogeneous anisotropic half-space under initial stress *Archive of Applied Mechanics* **83** 357-366

Torsional vibrations of circular poroelastic plates

S Ahmed Shah[1] and C Nageswara Nath[2]

[1] Department of Mathematics, Deccan College of Engineering and Technology, Hyderabad 500 001, India

[2] Department of Mathematics, CMR Technical Campus, Hyderabad - 501 401, India

E-mail: ahmed_shah67@yahoo.com, nagesh.nath@rediffmail.com

Abstract. Torsional vibrations of annular poroelastic circular plates are studied in the framework of Biot's theory of wave propagation in porous solids. The frequency equation of torsional vibrations is same for pervious and impervious surfaces. The frequency equation of torsional vibrations is obtained by using the traction free boundaries and edges of the annular circular plate. Non-dimensional frequency of annular plate is computed as a function of aspect ratio. The frequency equation is discussed for first two modes. Frequency equation of torsional vibrations of uniform circular poroelastic plate is obtained as a particular case of the annular plate. Resonant frequency of infinite poroelastic plate is obtained as a limiting case of annular plate. The expressions for phase velocity and attenuation are obtained and these are computed for two different poroelastic materials as a function of frequency in presence of dissipation. Phase velocity is almost same for the considered poroelastic materials and dissipations. Results of previous study are obtained as a particular case of the present investigation.

1. Introduction

Propagation of waves and vibrations in elastic plates has been extensively studied in monograph prepared by Mindlin [8]. The study of torsional vibrations of an elastic solid is important in several fields of engineering, for example, soil mechanics, transmission of power through shafts with flange at the end as integral part of the shaft. It is now recognized that virtually no high-speed equipment can be properly designed without obtaining solution to torsional vibrations problems. Examples of torsional vibrations are vibrations in gear train and motor-pump shafts. Thus, from engineering point of view the study of torsional vibrations has greater interest. Such vibrations, for example, are used in delay lines. The least mode of torsional vibration is dispersionless and is used for storing information in form of pulses. It receives a pulse from a transmitter and returns the pulse to the receiving end, essentially undistorted but delayed by a time interval which is determined by the path length and velocity. The memory properties of delay lines are used extensively in computers. Further, based on reflections and refractions during the propagation of a pulse imperfection can be identified. Other use of torsional vibrations is the measurement of the shear modulus of a crystal. These modes find applications in the design of resonators, transducers and sensors. Tajuddin and Ahmed Shah [9] studied torsional vibrations of hollow poroelastic cylinders in presence of dissipation. They obtained phase velocity, group velocity and attenuation of torsional vibrations for different dissipations. Ahmed Shah [1,2] studied the axially symmetric vibrations of fluid-filled poroelastic circular cylindrical shells and spherical shells of various wall-thicknesses. Gupta et al.[6] investigated the effect of presence of rigid boundary on the propagation of torsional surface wave in a homogeneous layer over a semi infinite heterogeneous half space with an irregularity at the interface. Ahmed Shah and Tajuddin [3] studied torsional vibrations of poroelastic spheroidal shells. Torsional vibrations of circular elastic plates with thickness steps are investigated by Kang et al.[7]. Zhou et al.[11] studied three dimensional vibrations analysis of circular and annular plates by using Chebyshev-Ritz method.

In the present analysis, the frequency equation of torsional vibrations of a homogeneous and isotropic poroelastic annular circular plate of uniform thickness is derived and then discussed. Let the boundaries of the annular plate are free from stress. The frequency equation is discussed for first two modes of vibrations. Frequency is computed as a function of aspect ratio for the considered poroelastic materials. The expressions for non-dimensional phase velocity and attenuation are obtained in presence of dissipation for an infinite poroelastic plate. Phase velocity and attenuation are computed as a function of frequency for two types of poroelastic materials and then discussed. The phase velocity is almost same for the considered poroelastic materials and different dissipations. Attenuation varies widely for the considered dissipations.

2. Basic equations

The equations of motion of a homogeneous, isotropic poroelastic solid [4] in presence of dissipation b are

$$\mathrm{N}\nabla^2\vec{\mathrm{u}} + (\mathrm{A}+\mathrm{N})\nabla \mathrm{e} + \mathrm{Q}\nabla \in = \frac{\partial^2}{\partial \mathrm{t}^2}(\rho_{11}\vec{\mathrm{u}} + \rho_{12}\vec{\mathrm{U}}) + \mathrm{b}\frac{\partial}{\partial \mathrm{t}}(\vec{\mathrm{u}} - \vec{\mathrm{U}}),$$

$$\mathrm{Q}\nabla \mathrm{e} + \mathrm{R}\nabla \in = \frac{\partial^2}{\partial \mathrm{t}^2}(\rho_{12}\vec{\mathrm{u}} + \rho_{22}\vec{\mathrm{U}}) - \mathrm{b}\frac{\partial}{\partial \mathrm{t}}(\vec{\mathrm{u}} - \vec{\mathrm{U}}). \quad (1)$$

where ∇^2 is the Laplace operator, $\vec{u}(u, v, w)$ and $\vec{U}(U, V, W)$ are displacements of solid and liquid respectively, e and ϵ are the dilatations of solid and liquid; A, N, Q, R are all poroelastic constants and ρ_{ij} (i,j=1,2) are the mass coefficients [4]. The poroelastic constants A, N correspond to familiar Lame constants in purely elastic solid. The coefficient N represents the shear modulus of the solid. The coefficient R is a measure of the pressure required on the liquid to force a certain amount of the liquid into the aggregate while total volume remains constant. The coefficient Q represents the coupling between the volume change of the solid to that of liquid. The stresses σ_{ij} and the liquid pressure s of the poroelastic solid are

$$\sigma_{\mathrm{ij}} = 2\mathrm{N}\mathrm{e}_{\mathrm{ij}} + (\mathrm{A}\mathrm{e} + \mathrm{Q} \in)\delta_{\mathrm{ij}}, (\mathrm{i},\mathrm{j} = \mathrm{r}, \theta, \mathrm{z}),$$

$$\mathrm{s} = (\mathrm{Q}\mathrm{e} + \mathrm{R} \in). \quad (2)$$

where δ_{ij} is the well-known Kronecker delta function and e_{ij} are the strain components of poroelastic solid.

3. Secular equation

Let (r, θ, z) be the cylindrical polar coordinates. Consider a homogeneous, isotropic annular circular poroelastic plate with inner and outer radii r_1 and r_2, respectively, whose axis is in the direction of z-axis and uniform thickness h_2. The flat surfaces of the plate lie at z=0 and z=h_2. The boundaries of the isotropic poroelastic plate are free from stress. The only non-zero displacement component of the solid and liquid media are v and V respectively. These displacements are functions of r, z and time, t. Then the equations of motion of poroelastic solid [4], that is equation (1) reduces to

$$\mathrm{N}(\nabla^2 - \frac{1}{\mathrm{r}^2})\mathrm{v} = \frac{\partial^2}{\partial \mathrm{t}^2}(\rho_{11}\mathrm{v} + \rho_{12}\mathrm{V}) + \mathrm{b}\frac{\partial}{\partial \mathrm{t}}(\mathrm{v} - \mathrm{V}),$$

$$0 = \frac{\partial^2}{\partial t^2}(\rho_{12}v + \rho_{22}V) - b\frac{\partial}{\partial t}(v - V). \tag{3}$$

Let the propagation mode shapes of solid and liquid v and V are

$$v = f_1(r)f_2(z)e^{i\omega t},$$

$$V = F_1(r)F_2(z)e^{i\omega t}, \tag{4}$$

where t is time, ω is circular frequency, and i is the complex unity. Substitution of equation (4) into equation (3) yield

$$N[f_1''f_2 + \frac{1}{r}f_1'f_2 + f_1f_2'' - \frac{1}{r^2}f_1f_2] = -\omega^2[K_{11}f_1f_2 + K_{12}F_1F_2],$$

$$, \qquad 0 = -\omega^2[K_{12}f_1f_2 + K_{22}F_1F_2]. \tag{5}$$

In equation (5), a dash over f_1 represents differentiation with respect to r, and a dash over f_2 represents differentiation with respect to z, and

$$K_{11} = \rho_{11} - \frac{ib}{\omega}, \quad K_{12} = \rho_{12} + \frac{ib}{\omega},$$

$$K_{22} = \rho_{22} - \frac{ib}{\omega}. \tag{6}$$

From second equation of (5), we get

$$F_1F_2 = -\frac{K_{12}}{K_{22}}f_1f_2. \tag{7}$$

Substitution of equation (7) into the first equation of (5) we obtain

$$f_1''f_2 + \frac{1}{r}f_1'f_2 + f_1f_2'' - \frac{1}{r^2}f_1f_2 + \xi_3^2f_1f_2 = 0, \tag{8}$$

where

$$\xi_3^2 = \frac{\omega^2}{V_3^2} = \frac{\omega^2(K_{11}K_{22} - K_{12}^2)}{NK_{22}}. \tag{9}$$

In equation (9), V_3 is shear wave velocity [4]. By using the method of separation of variables, equation (8) is simplified to

$$\frac{f_1'' + \frac{1}{r}f_1' - \frac{1}{r^2}f_1 + \xi_3^2f_1}{f_1} = -\frac{f_2''}{f_2} = k^2(\text{let}). \tag{10}$$

Then equation (10) is reduced to two separate equations given below

$$[\frac{d^2}{dr^2} + \frac{1}{r}\frac{d}{dr} + (\alpha_3^2 - \frac{1}{r^2})]f_1 = 0, \tag{11}$$

$$[\frac{d^2}{dz^2} + k^2]f_2 = 0. \tag{12}$$

In equation (11), α_3^2 is defined as

$$\alpha_3^2 = \xi_3^2 - k^2. \tag{13}$$

Solution of equation (11) is [bounded solution in case of $\alpha_3 = 0$]

$$f_1(r) = \begin{cases} [C_1J_1(\alpha_3 r) + C_2Y_1(\alpha_3 r)], \text{when } \alpha_3 \neq 0, \\ [C_1 r], \text{when } \alpha_3 = 0. \end{cases} \tag{14}$$

where C_1 and C_2 are the constants, J_1 and Y_1 are Bessel functions of first and second kind respectively each of order one. The solution of equation (12) is

$$f_2(z) = \begin{cases} [D_1\cos(kz) + D_2\sin(kz)], \text{when } k \neq 0, \\ [D_1 z + D_2], \text{when } k = 0. \end{cases} \tag{15}$$

where k, D_1 and D_2 are the constants. Substituting equations (14) and (15) into the first equation of (4), the propagation modes shapes are given by

$$v(r,z,t) = [C_1J_1(\alpha_3 r) + C_2Y_1(\alpha_3 r)][D_1\cos(kz) + D_2\sin(kz)]e^{i\omega t}, \tag{16}$$

for $\alpha_3 \neq 0$ and $k \neq 0$. From equation (4), it can be seen that the normal strains e_{rr},$e_{\theta\theta}$,e_{zz} all are zero. Therefore the dilatations of solid and liquid media each is zero. Hence the liquid pressure s is identically zero. Thus for torsional vibrations no distinction between a pervious and an impervious surface is made. Considering the edges and surfaces of the poroelastic plate to be stress free, the frequency equation obtained for torsional vibrations is same for both pervious and impervious surfaces. Following equation (2), the only non-zero stresses are

$$\sigma_{r\theta} = N(\frac{\partial v}{\partial r} - \frac{v}{r}),$$

$$\sigma_{z\theta} = N\frac{\partial v}{\partial z}. \tag{17}$$

Substituting equation (16) into equation (17), these stresses are simplified to

$$\sigma_{r\theta} = -N[C_1J_2(\alpha_3 r) + C_2Y_2(\alpha_3 r)][D_1\cos(kz) + D_2\sin(kz)]e^{i\omega t}, \tag{18}$$

$$\sigma_{z\theta} = Nk[C_1J_1(\alpha_3 r) + C_2Y_1(\alpha_3 r)][-D_1\sin(kz) + D_2\cos(kz)]e^{i\omega t}. \tag{19}$$

4. Boundary conditions and frequency equation

The stress free boundary conditions for torsional vibrations at the inner and outer edges and flat surfaces of the poroelastic annular circular plate are

$$\sigma_{r\theta} = 0, \text{at } r = r_1, r_2, \tag{20}$$

$$\sigma_{z\theta} = 0, \text{at } z = 0, h_2, \tag{21}$$

$$s = 0, \frac{\partial s}{\partial r} = 0, \text{at } r = r_1, r_2 \text{ and } z = 0, h_2. \tag{22}$$

Equations (20) and (21) together with first equation of (22) are to be satisfied for a pervious surface while equations (20) and (21) together with second equation of (22) are to be satisfied for an impervious surface. Since the considered vibrations are shear vibrations, the dilatations of solid and liquid media each is zero, thereby liquid pressure s developed in solid-liquid aggregate will be identically zero and no distinction between pervious and impervious surface is made. Thus equation (22) is satisfied identically. Equations (18) and (20) together yield the frequency equation of torsional vibrations

$$J_2(\alpha_3 r_1)Y_2(\alpha_3 r_2) - J_2(\alpha_3 r_2)Y_2(\alpha_3 r_1) = 0, \tag{23}$$

by eliminating the constants C_1 and C_2. Equation (23) is the frequency equation of torsional vibrations of hollow poroelastic cylinder studied by Tajuddin and Ahmed Shah [9]. Similarly, employing equation (19) into equation (21), after elimination of the constants D_1 and D_2, we get the equation $\sin(kh_2) = 0$.

Now we consider $\sin(kh_2) = 0$, or $kh_2 = q\pi, q = 1, 2, 3, \ldots\ldots$ then we have

$$k = \frac{q\pi}{h_2}, q = 1, 2, 3, \ldots.. \tag{24}$$

Let $r_2 - r_1 = h_1$ and $g = r_2/r_1$ sothat $h_1/r_1 = g - 1$ and $h_1/r_2 = (g-1)/g$. Also let the aspect ratio be $h_2/h_1 = h$. Let

$$R_n^2 = \alpha_3^2 h_1^2, \tag{25}$$

then equation (23) is reduced to

$$J_2(\frac{R_n g}{g-1})Y_2(\frac{R_n}{g-1}) - J_2(\frac{R_n}{g-1})Y_2(\frac{R_n g}{g-1}) = 0. \tag{26}$$

Substituting equation (24) into equation (13) and after necessary simplification, the resonant frequencies of the torsional vibrations of poroelastic annular plate are given by the equation

$$\omega = V_3\sqrt{(\frac{R_n^2}{h_1^2} + \frac{q^2\pi^2}{h_2^2})}. \tag{27}$$

In equation (27), R_n is the nth root of equation (26) for different values of g. By eliminating liquid effects from equation (27), the results of Kang et al.[7] are recovered as a particular case. Under the limiting conditions when $r_1 \to 0$ then $r_2 \to h_1$ and the poroelastic annular circular plate reduces to circular uniform poroelastic plate of thickness h_2. Under these conditions, equation (23) reduce to

$$J_2(\alpha_3 h_1) = 0. \tag{28}$$

And the resonant frequencies are given by the equation

$$\omega = V_3\sqrt{(\frac{R_n^2}{r_2^2} + \frac{q^2\pi^2}{h_2^2})}, \tag{29}$$

where Rn is the nth root of equation (28). When $r_2 \to \infty$ the uniform circular poroelastic plate reduces to infinite poroelastic plate of thickness h_2. Thus frequency equation (29) is reduced to

$$\omega = \frac{V_3 q\pi}{h_2}, \tag{30}$$

which is independent of R_n. Equation (30) gives the resonant frequencies of torsional vibrations of an infinite poroelastic plate of thickness h_2. From equation (24), it is clear that k is wavenumber. Thus substituting equation (24) into the equation (30), it reduces to

$$\omega = kV_3. \tag{31}$$

To analyze the frequency equations (27) and (31), it is convenient to introduce the following non-dimensional parameters:

$$a_4 = \frac{N}{H}, \tilde{z} = (\frac{V_0}{V_3})^2, h = \frac{h_2}{h_1},$$

$$m_{11} = \frac{\rho_{11}}{\rho}, \quad m_{12} = \frac{\rho_{12}}{\rho}, \quad m_{22} = \frac{\rho_{22}}{\rho}, \quad b_1 = \frac{bh_2}{\rho C_0}, \quad \Omega = \frac{\omega h_2}{C_0}, \tag{32}$$

where b_1, ω and h are non-dimensional dissipation, frequency and aspect ratio respectively, $V_0^2 = \frac{H}{\rho}$, $C_0^2 = \frac{N}{\rho}$ are the reference velocities, $\rho = \rho_{11} + 2\rho_{12} + \rho_{22}$ and H=P+2Q+R with P=A+2N. By using equation (32) into equation (27), it is simplified to

$$\Omega = \frac{(R_n^2 h^2 + q^2\pi^2)^{\frac{1}{2}}}{(a_4\tilde{z})^{\frac{1}{2}}}. \tag{33}$$

Now we employ equation (32) into equation (31) and with the help of equations (6) and (9), simplify it in the form

$$\frac{Nk^2}{\rho\omega^2} = E_r - iE_i, \tag{34}$$

where E_r and E_i are

$$E_r = \frac{\Omega^2 m_{22}(m_{11}m_{22} - m_{12}^2) + b_1^2}{\Omega^2 m_{22}^2 + b_1^2},$$

$$E_i = \frac{\Omega b_1 (m_{12} + m_{22})^2}{\Omega^2 m_{22}^2 + b_1^2}. \tag{35}$$

5. Phase velocity and attenuation

Poroelastic medium is dissipative in nature and thus the wavenumber k is complex. The waves generated obey diffusion type process and therefore get attenuated. Let $k=k_r+ik_i$, where k_r is real and k_i is the imaginary part of the wavenumber k. The real and imaginary part of the wavenumber corresponds to propagation and attenuation of waves. Hence the phase velocity C_p and attenuation X_h [4] are

$$C_p = \frac{\omega}{|k_r|}, \quad X_h = \frac{1}{|k_i|}. \tag{36}$$

By substituting $k=k_r+ik_i$ in equation (34) and using equation (32), we can separate the real and imaginary parts of the wavenumber as

$$k_r = \frac{1}{\sqrt{2}h_2}(B_3 + B_4)^{\frac{1}{2}}, \quad k_i = \frac{1}{\sqrt{2}h_2}(B_3 - B_4)^{\frac{1}{2}}, \tag{37}$$

where the expressions for B_3 and B_4 are

$$B_3 = \Omega^2(E_r^2 + E_i^2)^{\frac{1}{2}}, \quad B_4 = \Omega^2 E_r. \tag{38}$$

In equation (38), E_r and E_i are defined in equation (35). Thus the non-dimensional phase velocity $\frac{C_p}{C_0}$ and attenuation $\frac{X_{h_2}}{h_2}$ are

$$\frac{C_p}{C_0} = \sqrt{2}\Omega(B_3 + B_4)^{-\frac{1}{2}}, \quad \frac{X_{h_2}}{h_2} = \sqrt{2}(B_3 - B_4)^{-\frac{1}{2}}, \tag{39}$$

where B_3 and B_4 are defined in equation (38).

6. Results and discussion

Two types of poroelastic materials are considered to carry out the computational work, one is sandstone saturated with kerosene, say Material-I [5], the other one is sandstone saturated with water, say Material-II [10], whose non-dimensional physical parameters are given in Table 1.For a given poroelastic material, the non-dimensional frequency is computed as a function of aspect ratio, phase velocity and attenuation are determined as a function of non-dimensional frequency.The different dissipation parameters (b_1) chosen are 0.01, 0.1 and 1. Three values of g have been considered for the purpose of computation. These values are g=1.034, 3 and g $\rightarrow \infty$. These three cases physically represent the results related to poroelastic thin annular plate, poroelastic thick annular plate and uniform circular poroelastic plate, respectively. The first two roots of equation (26) for thin annular plate, thick annular plate and uniform circular plate are 3.1423, 6.2835; 3.736, 6.6477; and 5.1356, 8.4172, respectively. The frequency is obtained for first two roots (modes) of the equation (26) for the considered poroelastic materials with q=1,2,3,4 and 5 and presented in figure 1 -figure 6. Phase velocity and attenuation of an infinite poroelastic plate are presented as a function of frequency in figure 7 and figure 8 respectively, for different dissipations.

Table 1. Material parameters

Material/Parameter	a_4	$\tilde{z}$	m_{11}	m_{12}	m_{22}
Material-I	0.234	3.851	0.901	-0.001	0.101
Material-II	0.412	2.129	0.877	0	0.123

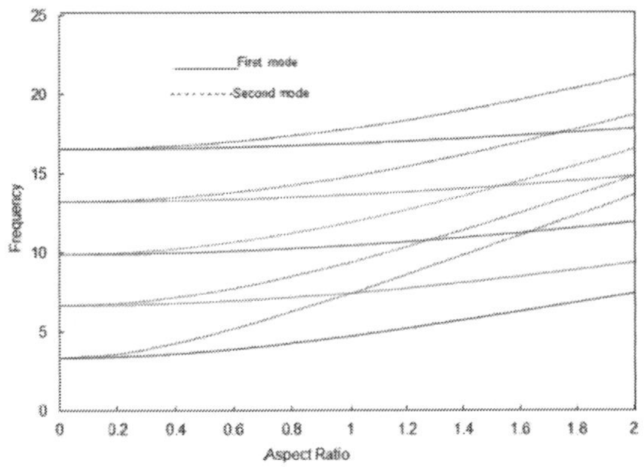

Figure 1. Frequency as a function of aspect ratio (Mat-I, Thin annular plate).

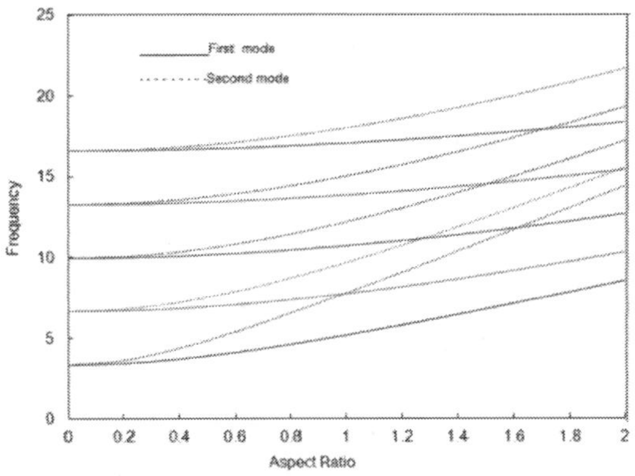

Figure 2. Frequency as a function of aspect ratio (Mat-I, Thick annular plate).

Figure 1 shows the frequency as a function of aspect ratio for thin annular poroelastic plate made of Material-I. The frequency is presented for first two modes of vibrations. From figure 1 it is clear that as the aspect ratio increases, the frequency increases gradually otherwise it is almost linear. The frequency for second mode is higher than that of the frequency of first mode. But it is interesting to note that frequency of first and second mode is same in for all the case. Variation of frequency in case of thick annular plate (Fig.2) and uniform circular plate

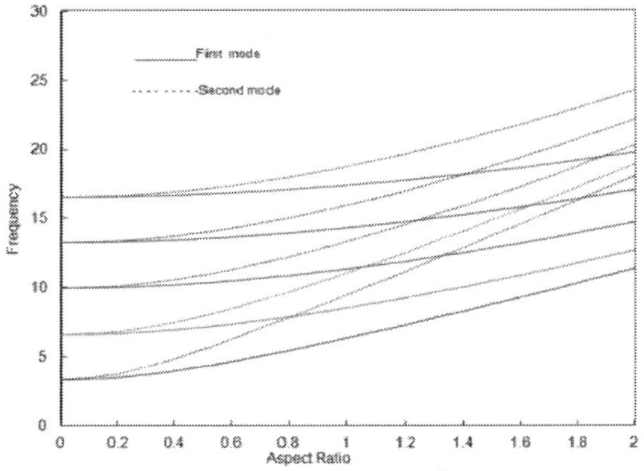

Figure 3. Frequency as a function of aspect ratio (Mat-I, Uniform circular plate).

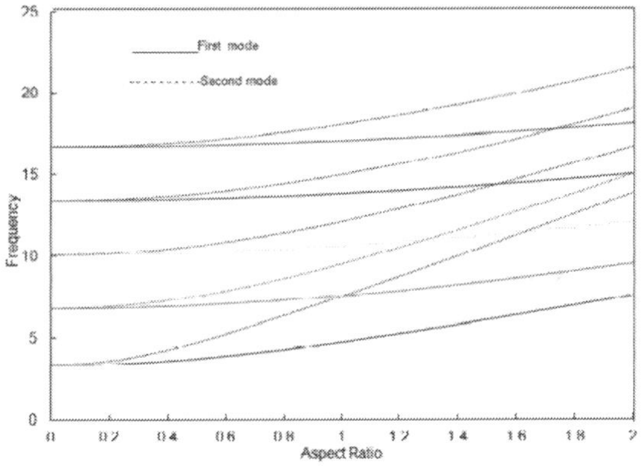

Figure 4. Frequency as a function of aspect ratio (Mat-II, Thin annular plate).

(Fig.3) made of Material-I is similar to that of thin annular plate. The variation of frequency for thin, thick annular plates and uniform circular plate made of Material-II (Fig.4-Fig.6) is same as that of respective plates made of Material-I. The frequency is same for considered plates made of Material-I and Material-II. Hence, it is noted that the presence of mass-coupling parameter does not affect the frequency for torsional vibrations.

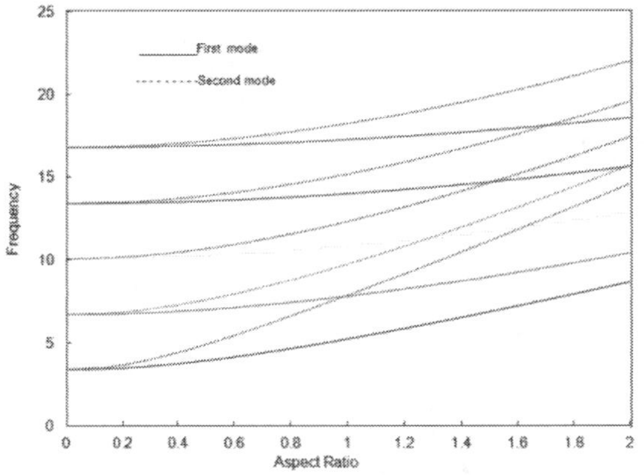

Figure 5. Frequency as a function of aspect ratio (Mat-II, Thick annular plate).

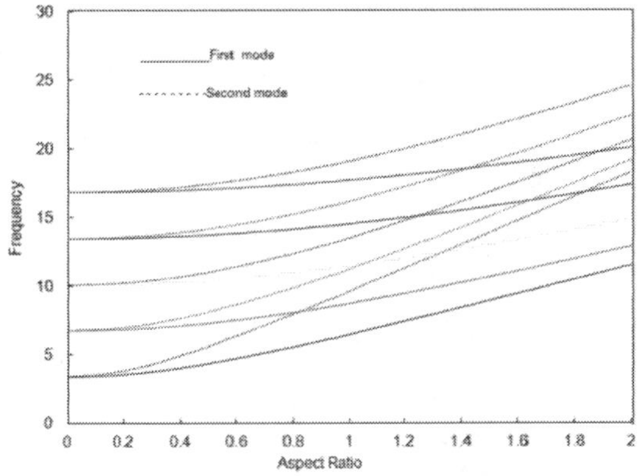

Figure 6. Frequency as a function of aspect ratio (Mat-II, Uniform circular plate).

The phase velocity of torsional vibrations of an infinite poroelastic plate for the Materials-I and II for different dissipations is presented in figure 7. It is clear that, the phase velocity in case of Material-I is slightly higher for the dissipation b_1=0.01. And as the dissipation increase, the phase velocity decrease to a small extent. The same is true in case of Material-II also. There is increase in phase velocity in (0, 2) and beyond it is almost constant. The phase velocity is same for Materials-I and II for different dissipations. Again here we see that the presence of mass-coupling parameter is not affecting the phase velocity. The attenuation of torsional vibrations of an infinite poroelastic plate for Materials-I and II is presented in figure 8 for different dissipations. The attenuation increase in (0, 2) that then it is almost linear in case of Material-I and b_1=0.01. The same trend is observed for b_1=0.1 and 1. But as the dissipation

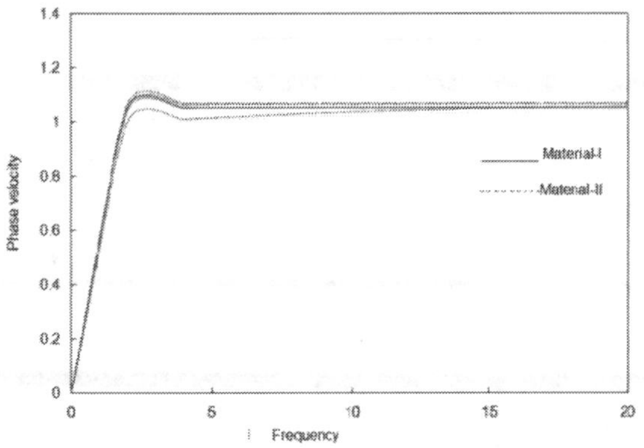

Figure 7. Phase velocity as a function of frequency (Infinite plate).

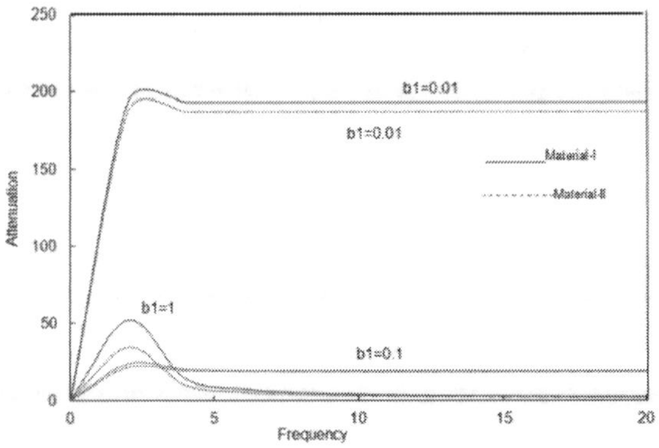

Figure 8. Attenuation as a function of frequency (Infinite plate).

increases, the attenuation decreases. A similar trend is seen in the variation of attenuation in case of Material-II for different dissipations. The attenuation is same for Materials-I and II for b_1=0.1. When b_1=0.01, the attenuation in poroelastic plate made of Material-I is higher than that of poroelastic plate made of Material-II for $\Omega \geq 2$. A similar behavior is observed in (0, 5) for b_1=1. Also we see that the presence of mass-coupling parameter is increasing the attenuation in particular range of frequency.

References

[1] Ahmed Shah S Axially symmetric vibrations of fluid-filled poroelastic circular cylindrical shells *Journal of Sound and Vibration* **318** 389-405

[2] Ahmed Shah S and Tajuddin M On axially symmetric vibrations of fluid filled poroelastic spherical shells *Open Journal of Acoustics* **1** 15-26

[3] Ahmed Shah S and Tajuddin M Torsional vibrations of poroelastic prolate spheroids *International Journal of Applied Mechanics and Engineering* **16** 521-29

[4] Biot M A Theory of propagation of elastic waves in fluid-saturated porous solid *Journal of the Acoustical Society of America* **28** 168-78

[5] Fatt I The Biot-Willis elastic coefficients for a sandstone *Journal of Applied Mechanics* **26** 296-97

[6] Gupta S, Chattopadhyay A and Kundu S Influence of irregularity and rigidity on the propagation of torsional wave *Applied Mathematical Sciences* **4** 805-16

[7] Kang Min K, Huang R and Knowles T Torsional vibrations of circular elastic plates with thickness steps *IEEE Transactions on Ultrasonics, Ferroelectrics, and Frequency Control* **53** 349-59

[8] Mindlin R D *An introduction to the Mathematical theory of vibrations of elastic plates*

[9] Tajuddin M and Ahmed Shah S On torsional vibrations of infinite hollow poroelastic cylinders *Journal of Mechanics of Materials and Structures* **2** 189-200

[10] Yew C H and Jogi P N Study of wave motions in fluid-saturated porous rocks *Journal of the Acoustical Society of America* **60** 2-8

[11] Zhou D, Au F T K, Cheung Y K and Lo S H Three dimensional vibration analysis of circular and annular plates via the Chebyshev-Ritz method *International Journal of Solids and Structures* **40** 3089-105

Effect of elastic foundations on vibrations of thick-walled hollow poroelastic cylinders

Srisailam Aleti and Manjula Ramagiri

Department of Mathematics, Kakatiya University, Warangal 506009, Telangana, India.

E-mail: asmou22@yahoo.com and manjularamagiri@gmail.com

Abstract. This paper deals with the effect of elastic foundations on the vibrations of thick-walled hollow poroelastic cylinders. The frequency equations of axially symmetric vibrations and non-axially symmetric vibrations, each for pervious and impervious surface are obtained using the analytical model based on Biot's theory. The plots of frequency versus ratio of thickness to inner radius of composite cylinder for different materials are presented graphically.

1. Introduction

The study of elastic foundations in poroelastic cylinder are fundamental and core structural elements in various fields of Engineering and Technology. Recently they have acquired a paramount and valuable place in practical applications of constructing gas cylinders, pressure vessels, boilers etc. Paliwal et. al [1] studied free vibrations of circular cylindrical shell on Wrinkler and Pasternak foundations. In the said paper, it is concluded that elastic foundation affects radial vibrations mode frequency while pertaining to torsional and longitudinal remains are unaffected. Goncalves and Batista [2] gave a theortical analysis for the free vibration of simply supported vertical shells partially filled with or submerged in a fluid and vibrations of the shells are examined using the sanders shell theory. Paliwal and Pandey [3] discussed the free vibrations of cylindrical shell on elastic foundation. In the said paper, the frequency equation of thin circular cylindrical shell resting on a elastic foundation is developed by using the first order shell theory of sanders and eigen frequencies are calculated. Natural frequency of a cylindrical panel on kerr foundation is studied by Cai et. al [4]. Exact analysis of plane-strain vibrations of thick-walled hollow elastic circular cylinder is discussed by Gazis [5]. In the said paper, approximate expressions for the frequencies of free modes are derived for moderately thick shells. Employing Biot's theory of poroelasticity [6], plane-strain vibrations of thick-walled hollow poroelastic cylinder is investigated by Malla Reddy and Tajuddin [7]. In the said paper, the free vibrations of an infinite thick-walled hollow poroelastic cylinder is studied both for axially symmetric and non-axially symmetric vibrations. Shankar et. al [8] investigated plane strain vibrations in a poroelastic composite hollow cylinder. In the said paper, frequency equation of plane strain vibrations of poroelastic composite hollow cylinder is obtained along with some particular cases such as poroelastic composite hollow cylinder with rigid casing, poroelastic composite solid cylinder with rigid casing, poroelastic composite bore and inner radius of core is discussed. Axially symmetric vibration of finite composite poroelastic cylinders is studied by Shah and Tajuddin [9]. Tajuddin [10] discussed torsional vibrations of finite composite cylinders. In the said paper, frequency equation is obtained for torsional vibrations of composite poroelastic cylinders either concentric or bonded end to end. Longitudinal shear vibrations of hollow poroelastic cylinder is investigated by Tajuddin and Ahmed shah [11]. In the said paper, frequency equation of longitudinal shear vibrations of a hollow poroelastic circular cylinders of infinite extent is derived and also the frequency is same for a thin poroelastic cylindrical shell in case of axially symmetric shear vibrations, flexural vibrations and non-axially symmetric shear vibrations. Malla Reddy and Tajuddin [12] investigated axially symmetric vibrations of a poroelastic composite cylinder in the context of fretting fatigue and the results are compared with that of Rule of Mixture (RoM). However, to the best of author's knowledge, effect of elastic foundations on vibrations of thick-walled hollow poroelastic cylinders is not yet investigated. Therefore, in this paper the same is investigated in the framework of Biot's theory. Non-dimensional frequency against the ratio of thickness to inner radius for different elastic foundations is studied for both axially symmetric and non-axially symmetric vibrations, each for a pervious and an impervious surface are computed for two types of materials.

This paper is organized as follows. In section 2, governing equations are given. In section 3, solutions of the problem are given. In section 4, boundary conditions and frequency equations are derived. Non-axially symmetric vibrations are presented in section 5. Axially symmetric vibrations are given in section 6. Numerical results are done in section 7. Finally, conclusion is given in section 8.

2. Governing equations

The equations of motion of a poroelastic solid [6] in presence of dissipation (b) which in terms of displacement vectors are

$$\begin{aligned} N\nabla^2\vec{u} + (A+N)\nabla e + Q\epsilon &= \tfrac{\partial^2}{\partial t^2}(\rho_{11}\vec{u} + \rho_{12}\vec{U}) + b\tfrac{\partial}{\partial t}(\vec{u} - \vec{U}), \\ \nabla(Qe + R\epsilon) &= \tfrac{\partial^2}{\partial t^2}(\rho_{12}\vec{u} + \rho_{12}\vec{U}) - b\tfrac{\partial}{\partial t}(\vec{u} - \vec{U}), \end{aligned} \tag{1}$$

where, ∇^2 is the Laplace operator, $\vec{u}(u, v, w)$ and $\vec{U}(U, V, W)$ are solid and fluid displacements, e and ϵ are the dilatations of solid and fluid respectively; the symbols A, N, Q, R are all poroelastic constants; ρ_{ij} are mass coefficients. The constitutive relations are

$$\begin{aligned} \sigma_{ij} &= 2Ne_{ij} + (Ae + Q\epsilon)\delta_{ij}, (i, j = 1, 2, 3), \\ s &= Qe + R\epsilon. \end{aligned} \tag{2}$$

In eq.(2), e_{ij}'s are strain displacements σ_{ij}'s are solid stresses and the fluid pressure s, δ_{ij} is the well known Kronecker delta function.

3. Solution of the problem

Let (r, θ, z) be the cylindrical polar coordinates. Consider a thick-walled homogenous isotropic infinite poroelastic cylinder with inner and outer radii r_1 and r_2 respectively, whose axis is in the direction of z-axis. Then the solid displacements $\vec{u}(u, v, o)$ which can be evaluated from eq. (1) are

$$\begin{aligned} u(r,\theta,t) = &-[C_1\xi_1(J_{n+1}(\xi_1 r) - \tfrac{n}{\xi_1 r}J_n(\xi_1 r)) + C_2\xi_1(Y_{n+1}(\xi_1 r) - \tfrac{n}{\xi_1 r}Y_n(\xi_1 r)) + C_3\xi_2(J_{n+1}(\xi_2 r) \\ &-\tfrac{n}{\xi_2 r}J_n(\xi_2 r)) + C_4\xi_2(Y_{n+1}(\xi_2 r) - \tfrac{n}{\xi_2 r}Y_n(\xi_2 r)) - A_1\tfrac{n}{r}J_n(\xi_3 r) \; - B_1\tfrac{n}{r}Y_n(\xi_3 r)]^{sin}_{cos}(n\theta)e^{i\omega t}, \\ \\ v(r,\theta,t) = &[C_1\tfrac{n}{r}J_n(\xi_1 r) + C_2\tfrac{n}{r}Y_n(\xi_1 r) + C_3\tfrac{n}{r}J_n(\xi_2 r) + C_4\tfrac{n}{r}Y_n(\xi_2 r) + A_1\xi_3(J_{n+1}(\xi_3 r) - \tfrac{n}{r}J_n(\xi_3 r)) \\ &+B_1\xi_3(Y_{n+1}(\xi_3 r) - \tfrac{n}{\xi_3 r}Y_n(\xi_3 r))]^{sin}_{cos}(n\theta)e^{i\omega t}. \end{aligned} \tag{3}$$

In eq. (3) $C_1, C_2, C_3, C_4, A_1, B_1$ are all arbitrary constants, ω is the frequency of wave, n is the number of waves and $\xi_i = \frac{\omega}{kV_i}, i = 1,2,3$. Here $V_i (i = 1,2,3)$ are the dilatational wave velocities of first and second kind and shear wave velocity, respectively. By substituting the displacements functions, u, v in eq. (2), the relevant stresses are

$$\sigma_{rr} + s - Ku - G\Delta u = (M_{11}(r)C_1 + M_{12}(r)C_2 + M_{13}(r)C_3 + M_{14}(r)C_4 + M_{15}(r)A_1 + M_{16}(r)B_1)_{cos}^{sin}(n\theta)e^{i\omega t},$$

$$\sigma_{r\theta} = (M_{21}(r)C_1 + M_{22}(r)C_2 + M_{23}(r)C_3 + M_{24}(r)C_4 + M_{25}(r)A_1 + M_{26}(r)B_1)_{cos}^{sin}(n\theta)e^{i\omega t},$$

$$s = (M_{31}(r)C_1 + M_{32}(r)C_2 + M_{33}(r)C_3 + M_{34}(r)C_4)_{cos}^{sin}(n\theta)e^{i\omega t},$$

$$\frac{\partial s}{\partial r} = (N_{31}(r)C_1 + N_{32}(r)C_2 + N_{33}(r)C_3 + N_{34}(r)C_4)_{cos}^{sin}(n\theta)e^{i\omega t}. \tag{4}$$

Where the coefficients M_{ij} and N_{ij} are

$M_{11}(r) = (2N(\frac{n(n-1)}{r^2} - \xi_1^2) - (A+Q)\xi_1^2 + (Q+R)\xi_1^2\delta_1^2 - \frac{Kn}{r})J_n(\xi_1 r) + (\frac{2N\xi_1}{r} + K\xi_1)J_{n+1}(\xi_1 r)$,

$M_{15}(r) = \frac{2nN\xi_3}{r}J_{n+1}(\xi_3 r) + (\frac{2Nn(1-n)}{r^2} - \frac{Kn}{r})J_n(\xi_3 r)$,

$M_{21}(r) = \frac{2Nn(n-1)}{r^2}J_n(\xi_1 r) - \frac{2Nn\xi_1}{r}J_{n+1}(\xi_1 r)$,

$M_{25}(r) = N(\frac{2n(1-n)}{r^2} + \xi_3^2)J_n(\xi_3 r) - \frac{2N\xi_3}{r}J_{n+1}(\xi_3 r)$,

$M_{31}(r) = (R\delta_1^2 - Q)\xi_1^2 J_n(\xi_1 r)$,

$N_{31}(r) = (\frac{R\delta_1^2 - Q}{r})\xi_1^2 n J_n(\xi_1 r) + (Q - R\delta_1^2)\xi_1^3 J_{n+1}(\xi_1 r)$,

$M_{12}(r)$=similar expression as $M_{11}(r)$ with J_n and J_{n+1} replaced by Y_n and Y_{n+1} respectively,

$M_{13}(r)$=similar expression as $M_{11}(r)$ with δ_1^2 and ξ_1 replaced by δ_2^2 and ξ_2 respectively,

$M_{14}(r)$=similar expression as $M_{11}(r)$ with δ_1^2 , ξ_1,J_n and J_{n+1} replaced δ_2^2, ξ_2, Y_n and Y_{n+1} respectively,

$M_{16}(r)$=similar expression as $M_{15}(r)$ with J_n and J_{n+1} replaced by Y_n and Y_{n+1} respectively,

$M_{22}(r)$=similar expression as $M_{21}(r)$ with J_n and J_{n+1} replaced by Y_n and Y_{n+1} respectively,

$M_{23}(r)$=similar expression as $M_{21}(r)$ with ξ_1 replaced by ξ_2,

$M_{24}(r)$=similar expression as $M_{21}(r)$ with J_n , J_{n+1} and ξ_1 replaced by Y_n ,Y_{n+1} and ξ_2 respectively,

$M_{26}(r)$=similar expression as $M_{25}(r)$ with J_n , J_{n+1} replaced by Y_n and Y_{n+1} respectively,

$M_{32}(r)$=similar expression as $M_{31}(r)$ with J_n replaced by Y_n ,

$M_{33}(r)$=similar expression as $M_{31}(r)$ with δ_1^2 and ξ_1 replaced by δ_2^2 and ξ_2 respectively,

$M_{34}(r)$=similar expression as $M_{31}(r)$ with δ_1^2 , ξ_1,and J_n replaced by δ_2^2, ξ_2, Y_n respectively,

$N_{32}(r)$=similar expression as $N_{31}(r)$ with J_n and J_{n+1} replaced by Y_n and Y_{n+1} respectively,

$N_{33}(r)$=similar expression as $N_{31}(r)$ with δ_1^2 ,ξ_1 replaced by δ_2^2, and ξ_2, respectively,

$N_{34}(r)$=similar expression as $N_{31}(r)$ with δ_1^2 , ξ_1,J_n and J_{n+1} replaced δ_2^2, ξ_2, Y_n and Y_{n+1} respectively.
In the above, δ_1^2 and δ_2^2 are

$\delta_1^2=\frac{(Rm_{11}-Qm_{12}-V_1^{-2}(PR-Q^2)}{Rm_{12}-Qm_{22}}$,

δ_2^2=similar expression as δ_1^2 with V_1^{-2} replaced by V_2^{-2},

and $m_{11}=\rho_{11}-ib\omega^{-1}$,$m_{12}=\rho_{12}+ib\omega^{-1}$, $m_{22}=\rho_{22}-ib\omega^{-1}$.

4. Boundary conditions and frequency equation

The boundary conditions for free vibrations in case of a pervious surface $r=r_1$ and $r=r_2$ is

$$\sigma_{rr}+s-Ku-G\triangle u=\sigma_{r\theta}=s=0, \tag{5}$$

The boundary conditions for free vibrations in case of a impervious surface $r=r_1$ and $r=r_2$ is

$$\sigma_{rr}+s-Ku-G\triangle u=\sigma_{r\theta}=\frac{\partial s}{\partial r}=0. \tag{6}$$

In eq. (5) and (6), $\triangle=\frac{\partial^2}{\partial r^2}+\frac{1}{r}\frac{\partial}{\partial r}+\frac{\partial^2}{\partial z^2}$, K is the foundation modulus. If shear modulus G goes to zero, the Pasternak foundation will reduce to Wrinkler foundation [13]. Eq. (5) and (6) gives a system of six homogeneous equations in $C_1, C_2, C_3, C_4, A_1, B_1$ each for a pervious surface and an impervious surface. A nontrivial solution can be obtained if the determinant of coefficient vanishes. Accordingly, the frequency equation in the case of pervious surface is

$$\begin{vmatrix} M_{11}(r_1) & M_{12}(r_1) & M_{13}(r_1) & M_{14}(r_1) & M_{15}(r_1) & M_{16}(r_1) \\ M_{21}(r_1) & M_{22}(r_1) & M_{23}(r_1) & M_{24}(r_1) & M_{25}(r_1) & M_{26}(r_1) \\ M_{31}(r_1) & M_{32}(r_1) & M_{33}(r_1) & M_{34}(r_1) & 0 & 0 \\ M_{11}(r_2) & M_{12}(r_2) & M_{13}(r_2) & M_{14}(r_2) & M_{15}(r_2) & M_{16}(r_2) \\ M_{21}(r_2) & M_{22}(r_2) & M_{23}(r_2) & M_{24}(r_2) & M_{25}(r_2) & M_{26}(r_2) \\ M_{31}(r_2) & M_{32}(r_2) & M_{33}(r_2) & M_{34}(r_2) & 0 & 0 \end{vmatrix}=0. \tag{7}$$

In the case of an impervious surface, the frequency equation is

$$\begin{vmatrix} M_{11}(r_1) & M_{12}(r_1) & M_{13}(r_1) & M_{14}(r_1) & M_{15}(r_1) & M_{16}(r_1) \\ M_{21}(r_1) & M_{22}(r_1) & M_{23}(r_1) & M_{24}(r_1) & M_{25}(r_1) & M_{26}(r_1) \\ N_{31}(r_1) & N_{32}(r_1) & N_{33}(r_1) & N_{34}(r_1) & 0 & 0 \\ M_{11}(r_2) & M_{12}(r_2) & M_{13}(r_2) & M_{14}(r_2) & M_{15}(r_2) & M_{16}(r_2) \\ M_{21}(r_2) & M_{22}(r_2) & M_{23}(r_2) & M_{24}(r_2) & M_{25}(r_2) & M_{26}(r_2) \\ N_{31}(r_2) & N_{32}(r_2) & N_{33}(r_2) & N_{34}(r_2) & 0 & 0 \end{vmatrix}=0. \tag{8}$$

5. Non-axially symmetric vibrations

If n is non-zero, the dilatational and shear vibrations are coupled as indicated by the frequency eqs (7) and (8). Using the eq. (7), we obtain the frequency equation of pervious surface in the following

$$|A_{ij}|=0, \quad (i,j=1,2,3,4,5,6). \tag{9}$$

Arguing in the similar lines, using eq. (8), we obtain the frequency equation of an impervious surface in the following form

$$| B_{ij} |= 0, \quad (i, j = 1, 2, 3, 4, 5, 6). \tag{10}$$

6. Axially symmetric vibrations

If we set $n = 0$, then the frequency eq. (7) for pervious surface degenerates into the product of two determinants each of second and fourth order respectively. The second determinant corresponds to shear vibrations and fourth determinant corresponds to extensional vibrations.

6.1. Shear vibrations

The frequency equation of shear vibrations for both pervious and impervious can be obtained from eq. (7) and (8) in the following form

$$\begin{vmatrix} M_{25}(r_1) & M_{26}(r_1) \\ M_{25}(r_2) & M_{26}(r_2) \end{vmatrix} = 0. \tag{11}$$

Eq. (11), corresponds to zero dilatational potential and is independent of nature of surface. By substituting $M_{ij}(r)$ into eq. (11), and using recurrence relations [14] the frequency equation simplifies to

$$J_2(\xi_3 r_1)Y_2(\xi_3 r_2) - J_2(\xi_3 r_2)Y_2(\xi_3 r_1 = 0. \tag{12}$$

6.2. Extensional vibrations

The frequency equation of extensional vibrations for pervious surface can be obtained from eq. (7) is

$$\begin{vmatrix} M_{11}(r_1) & M_{12}(r_1) & M_{13}(r_1) & M_{14}(r_1) \\ M_{31}(r_1) & M_{32}(r_1) & M_{33}(r_1) & M_{34}(r_1) \\ M_{11}(r_2) & M_{12}(r_2) & M_{13}(r_2) & M_{14}(r_2) \\ M_{31}(r_2) & M_{32}(r_2) & M_{33}(r_2) & M_{34}(r_2) \end{vmatrix} = 0. \tag{13}$$

Similarly, it can be seen the frequency eq. (8) for an impervious surface yields the product of two determinants each of second and fourth order respectively. The second order determinant is same as in eq. (11), while the fourth order determinant yields the extensional vibrations of an impervious surface is

$$\begin{vmatrix} M_{11}(r_1) & M_{12}(r_1) & M_{13}(r_1) & M_{14}(r_1) \\ N_{31}(r_1) & N_{32}(r_1) & N_{33}(r_1) & N_{34}(r_1) \\ M_{11}(r_2) & M_{12}(r_2) & M_{13}(r_2) & M_{14}(r_2) \\ N_{31}(r_2) & N_{32}(r_2) & N_{33}(r_2) & N_{34}(r_2) \end{vmatrix} = 0. \tag{14}$$

7. Numerical results

For the sake of numerical work, the dissipative coefficient b is taken zero and hence we obtained only real frequency Ω. To analyze the frequency equations, it is convenient to introduce the following non-dimensional parameters.

$$
\begin{aligned}
&a_1 = \tfrac{P}{H_1}, \quad a_2 = \tfrac{Q}{H_1}, \quad a_3 = \tfrac{R}{H_1}, \quad a_4 = \tfrac{N}{H_1},\\
&d_1 = \tfrac{\rho_{11}}{\rho_1}, \quad d_2 = \tfrac{\rho_{12}}{\rho_1}, \quad d_3 = \tfrac{\rho_{22}}{\rho_1},\\
&\tilde{x} = (\tfrac{V_0}{V_1})^2, \quad \tilde{y} = (\tfrac{V_0}{V_2})^2, \quad \tilde{z} = (\tfrac{V_0}{V_3})^2,\\
&b_1 = \tfrac{P_1}{H_1}, \quad b_2 = \tfrac{Q_1}{H_1}, \quad b_3 = \tfrac{R_1}{H_1}, \quad b_4 = \tfrac{N_1}{H_1},\\
&g_1 = \tfrac{(\rho_{11})_1}{\rho_1}, \quad g_2 = \tfrac{(\rho_{12})_1}{\rho_1}, \quad g_3 = \tfrac{(\rho_{22})_1}{\rho_1}, \quad m = \tfrac{C}{C_0},\\
&\tilde{x_1} = (V_0/(V_1)_1)^2, \quad \tilde{y_1} = (V_0/(V_2)_1)^2, \quad \tilde{z_1} = (V_0/(V_3)_1)^2.
\end{aligned}
\tag{15}
$$

In eq. (15), C is the phase velocity, Ω is the non-dimensional frequency, C_0 and V_0 are references velocities $C_0^2 = \frac{N_1}{\rho_1}$, $V_0^2 = \frac{H_1}{\rho_1}$ and m is the non-dimensional phase velocity, $\rho_1 = (\rho_{11})_1 + 2(\rho_{12})_1 + (\rho_{22})_1$, $H_1 = P_1 + 2Q_1 + R_1$. Let $g = \frac{r_2}{r_1}$ so that and $\frac{h}{r_1} = g - 1$ and $\Omega = \frac{\omega h}{C_0}$. In all the above subscript '1' stand for the quantities related to inner cylinder. Employing the non-dimensional quantities in the frequency equation we obtain a implicit relation between frequency and ratio of wall thickness to inner radius. Non-dimensional frequency is calculated for two types of composite cylinders, namely composite cylinder-I and composite cylinder-II, for each pervious and impervious surface using numerical process bisection method performed in MATLAB. The composite cylinder consists of two cylinders one cylinder is made up of sandstone saturated with kerosene[15], and the other is made up of sandstone saturated with water[16], while in composite cylinder-II the materials are reversed. The physical parameters of composite cylinders are given in Table 1. Figures 1-8 plots the non-dimensional frequency against the ratio of thickness to inner cylinder pertaining to elastic foundations 10, 50. From the figures 1-8, it is observed that frequency of cylinder-I is more than the frequency of cylinder-II. From figures 1, 3 it is clear that frequency is greater for impervious surface than that of pervious surface in the case of cylinder-I. Also the frequency is steady and constant beyond the ratio of thickness to inner radius 7 for pervious surface. From figures 2, 4 it is observed that frequency is greater for pervious surface than that of an impervious surface in the case of cylinder-II. Also, the frequency is steady and constant beyond the ratio of thickness to inner radius 13 for both pervious and impervious surfaces. From figures 5, 7 it is seen that frequency is more for pervious surface than that of an impervious surface in the case of cylinder-I. Also, from figure7 it is clear that frequency is steady and constant beyond the ratio of thickness to inner radius 7 for pervious surface and frequency is constant beyond 11 for an impervious surface. From figures 6, 8 it is clear that frequency is greater for impervious surface than that of pervious surface in the case of cylinder-II. Also, from figure 6 it is clear that frequency is steady and constant beyond the ratio of thickness to inner radius 11 for pervious surface and frequency is constant beyond 7 for an impervious surface. The trend is reversed in the cases of cylinders this is due to the influence of elastic foundation present in materials.

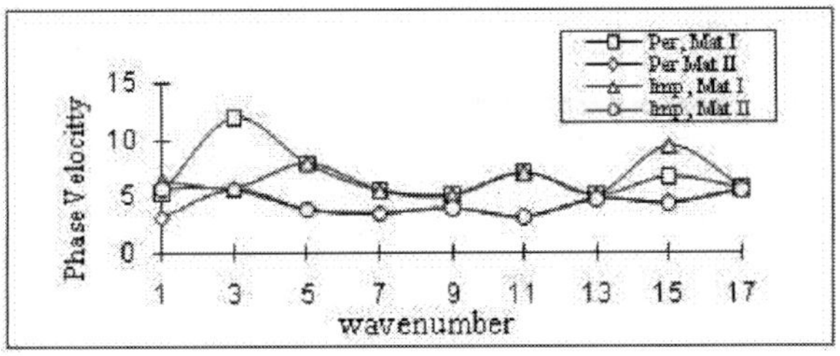

Figure 1. Non-axially symmetric vibrations: variation of frequency with $\frac{h}{r_1}$ at $K = 10$ (composite cylinder-I).

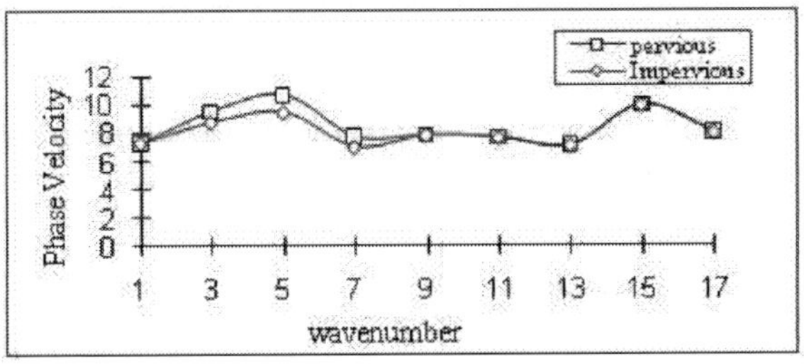

Figure 2. Non-axially symmetric vibrations: variation of frequency with $\frac{h}{r_1}$ at elastic foundation at $K = 10$ (composite cylinder-II).

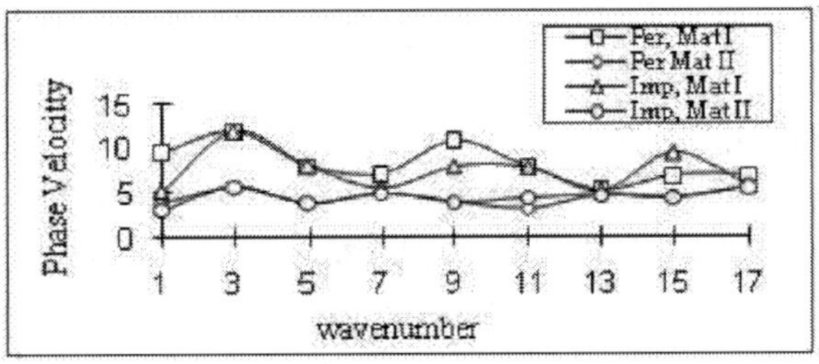

Figure 3. Non-axially symmetric vibrations: variation of frequency with $\frac{h}{r_1}$ at elastic foundation at $K = 50$(composite cylinder-I) .

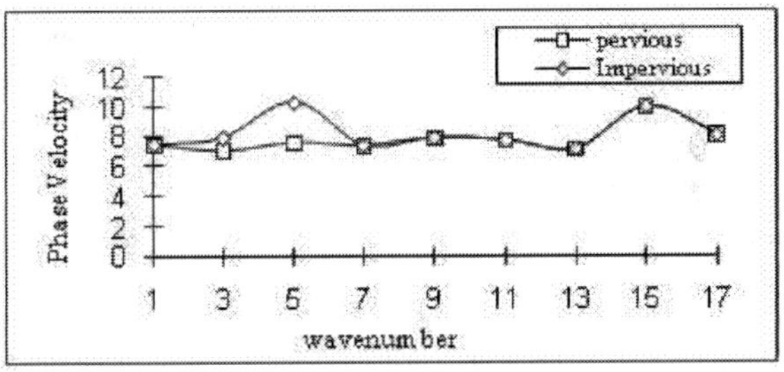

Figure 4. Non-axially symmetric vibrations: variation of frequency with $\frac{h}{r_1}$ at elastic foundation at $K = 50$ (composite cylinder-II).

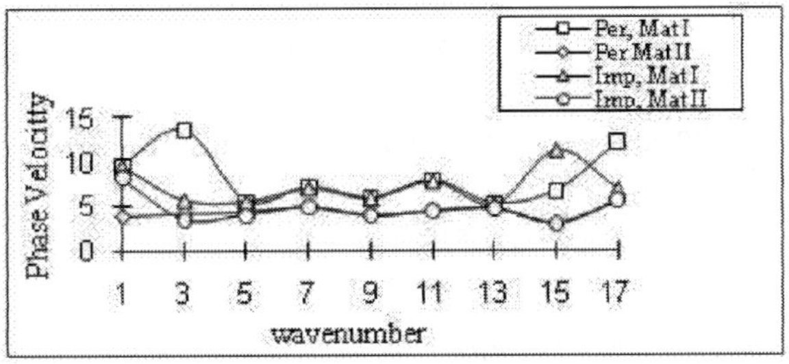

Figure 5. Axially symmetric vibrations: variation of frequency with $\frac{h}{r_1}$ at elastic foundation $K = 10$ (composite cylinder-I).

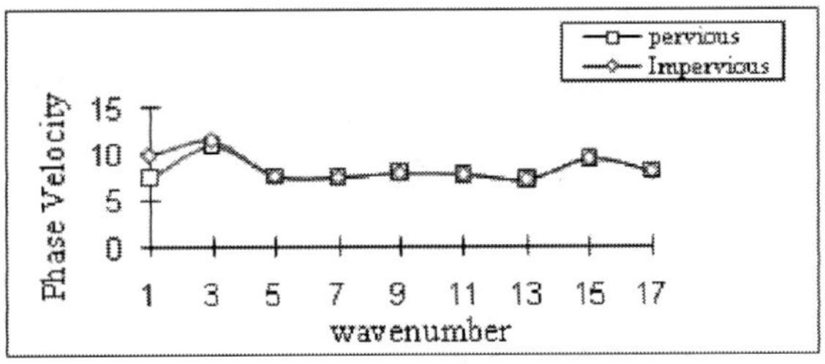

Figure 6. Axially symmetric vibrations: variation of frequency with $\frac{h}{r_1}$ at elastic foundation at $K = 10$ (composite cylinder-II) .

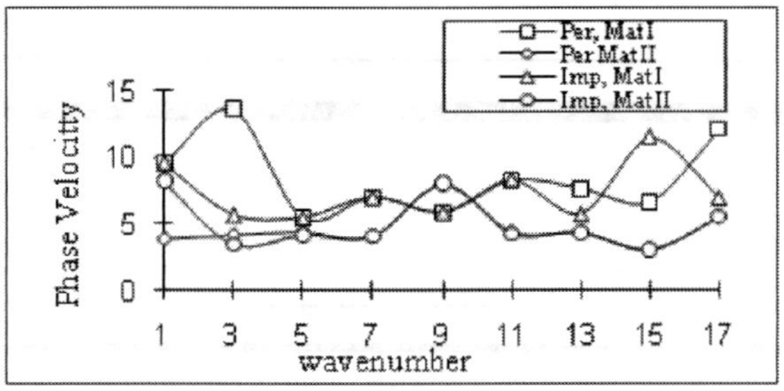

Figure 7. Axially symmetric vibrations: variation of frequency with $\frac{h}{r_1}$ at elastic foundation at $K = 50$ (composite cylinder-I).

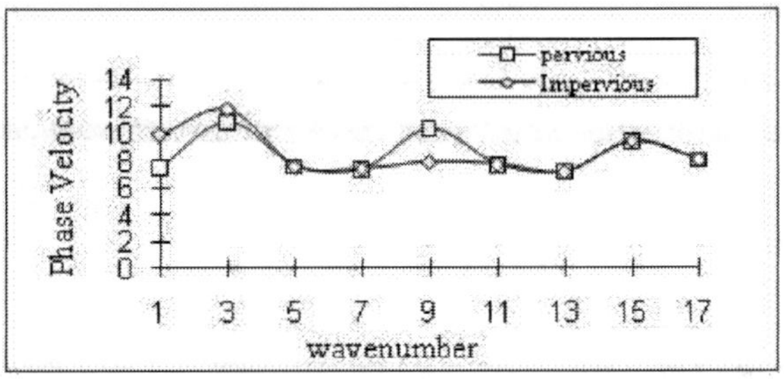

Figure 8. Axially symmetric vibrations: variation of frequency with $\frac{h}{r_1}$ at elastic foundation at $K = 50$ (composite cylinder-II).

Table 1. Material parameters

Material parameters	cylinder-I	cylinder-II
a_1	0.445	1.819
a_2	0.034	0.011
a_3	0.015	0.054
a_4	0.123	0.780
d_1	0.887	0.891
d_2	0.001	0
d_3	0.1	0.124
$\tilde{x}$	1.863	0.489
$\tilde{y}$	8.884	2.330
$\tilde{z}$	7.183	1.142
b_1	0.96	0.843
b_2	0.006	0.065
b_3	0.028	0.028
b_4	0.142	0.234
g_1	0.877	0.901
g_2	0	-0.001
g_3	0.123	0.101
$\tilde{x_1}$	0.913	0.999
$\tilde{y_1}$	4.347	4.763
$\tilde{z_1}$	2.129	3.851

8. Conclusion

Effect of elastic foundations on the vibrations of thick-walled hollow poroelastic cylinder are investigated in the framework of Biot's theory. Non dimensional frequency against ratio of thickness to inner radius is computed for two types of materials. In the case of non-axially symmetric vibrations, extensional and shear vibrations are coupled. The frequency of impervious surface values are greater than that of pervious surface in the case of cylinder-I. The frequency of pervious surface values are greater than that of an impervious surface in the case of cylinder-II. In the case of axially symmetric vibrations, the extensional and shear vibrations exist uncoupled. The frequency of pervious surface values are greater than that of an impervious surface in case of cylinder-I. The frequency of impervious surface values are greater than that of pervious surface in case of cylinder-II. This kind of analysis can be made if pertinent materials are available.

Acknowledgement
The first author would like to acknowledge University Grants Commission (UGC), Government of India for the funding through Dr D.S.Kotary Postdoctoral Fellowship with grant no.F.4-2/2006 (BSR)/13-661/2012(BSR).

References
[1] DN Paliwal, RK Pandey and T Nath 1996 Free vibrations of circular cylindrical shell on Wrinkler and Pasternak foundation *Int. J. Press. Vessel Piping* **69** 79-89
[2] Gonclaves PB, Batista RC 1987 Frequency response of cylindrical shells partially submerged or filled with liquid, *J. Sound and vib* **113** 59-70
[3] DN Paliwal and RK Pandey 1998 The free vibration of a cylindrical shell on elastic foundation *J. Vib. Acoust* **120** 63-71
[4] Cai JB, Chen WQ, GR Ye and HJ Ding 2000 On natural frequencies of a transversely isotropic cylindrical panel on a Kerr foundation*J. Sound and Vib* **232** 997-1004
[5] Gazis DC 1985 Exact analysis of plane-strain vibrations of thick-walled hollow cylinders *J. Acoust. Soc. Am.* **30** 786-794
[6] Biot MA 1956 The theory of propagation of elastic wave in fluid-saturated porous solid *J. Acoust. Soc. Am.* **28** 168-191
[7] Malla Reddy and Tajuddin 2000 Exact analysis of the plane strain vibrations of thick walled hollow poroelastic cylinders *Int. J. Solids. Struct* **37** 3439-56
[8] B Shanker, Manoj Kumar, Ahmed Shah and Nageswaranath 2013 A Study on plane strain vibrations in a poroelastic composite hollow cylinder *J.Vib. Analysis Measurement and Control* **1** 1-29
[9] Shah SA and Tajuddin M 2009 Axially symmetric vibrations of finite composite poroelastic cylinders *Int. J. Appl. Mech. Eng* **14** 865-877
[10] Tajuddin M and Sarma KS 1982 Torsional vibrations of finite composite poroelastic cylinders *Indian. J. Pure. App. Math* **13** 375-381
[11] Tajuddin M and Shah SA 2010 Longitudinal shear vibrations of hollow poroelastic cylinders *Bull. Calcutta Math Soc* **102** 289-298
[12] MallaReddy P and Tajuddin M 2010 Axially symmetric vibrations of poroelastic composite cylinder in the context of fretting fatigue *Spec Top Revi. Porous Media* **1** 311-320
[13] Pronk C and Marion E VandenBol 1998 Wrinkler-Pasternak-Foundations, Notes on boundary conditions
[14] Watson GN 1962 *Theory of Bessel Functions* (London; Cambridge University Press)
[15] Yew CH and Jogi PN 1976 Study of wave motions in fluid-saturated porous rocks*J. Acou. Soc. Am* **60** 2-8
[16] Fatt I 1957 The Biot-Willis elastic coefficient for a sand stone *J. App. Mech* 296-297

Edge waves in an initially stressed visco-elastic plate

N Kumari, A Chattopadhyay, S A Sahu and A K Singh

Department of Applied Mathematics, Indian School of Mines, Dhanbad, 826004, India.

E-mail: nirmala.ism@gmail.com, amares.c@gmail.com, ism.sanjeev@gmail.com, abhi.5700@gmail.com

Abstract. The present study analytically investigates the propagation of edge waves in a pre-stressed homogeneous visco-elastic plate. The plate is initially stressed in horizontal direction only. The solution for potential function and displacement components of the edge wave propagation have been obtained separately. Dispersion equation of edge wave has been established in closed form.The horizontal tensile and compressive initial stress parameters give the remarkable effect on the phase velocity of edge wave propagation.

1. Introduction

Viscoelastic materials are considered in the broader context of physical properties. Elastic solids support both shear and hydrostatic stresses and their properties are independent of time or frequency, whereas viscoelastic materials exhibit time and frequency dependence. Isotropic elastic solids are describable by two elastic constants, (viz. shear and bulk modulus). By contrast, viscoelastic materials require a function of time or frequency to describe the behavior. Therefore, a rich set of physical phenomena can occur. For an elastic wave in a body, the velocity of propagation at an edge differs from that at a point far from it. This is due to reflection and refraction of a wave at the edge. The waves propagating at the edge of a plate of finite thickness are referred as Edge wave. The stresses, which exist in an elastic body even though external forces are absent, are termed as initial stresses and the body is said to be initially stressed. Biot [1] mentioned that an initial stress has remarkable effect on the propagation of elastic waves in body. He showed that the propagation of waves in an initially stressed body is fundamentally different from the classical theory of elasticity to a great extent. Ewing et al. [2] discussed that manufacturing process, body forces and unequal rates of heating and cooling, creep and inelastic deformation cause enormous initial stresses in a body. Love [3] predicted that the earth is in a state of high initial stress.For example, due to the atmospheric pressure, gravity, difference in temperature, the large amount of initial stresses may exist inside the earth, so the earth is considered to be an initially stressed.These stresses might exert significant influence on the elastic waves produced by earthquakes, explorations or impacts. Thus, it is imperative to deal with the properties of wave propagation under initial stress. It was Biot [4] who first pointed out that the initial stress influences elastic waves to a great extent. Being motivated by these facts, we have also taken initially stressed plate into account. Also the propagation of edge waves in plates was studied by Dey [5] and Kumar [6]. Dey and De [7] have studied the edge wave propagation in an incompressible anisotropic initially stressed plate of finite thickness. Gupta et al. [8] have deduced the propagation of S-waves in a non-homogeneous anisotropic incompressible and initially stressed medium. Chattopadhyay et al. [9] have studied the propagation of shear waves

Journal of Physics: Conference Series **662** (2015) 012011 doi:10.1088/1742-6596/662/1/012011

in visco-elastic medium at irregular boundaries. Singh et al.[10] discussed the smooth moving punch in an initially stressed transversely isotropic magneto-elastic medium due to shear wave. To study the effect of viscoelasticity in wave propagation,some attempts have been done earlier but no attempts have been made to study edge wave propagation in a viscoelastic medium. In this paper we investigate the propagation of edge waves in a homogeneous pre-stressed viscoelastic plate of finite thickness. Dispersion equation of edge wave has been obtained in closed form. It has been observed that compressive initial stress and tensile initial stress have their substantial effect on phase velocity of edge wave.

2. Formulation and solution of the problem

To formulate the edge wave propagation in an initially stressed uniformly homogeneous visco-elastic plate of thickness H and infinite length, we consider x-axis is along the direction of wave propagation and y-axis is positive pointing vertically upward. The plate is horizontally initially stressed S_{11} along the x-direction respectively.

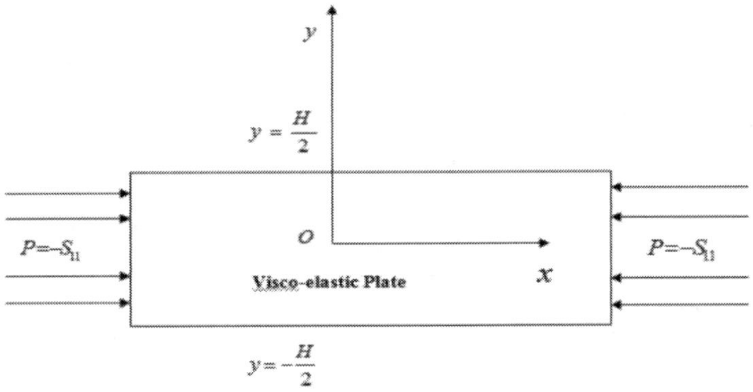

Figure 1. A homogeneous viscoelastic plate of thickness H and of infinite length under uniform initial stress S_{11}.

The dynamical equations of motion for the propagation of edge wave are

$$(\lambda + 2\bar{\mu})\frac{\partial^2 u}{\partial x^2} + \left(\bar{\mu} + \frac{P}{2}\right)\frac{\partial^2 u}{\partial y^2} + \left(\lambda + \bar{\mu} - \frac{P}{2}\right)\frac{\partial^2 v}{\partial x \partial y} = \rho\frac{\partial^2 u}{\partial t^2}, \tag{1}$$

and

$$\left(\bar{\mu} - \frac{P}{2}\right)\frac{\partial^2 v}{\partial x^2} + (\lambda + 2\bar{\mu})\frac{\partial^2 v}{\partial y^2} + \left(\lambda + \bar{\mu} + \frac{P}{2}\right)\frac{\partial^2 u}{\partial x \partial y} = \rho\frac{\partial^2 v}{\partial t^2}, \tag{2}$$

where $\bar{\mu} = \mu + \mu'\frac{\partial}{\partial t}$, μ and μ' are stiffness of the material due to elastic and viscoelastic properties. $P(= -S_{11})$ is the horizontal initial stress along x-direction, u and v are the displacement components along x and y directions respectively. The condition of incompressibility is

$$e_{xx} + e_{yy} = 0, \tag{3}$$

where $e_{xx} = \frac{\partial u}{\partial x}$ and $e_{yy} = \frac{\partial v}{\partial y}$ are the incremental normal strain tensors.
We introduce potential function $\varphi(x, y, t)$ as follows:

$$u = -\frac{\partial \varphi(x, y, t)}{\partial y} \text{ and } v = \frac{\partial \varphi(x, y, t)}{\partial x}, \tag{4}$$

where $\varphi(x, y, t)$ is a differentiable function.
Using eqns.(1),(2) and (3), we get

$$(1-\bar{\xi})\frac{\partial^4\varphi}{\partial x^4}+2\frac{\partial^4\varphi}{\partial x^2\partial y^2}+(1+\bar{\xi})\frac{\partial^4\varphi}{\partial y^4}=\frac{1}{\bar{\beta}^2}\frac{\partial^2}{\partial t^2}\left(\frac{\partial^2\varphi}{\partial x^2}+\frac{\partial^2\varphi}{\partial y^2}\right), \quad (5)$$

where $\bar{\xi}=\frac{P}{2\bar{\mu}}$ and $\bar{\beta}^2=\bar{\mu}/\rho$.
We have considered the solution of eq.(5)as

$$\varphi(x, y, t)=\frac{1}{l^2}f(ly)\exp\{i(lx-\alpha t)\}, \quad (6)$$

where $\alpha(=lc)$ and l are frequency of oscillation and wave number.
With the help of eqns. (5) and (6), we obtain the potential function $\varphi(x, y, t)$ as

$$\varphi(x, y, t)=\frac{1}{l^2}\{c_1\cosh(\beta_1 ly)+c_2\cosh(\beta_2 ly)\}\sin(lx-\alpha t), \quad (7)$$

where c_1 and c_2 are arbitrary constants,
$\beta_1^2=R+\sqrt{R^2-S^2}$, $\beta_2^2=R-\sqrt{R^2-S^2}$, $S^2=\frac{1-\bar{\xi}}{1+\bar{\xi}}-\frac{\rho\alpha^2}{\bar{\mu}l^2(1+\bar{\xi})}$ and $R=\frac{1}{1+\bar{\xi}}-\frac{\rho\alpha^2}{2\bar{\mu}l^2(1+\bar{\xi})}$.
with the help of eqns. (4) and (7), we get the displacement components as

$$u=-\frac{1}{l}\left[c_1\beta_1\sinh(\beta_1 ly)+c_2\beta_2\sinh(\beta_2 ly)\right]\sin(lx-\alpha t), \quad (8)$$

and

$$v=\frac{1}{l}\left[c_1\cosh(\beta_1 ly)+c_2\cosh(\beta_2 ly)\right]\cos(lx-\alpha t). \quad (9)$$

2.1. Boundary conditions and dispersion equation

The bounding planes $y=\pm\frac{H}{2}$ are supposed to be free from tractions, hence the boundary conditions are

$$\left.\begin{array}{l}\Delta f_x=0,\\ \Delta f_y=0,\end{array}\right\} \text{ at } y=\pm\frac{H}{2}, \quad (10)$$

where

$$\Delta f_x=\bar{\mu}(1+\bar{\xi})\left(\frac{\partial u}{\partial y}+\frac{\partial v}{\partial x}\right), \quad (11)$$

and

$$\Delta f_y=\lambda\frac{\partial u}{\partial x}+(\lambda+2\bar{\mu})\frac{\partial v}{\partial y}, \quad (12)$$

are incremental forces per unit initial area at the boundary.
With the help of eqns. (8), (9) and (10) we get the following two equations as

$$c_1\left(1+\beta_1^2\right)\cosh(\Lambda\beta_1)+c_2\left(1+\beta_2^2\right)\cosh(\Lambda\beta_2)=0, \quad (13)$$

and

$$c_1\beta_1\sinh(\Lambda\beta_1)+c_2\beta_2\sinh(\Lambda\beta_2)=0, \quad (14)$$

where $\Lambda(=lH/2)$ is the wavelength.
Eliminating arbitrary constants c_1 and c_2 from eqns. (13) and (14), we get

$$\beta_2\left(1+\beta_1^2\right)\tanh(\Lambda\beta_2)-\beta_1\left(1+\beta_2^2\right)\tanh(\Lambda\beta_1)=0. \quad (15)$$

This is the dispersion equation of edge wave.

 doi:10.1088/1742-6596/662/1/012011

3. Numerical examples and discussion

For the numerical computation of phase velocity of edge wave propagation at the edge of a pre-stressed viscoelastic plate, the following data have been considered [11]:
$\mu = 6.77 \times 10^{10} N/m^2$, $\rho = 3323\, kg/m^3$.
Moreover, we consider following data for computation
$P/2\mu = -0.2, -0.1, 0.0, 0.2, 0.1;\ \alpha\mu'/\mu = 0.05$.

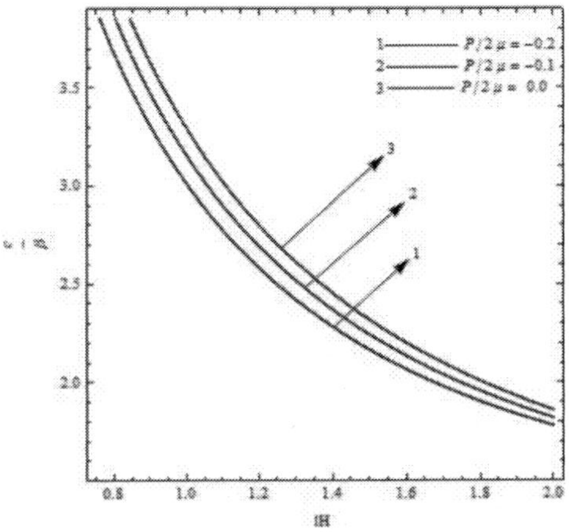

Figure 2. Variation in dimensionless phase velocity (c/β) against dimensionless wave number (lH) for different values of horizontal tensile initial stress when $\alpha\mu'/\mu = 0.05$.

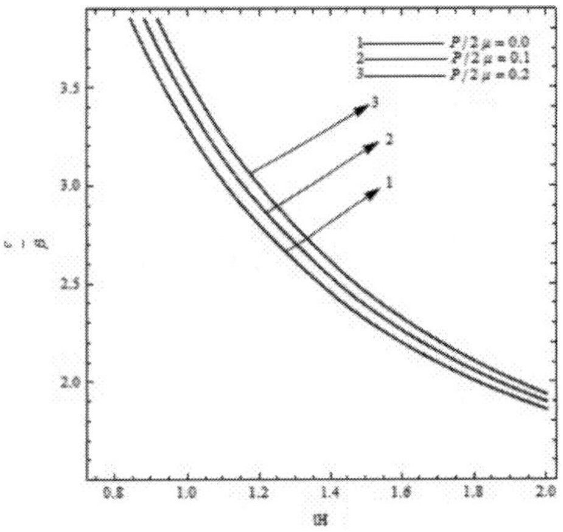

Figure 3. Variation in dimensionless phase velocity (c/β) against dimensionless wave number (lH) for different values of horizontal compressive initial stress when $\alpha\mu'/\mu = 0.05$.

The dispersion equation for the propagation of edge wave in a visco-elastic plate of finite width under horizontal initial stress is obtained. It relates the dimensionless phase velocity (c/β) with dimensionless wave number (lH). The variations of dimensionless phase velocity against dimensionless wave number for different values of initial stress parameters have been shown in Figures 2 and 3. Figure 2 gives the effect of horizontal tensile initial stress parameter on the dispersion curve when the values of tensile initial stress parameter are taken -0.2, -0.1 0.0. However, Figure 3 shows the curves for different values of horizontal compressive initial stress parameter and the values are taken as 0.0, 0.1, 0.2. In both the cases viscosity parameter $\alpha\mu'/\mu$ kept fixed at 0.05. It is clear from these figures that the dimensionless phase velocity increases with dimensionless wave number when the values of dimensionless horizontal initial stress (tensile and compressive) parameters increase.

4. Conclusion

The propagation of edge wave at the edge of a pre-stressed homogeneous viscoelastic plate has been studied. The dispersion equation for the propagation of edge wave has been obtained in closed form. The dispersion equation relates the phase velocity with wave number. The tensile and compressive initial stress parameters have a significant favoring effect on phase velocity of edge wave. Also it has been observe that the phase velocity increases with wave number when the initial stress parameters increase.

Acknowledgments
The authors convey their sincere thanks to Indian School of Mines, Dhanbad for providing JRF to Ms. Nirmala Kumari and also facilitating us with its best facility.

References

[1] Biot M A 1965 *Mechanics of incremental deformations. (New York: John Wiley and Sons Inc)*
[2] Ewing W M, Jardetzky W S and Press F 1957 *Elastic waves in layered media. (New York: McGraw Hill Book Company Inc)*
[3] Love A E H 1944 *A treatise on the mathematical theory of elasticity. (New York: Fourth ed. Dover)*
[4] Biot M A 1940 *The influence of initial stress on elastic waves. J. Appl. Phys.* **11(8)** 522-530
[5] Das S C and Dey S 1998 *Edge wave under initial stress. Appl. Sci. Res.***22** 382-389
[6] Kumar S 1959 *Edge waves in plates, Int. Symposium on Stress Wave Propagation in Materials. (Pennsylvania State University (Penn.) U.S.A)*
[7] Dey S and De P K 2009 *Edge wave propagation in an incompressible anisotropic initially stressed plate of finite thickness. Int. J. Comp. Cong.n***7(3)** 55-60
[8] Gupta S, Kundu S, Verma A K and Verma R 2010 *Propagation of S-waves in a non-homogeneous anisotropic incompressible and initially stressed medium.Int.J.Eng., Sci.Technol.***2(2)** 31-42
[9] Chattopadhyay A, Gupta S, Sharma V K and Kumari P 2009 *Propagation of shear waves in visco-elastic medium at irregular boundaries. Acta Geophys.* **58(2)** 195-214
[10] Singh A K, Kumari N, Chattopadhyay A and Sahu S A 2015 *Smooth moving punch in an initially stressed transversely isotropic magnetoelastic medium due to shear wave. Mech. Adv. Mater. Struct.***DOI :10.1080/15376494.2015.1029161**
[11] Gubbins D 1990 *Seismology and plate tectonics. (London: Cambridge University Press)*

Unsteady MHD free convection and mass transfer flow past a porous vertical plate in presence of viscous dissipation

D Mahender and P Srikanth Rao

Department of Mathematics, B.V.R.I.T, Narsapur, Medak, Telangana State, India.

E-mail: mahendar.dasari280979@gmail.com

Abstract. This study examines the influence of unsteady magneto hydrodynamic free convection and mass transfer flow past a vertical porous plate in presence of viscous dissipation and heat source. The governing non linear partial differential equations are transformed into a system of coupled non linear partial differential equations and solved using a Finite element technique. The effects of various physical parameters on the dimensionless velocity, temperature and concentration profiles are depicted graphically and analyzed in detail.

1. Introduction

The phenomenon of hydromagnetic flow with heat and mass transfer in an electrically conducting fluid past a porous plate embedded in a porous medium has attracted the attention of a good number of investigators because of its varied applications in many engineering problems such as MHD generators, plasma studies, nuclear reactors, oil exploration, geothermal energy extractions and in the boundary layer control in the field of aerodynamics. Heat transfer in laminar flow is important in problems dealing with chemical reactions and in dissociating fluids. In view of its wide applications, Hasimoto [1] initiated the boundary layer growth on a flat plate with suction or injection. Yamamoto and Iwamura [2] explained the flow of a viscous fluid with convective acceleration through a porous medium. Mansutti et al. [3] have discussed the steady flow of a non Newtonian fluid past a porous plate with suction or injection. Jha [4] analyzed the effect of applied magnetic field on transient free convective flow in a vertical channel. Chandran et al. [5] have discussed the unsteady free convection flow of an electrically conducting fluid with heat flux and accelerated boundary layer motion in presence of a transverse magnetic field. Acharya et al. [6] have reported the problem of heat and mass transfer over an accelerating surface with heat source in presence of suction and blowing. The unsteady free convective MHD flow with heat transfer past a semi infinite vertical porous moving plate with variable suction has been studied by Kim [7]. Singh and Thakur [8] have given an exact solution of a plane unsteady MHD flow of a non Newtonian fluid. Sharma and Pareek [9] explained the behavior of steady free convective MHD flow past a vertical porous moving surface. Das et al. [10] estimated the mass transfer effects on unsteady flow past an accelerated vertical porous plate with suction employing finite difference analysis. Das et al. [11] investigated numerically the unsteady free convective MHD flow past an accelerated vertical plate with suction and heat flux. Das and Mitra [12] discussed the unsteady mixed convective MHD flow and mass transfer past an accelerated infinite vertical plate with suction. Recently, Das et al. [13] analyzed the effect of mass transfer on MHD flow and heat transfer past a vertical porous plate through a porous medium under oscillatory suction and heat source. More recently, Das et al. [14] investigated the hydro magnetic convective flow past a vertical porous plate through a porous medium with suction and heat source. The study of stellar structure on solar surface is connected with mass transfer phenomena. Its origin is attributed to difference in temperature caused by the non homogeneous production of heat which in many cases can rest not only in the formation of convective currents but also in violent explosions. Mass transfer certainly occurs within the mantle and cores of planets of the size of or larger than the earth.

In the present study we therefore, propose to analyze the effect of mass transfer on unsteady free convective flow of a viscous incompressible electrically conducting fluid past an infinite vertical porous plate with viscous dissipation and heat source. The present investigations can be utilized as a basis for studying more complex systems that arise in engineering and industrial applications.

2. Mathematical formulation

Consider the unsteady free convective mass transfer flow of a viscous incompressible electrically conducting fluid past an infinite vertical porous plate in presence of constant suction and heat source and transverse magnetic field. We made the following assumptions. In Cartesian coordinate system, let x'- axis is taken to be along the plate and the y'- axis normal to the plate. Since the plate is considered infinite in x'- direction, hence all physical quantities will be independent of x'- direction. The wall is maintained at constant temperature $T_w^{'}$ and concentration $C_w^{'}$ higher than the ambient temperature $T_\infty^{'}$ and concentration $C_\infty^{'}$ respectively. A uniform magnetic field of magnitude B_0 is applied normal to the plate. The transverse applied magnetic field and magnetic Reynolds number are assumed to be very small, so that the induced magnetic field is negligible. The homogeneous chemical reaction is of first order with rate constant $\bar{K}$ between the diffusing species and the fluid is neglected. It is assumed that there is no applied voltage which implies the absence of an electric field. The fluid has constant kinematic viscosity and constant thermal conductivity and the boussinesq's approximation have been adopted for the flow. The concentration of the diffusing species in the binary mixture is assumed to be very small in comparison with the other chemical species, which are present and hence soret and dufour effects are negligible. The magneto hydrodynamic unsteady mixed convective boundary layer equations under the boussinesq's approximations are:

Continuity Equation:

$$\frac{\partial v^{'}}{\partial y^{'}} = 0 \Rightarrow v^{'} = -v_0^{'}(constant), \tag{1}$$

Momentum Equation:

$$\frac{\partial u^{'}}{\partial t^{'}} + v^{'}\frac{\partial u^{'}}{\partial y^{'}} = g\beta(T^{'} - T_\infty^{'}) + g\beta^{*}(C^{'} - C_\infty^{'}) + v\frac{\partial^2 u^{'}}{\partial y^{2'}} - \frac{\sigma B_0^2}{\rho}u^{'} - \frac{v}{k^{'}}u^{'}. \tag{2}$$

Energy Equation:

$$\frac{\partial T^{'}}{\partial t^{'}} + v^{'}\frac{\partial T^{'}}{\partial y^{'}} = \kappa\frac{\partial^2 T^{'}}{\partial y^{2'}} + \frac{v}{C_p}\left(\frac{\partial u^{'}}{\partial t^{'}}\right)^2 + S^{'}(T^{'} - T_\infty^{'}). \tag{3}$$

Concentration Equation:

$$\frac{\partial C^{'}}{\partial t^{'}} + v^{'}\frac{\partial C^{'}}{\partial y^{'}} = D\frac{\partial^2 C^{'}}{\partial y^{'}2}. \tag{4}$$

The boundary conditions of the problem are :

$$u' = 0, v' = -v_0', T' = T_w' + \epsilon(T_w' - T_\infty')e^{iw't'}, C' = C_w' + \epsilon(C_w' - C_\infty')e^{iw't'} \ at \ y' = 0, \tag{5}$$

$$u' \to 0, T' \to T_\infty', C' \to C_\infty' \ as \ y' \to \infty. \tag{6}$$

Introducing the following non dimensional variables and parameters,

$$y = \frac{y'v_0'}{v}, t = \frac{t'v_0'^2}{4v}, \omega = \frac{4v\omega'}{v_0'^2}, u = \frac{u'}{v_0'}, v = \frac{\eta_0}{\rho}, \tag{7}$$

$$M = \left(\frac{\sigma B_0^2}{\rho}\right)\frac{v}{v_0'^2}, K = \frac{v_0'^2 K'}{v^2}, \theta = \frac{(T' - T_\infty')}{(T_w - T_\infty')}, C = \frac{(C' - C_\infty')}{(C_w - C_\infty')}, \tag{8}$$

$$Pr = \frac{\nu}{k}, Gr = \frac{vg\beta(T_w' - T_\infty')}{v_0'^3}, Gc = \frac{vg\beta^*(C_w' - C_\infty')}{v_0'^3}, Sc = \frac{v}{D}, S = \frac{4vS'}{v_0'^2}, Ec = \frac{v_0'^2}{Cp(T_w' - T_\infty')}, \tag{9}$$

$$Sc = \frac{v}{D}, S = \frac{4vS'}{v_0'^2}, Ec = \frac{v_0'^2}{Cp(T_w' - T_\infty')}. \tag{10}$$

Substituting (7), (8), (9) and (10) in equations (2), (3) and (4) under boundary conditions (5) and (6), we get:

$$\frac{1}{4}\frac{\partial u}{\partial t} - \frac{\partial u}{\partial y} = (Gr)\theta + (Gc)C + \frac{\partial^2 u}{\partial y^2} - \left(M + \frac{1}{K}\right)u, \tag{11}$$

$$\frac{1}{4}\frac{\partial \theta}{\partial t} - \frac{\partial \theta}{\partial y} = \frac{1}{Pr}\frac{\partial^2 \theta}{\partial y^2} + \frac{1}{4}S\theta + (Ec)\left(\frac{\partial u}{\partial y}\right)^2, \tag{12}$$

$$\frac{1}{4}\frac{\partial C}{\partial t} - \frac{\partial C}{\partial y} = \frac{1}{Sc}\frac{\partial^2 C}{\partial y^2}. \tag{13}$$

The corresponding boundary conditions are:

$$u = 0, \theta = 1 + \epsilon e^{i\omega t}, C = 1 + \epsilon e^{i\omega t} \ at \ y = 0, \tag{14}$$

$$u \to 0, \theta \to 0, C \to 0 \ as \ y \to \infty, \tag{15}$$

where B_0, magnetic field component along y'-axis, C_p, specific heat at constant pressure, G_r, grashof number,G_c, modified trashof number, g, acceleration of gravity, K', the permeability of medium, K, the permeability parameter, M, hartmann number,P_r, prandtl number,S_c, schmidt number,S, heat source parameter,E_c, eckert number, D, chemical molecular diffusivity, T', temperature of fluid near the plate, T_w', temperature of the fluid far away of the fluid from the plate, T_∞', temperature of the fluid at infinity, C, concentration of the fluid, C', concentration of fluid near the plate, C_w', concentration of the fluid far away of the fluid from the plate, C_∞', concentration of the fluid at infinity, t', time in x',y'-, coordinate system, t, time in dimensionless coordinates, u', velocity component in x'- direction, u, dimensionless velocity component in x'- direction, v', velocity component in y'- direction, v_0', dimensionless velocity component in y'- direction,x',y', coordinate system, x,y, dimensionless coordinates, β, coefficient of volume expansion for heat transfer,β^*, coefficient of volume expansion for mass transfer, κ, thermal conductivity of the fluid, τ, electrical conductivity of the fluid, v, kinematic viscosity, θ, non dimensional temperature, ρ, density of the fluid, ω, angular frequency, ωt, phase angle.

3. Method of solution

The systems of coupled nonlinear equations (11), (12) and (13) with the relevant boundary conditions (14) and (15) are solved numerically for the velocity, temperature and concentration distributions. The finite element method is used to obtain an accurate and efficient solution to the boundary value problem under consideration. The fundamental steps comprising the method are as follows:

3.1. Step 1: Discretization of the domain into elements:

The whole domain is divided into finite number of sub-domains, a process known as discretization of the domain. Each sub-domain is termed a finite element. The collection of elements is designated the finite element mesh.

3.2. Step 2: Derivation of the element equations:

The derivation of finite element equations i.e.algebraic equations among the unknown parameters of the finite element approximation, involves the following three steps: a) Construct the variational formulation of the differential equation. b) Assume the form of the approximate solution over a typical finite element. c) Derive the finite element equations by substituting the approximate solution into variational formulation.

3.3. Step 3: Assembly of element equations:

The algebraic equations so obtained are assembled by imposing the inter-element continuity conditions. This yields a large number of algebraic equations, constituting the global finite element model, which governs the whole flow domain.

3.4. Step 4: Impositions of boundary conditions:

The physical boundary conditions defined in equation (14) and (15) are imposed on the assembled equations.

3.5. Step 5: Solution of the assembled equations:

The final matrix equation can be solved by a direct or indirect (iterative) method. Numerical solutions for these equations are obtained by MATLAB. In order to prove the convergence and stability of finite element method, the same MATLAB was run with slightly changed values of the previous and no significant change was observed in the values of u, θ and C. This process is repeated until the desired accuracy of 0.0005 is obtained. Hence, the finite element method is stable and convergent.

4. Results and discussion

In order to get physical insight into the problem, we have carried out numerical calculations for nondimensional velocity field, temperature field, species concentration field and by assigning some specific values to the parameters entering into the problem and the effects of these values on the above fields are demonstrated graphically.In the present study we adopted the following default parameter values of finite element computations: $Gr = 5.0$, $Gc = 5.0$, M = 1.0, K = 1.0,$Pr = 0.71$,$E_c = 0.001$, $Sc = 0.6$,$\epsilon = 0.002$, $\omega = 1.0$, t = 1.0. All graphs therefore correspond to these values unless specifically indicated on the appropriate graph.

Figure 1 shows the typical velocity profiles in the boundary layer for various values of the magnetic parameter M. Due to the damping effect of the magnetic field, increase in the values of M have a tendency to slow the motion of the fluid and make it warmer as it moves along the vertical plate causing the tangential velocity to decrease. Figures 2 and 4 show the effects of the heat source parameter S on the velocity and temperature profiles respectively. It is clear that

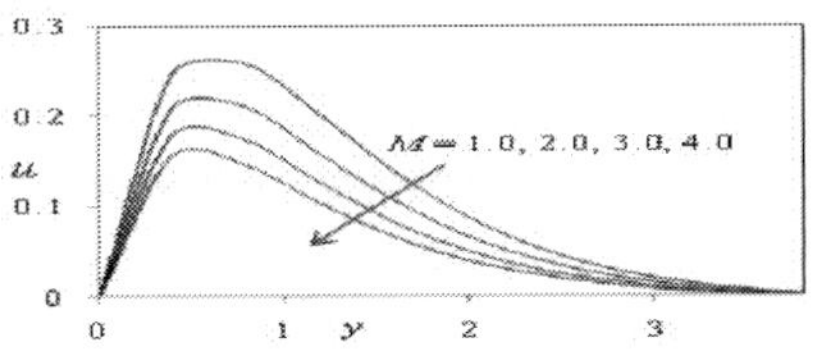

Figure 1. Effect of hartmann number on velocity profiles u.

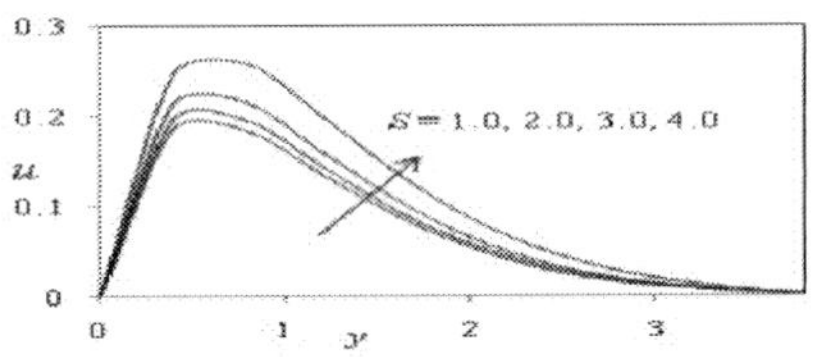

Figure 2. Effect of heat source parameter on velocity profiles u.

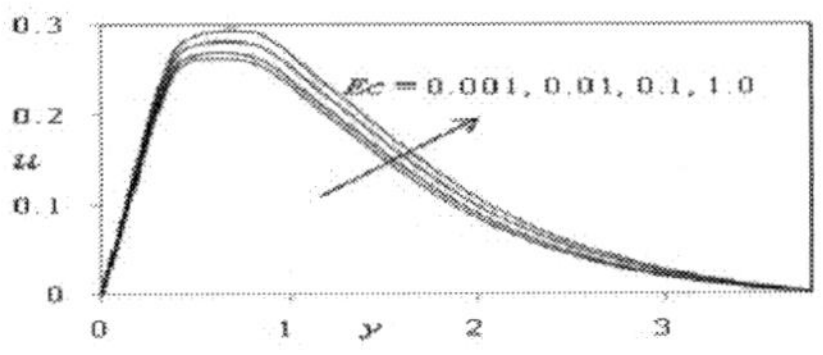

Figure 3. Effect of eckert number on primary velocity profiles u.

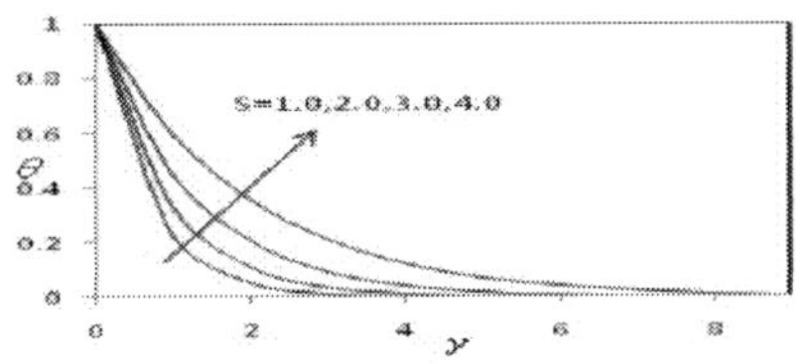

Figure 4. Effect of heat source parameter on temperature profiles θ.

as heat source parameter increases, an increase in the fluid temperature occurs and through the thermal buoyancy effect, the velocity increases. The tangential velocity and temperature profiles for various values of eckert numbers are shown in figure 3 and figure 4. From figure 3 we may conclude that the presence of viscous dissipation and joule heating increase in the tangential velocity and the temperature. This rise in temperature is due to the heat created by viscous dissipation and compression work ($E_c \neq 0$).

Figure 6 shows the effect of the schmidt number Sc on the concentration profile. As the schmidt number Sc increases, the concentration decreases. This causes the concentration buoyancy effects to decrease yielding a reduction in the fluid velocity and also show that the reductions the concentration profiles are accompanied by the simultaneous reductions in the concentration boundary layers.

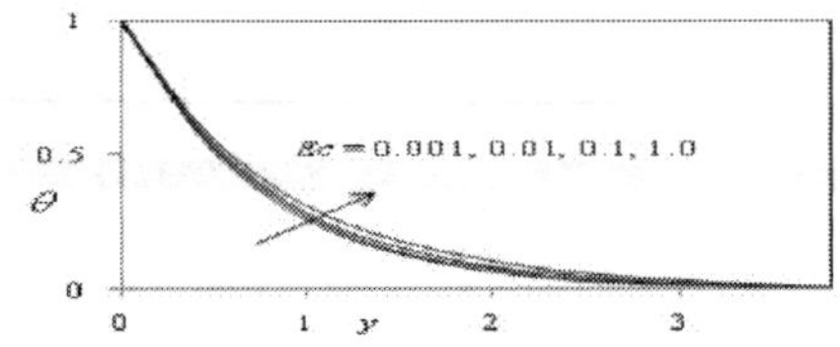

Figure 5. Effect of eckert number on temperature profiles θ.

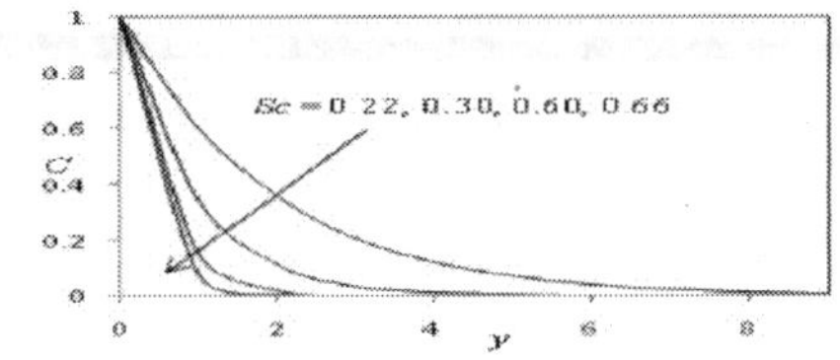

Figure 6. Concentration profiles for different values of Sc.

5. Conclusion

In this study, the governing non-linear partial differential equations are transformed into system of algebraic non-linear equations by using finite element method.The numerical results conclude that as magnetic parameter increases, the value of the velocity decreases. This conclusion meets the logic of magnetic field exerting a retarding force on the free convection flow. It is notice that there is a rise in the temperature and the velocity due to the heat created by the viscous dissipation, free convection and heat source. The effect of increasing schmidt number is to reduce the concentration boundary layer thickness of the flow field at all points.

References

[1] Hasimoto H 1957 *J. Phys. Soc. Japan.* **12** 68-72
[2] Yamamoto K and Iwamura N 1976 *Engg. Math.* **10** 41-54
[3] Mansutti D Pontrelli G and Rajagopal K R 1993 *Int. J. Num. Methods Fluids.* **17** 927-941
[4] Jha B K 1998 *Ind. J. Pure Appl. Math.* **29** 441-445
[5] Chandran P Sacheti N C and Singh A K 1998 *J. Phys. Soc. Japan.* **67** 124-129
[6] Acharya M Singh L P and Dash G C 1999 *Int. J. Engng. Sci.* **37** 1-18
[7] Kim Y J 2000 *Int. J.Engng. Sci.* **38** 833-845
[8] Singh B and Thakur C 2002 *Ind. J. Pure Appl. Math.* **33** 993-1001
[9] Sharma P R and Pareek D 2002 *Ind. J. Theo. Phys.* **50** 5-13
[10] Das S S Sahoo S K and Dash G C 2006 Bull. *Malays. Math. Sci. Soc.* **29** 33-42
[11] Das S S Satapathy A, Das J K and Sahoo S K 2007 *J. Ultra Sci. Phys. Sci.* **19** 105-112
[12] Das S S and Mitra M 2009 *Ind. J. Sci. Tech.* 2 18-22
[13] Das S S Satapathy A, Das J K and Panda J P 2009 *Int. J. Heat Mass Transfer.* **52** 5962-69
[14] Das S S Tripathy U K and J. K. Das 2010 *International Journal of Energy and Environment.***1** 467-478.

Heat and mass transfer effect on MHD natural convection flow past a moving vertical plate

Siva Reddy Sheri

Department of Mathematics, GITAM University, Hydrabad Campus, Telangana State, India.

E-mail: sreddy7@yahoo.co.in

J Anand Rao

Department of Mathematics, Osmania University, Hydrabad , Telangana State, India.

Abstract. The present study considers the effect of heat and mass transfer on MHD natural convective flow of an incompressible electrically conducting fluid past a moving vertical plate. The dimensionless governing non-linear partial differential equations for this investigation are solved numerically using finite element method. The effect of the various dimensionless parameters entering into the problem on the velocity, temperature and concentration profiles across the boundary layer is investigated through graphs.

1. Introduction

The study of heat and mass transfer with chemical reaction is of great practical importance to engineers and scientists because of its almost universal occurrence in many branches of science and engineering. Possible applications of this type of flow can be found in many industries like power industry and chemical process industries. Many natural phenomena and engineering applications are susceptible to magnetohydrodynamic (MHD) analysis. From technological point of view, magnetohydrodynamic flow finds application in the fields of stellar and planetary magneto-spheres, aeronautics, meteorology, solar physics, cosmic fluid dynamics, chemical engineering, electronics and induction flow metry, MHD generators, MHD accelerators, construction of turbine and other centrifugal machines. Very often, along with the free convection currents caused by the temperature difference, the flow is also affected by the difference in concentration of material constituents. In engineering application, the concentration differences are created either by injecting foreign gases or by coating a substrate with a material, and subsequently heating it, so that the material evaporates. Combined heat and mass transfer problems with chemical reaction are of importance in many processes and have, therefore, received a considerable amount of attention in recent years.

The problem of free convection under the influence of the magnetic field has attracted the interest of many researchers in view of its applications in geophysics and astrophysics. The problem under consideration has important applications in the study of geophysical formulations, in the explorations and thermal recovery of oil, and in the underground nuclear waste storage sites. The unsteady natural convection flow past a semi infinite vertical plate was first solved by Soundalgekar and Ganesan [1]. Elbashbeshy [2] studied the heat and mass transfer along a vertical plate under the combined buoyancy effects of thermal and species diffusion, in the presence of the magnetic field. Ganesan and Palani [3] obtained a numerical solution of the unsteady MHD flow past a semi infinite isothermal vertical plate using the finite difference method. Das et al. [4] investigated the hydromagnetic convective flow past a vertical porous plate through a porous medium with suction and heat source. Palani and Kim [5] analyzed the effect of thermal radiation on convection flow past a vertical cone with surface heat flux. Sivaiah et al [6-8] studied the finite element solution of heat and mass transfer past an impulsively started infinite vertical plate.

In this paper we studied that the heat and mass transfer effect on magneto hydrodynamics natural convection flow past a moving vertical plate. The problem is governed by system of coupled non linear partial differential equations whose exact solution is difficult to obtain, so, the problem is solved by using Galerkin finite element method which is more economical from computational view point.

2. Mathematical formulation

An unsteady two-dimensional hydro magnetic laminar mixed convective boundary layer flow of a viscous, incompressible, electrically conducting and chemically reacting fluid in an optically thin environment, past a semi infinite vertical permeable moving plate embedded in a uniform porous medium, in the presence of thermal radiation is considered. The x' axis is taken in the upward direction along the plate and y' axis normal to it. A uniform magnetic field is applied in the direction perpendicular to the plate. Now, under the usual Boussinesqs approximation, the governing boundary layer equations are:

Continuity Equation:

$$\frac{\partial v'}{\partial y'} = 0 \Rightarrow v' = -v_0'(constant). \tag{1}$$

Momentum Equation:

$$\frac{\partial u'}{\partial t'} + v'\frac{\partial u'}{\partial y'} = g\beta(T' - T_\infty') + g\beta^*(C' - C_\infty') + v\frac{\partial^2 u'}{\partial y'^2} - \frac{\sigma B_0^2}{\rho}u' - \frac{v}{k'}u'. \tag{2}$$

Energy Equation:

$$\frac{\partial T'}{\partial t'} + v'\frac{\partial T'}{\partial y'} = \frac{\kappa}{\rho C_p}\frac{\partial^2 T'}{\partial y'^2}. \tag{3}$$

Concentration Equation:

$$\frac{\partial C'}{\partial t'} + v'\frac{\partial C'}{\partial y'} = D\frac{\partial^2 C'}{\partial y'^2}. \tag{4}$$

The boundary conditions of the problem are:

$$t' \leq 0 : u' = 0, v' = 0, T' = 0, C' = 0, \quad for \ \ all \ \ y' \tag{5}$$

$$t' > 0 : u' = 0, v' = -v_0', T' = T_w' + \epsilon(T_w' - T_\infty')e^{in't'}, C' = C_w' + \epsilon(C_w' - C_\infty')e^{in't'} \ \ at \ \ y' = 0, \tag{6}$$

$$u' \to 0, T' \to T_\infty', C' \to C_\infty' \ \ as \ \ y' \to \infty. \tag{7}$$

Introducing the following non dimensional variables and parameters,

$$y = \frac{y' v_0'}{v}, t = \frac{t' v_0'^2}{v}, n = \frac{vn'}{v_0'^2}, u = \frac{u'}{v_0'}, M = \left(\frac{\sigma B_0^2}{\rho}\right)\frac{v}{v_0'^2}, K = \frac{K' v_0'^2}{v^2}, Sc = \frac{v}{D}, Pr = \frac{\nu}{k}, \tag{8}$$

$$Gr = \frac{vg\beta(T_w' - T_\infty')}{v_0'^3}, Gc = \frac{vg\beta^*(C_w' - C_\infty')}{v_0'^3}, \theta = \frac{(T' - T_\infty')}{(T_w' - T_\infty')}, C = \frac{(C' - C_\infty')}{(C_w' - C_\infty')}. \tag{9}$$

Substituting (8) and (9) in equations (2), (3) and (4) under boundary conditions (5) to (7), we get

$$\frac{\partial u}{\partial t} - \frac{\partial u}{\partial y} = (Gr)\theta + (Gc)C + \frac{\partial^2 u}{\partial y^2} - \left(M + \frac{1}{K}\right)u, \tag{10}$$

$$\frac{\partial \theta}{\partial t} - \frac{\partial \theta}{\partial y} = \frac{1}{Pr}\frac{\partial^2 \theta}{\partial y^2}, \tag{11}$$

$$\frac{\partial C}{\partial t} - \frac{\partial C}{\partial y} = \frac{1}{Sc}\frac{\partial^2 C}{\partial y^2}. \tag{12}$$

The corresponding boundary conditions are:

$$u = 0, \theta = 1 + \epsilon e^{nt}, C = 1 + \epsilon e^{nt} aty = 0, \tag{13}$$

$$u \to 0, \theta \to 0, C \to 0 asy \to \infty. \tag{14}$$

3. Method of solution

The systems of coupled nonlinear equations (10), (11) and (12) with the relevant boundary conditions (13)and (14) are solved numerically for the velocity, temperature and concentration distributions, by Bathe [9] and Reddy[10]. The finite element method is used to obtain an accurate and efficient solution to the boundary value problem under consideration. The fundamental steps comprising the method are as follows:

3.1. Step 1: Discretization of the domain into elements:

The whole domain is divided into finite number of sub-domains, a process known as discretization of the domain. Each sub-domain is termed a finite element. The collection of elements is designated the finite element mesh.

3.2. Step 2: Derivation of the element equations:

The derivation of finite element equations i.e. algebraic equations among the unknown parameters of the finite element approximation, involves the following three steps: a) Construct the variational formulation of the differential equation. b) Assume the form of the approximate solution over a typical finite element. c) Derive the finite element equations by substituting the approximate solution into variational formulation.

3.3. Step 3: Assembly of element equations:

The algebraic equations so obtained are assembled by imposing the inter-element continuity conditions. This yields a large number of algebraic equations, constituting the global finite element model, which governs the whole flow domain.

3.4. Step 4: Impositions of boundary conditions:

The physical boundary conditions defined in equation (13) and (14) are imposed on the assembled equations.

3.5. Step 5: Solution of the assembled equations:

The final matrix equation can be solved by a direct or indirect (iterative) method. Numerical solutions for these equations are obtained by MATLAB. In order to prove the convergence and stability of finite element method, the same MATLAB was run with slightly changed values of u, θ and C, no significant change was observed in the values of u, θ and C. This process is repeated until the desired accuracy of 0.0005 is obtained. Hence, the finite element method is stable and convergent.

4. Results and discussion

To understand the physical meaning of the problem, we have computed the expression for u, θ and C for different values of thermal Grashof number Gr, solutal Grashof number Gc , magnetic field parameter M, permeability parameter K , Prandtl number Pr , Schmidt number Sc. The purpose of the numerical result given here is to assess the effects of different parameters upon the nature of the flow, temperature and concentration. The values of the Prandtl number are chosen $Pr = 7$ (water) and $Pr= 0.71$(air). The values of the Schmidt number are chosen to represent the presence of species by hydrogen (0.22), water vapor (0.60) and ammonia (0.78). In the present study we have chosen Gr=2.0,Gc=2.0, M=1.0,K=0.5,Pr=0.71, Sc=0.22,ε=0.2,n=1.0,t=1.0 while Gr, Gc, M, K, Pr, Sc are varied over a range, which are listed in the figures.

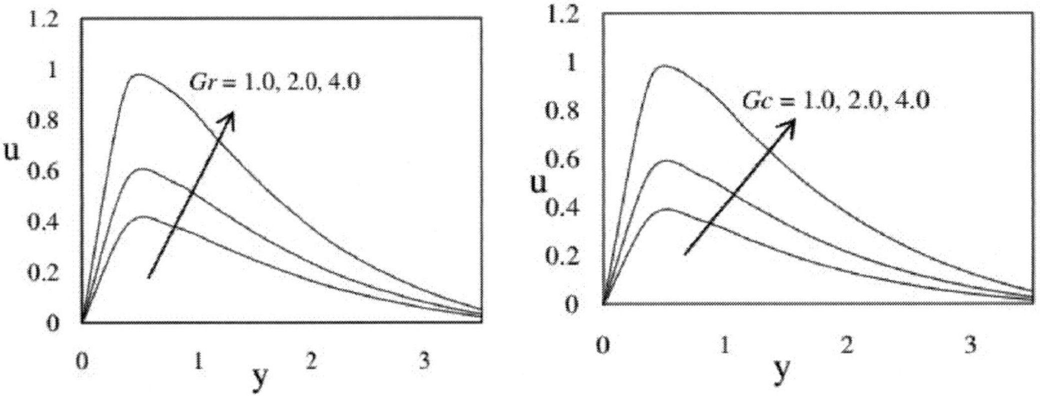

(a) Velocity profiles for different values of Gr **(b)** Velocity profiles for different values of Gc

Figure 1: Velocity profiles for different values of Gr and Gc.

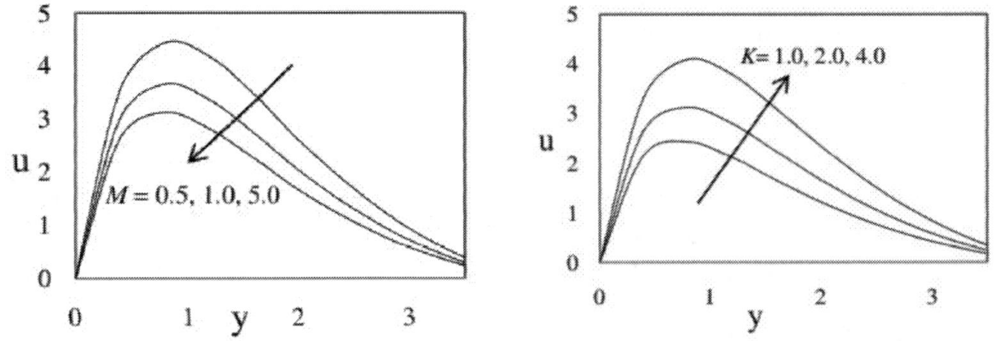

(a) Velocity profiles for different values of M **(b)** Velocity profiles for different values of K

Figure 2: Velocity profiles for different values of M and K.

The temperature and the species concentration are coupled to the velocity via thermal Grashof number Gr and solutal Grashof numberGc as seen in equation (10). For various values of thermal Grashof number and solutal Grashof number, the velocity profiles are plotted in figures 1.(a) and (b). The thermal Grashof numberGr signifies the relative effect of the thermal buoyancy force to the viscous hydrodynamic force in the boundary layer. As expected, it is

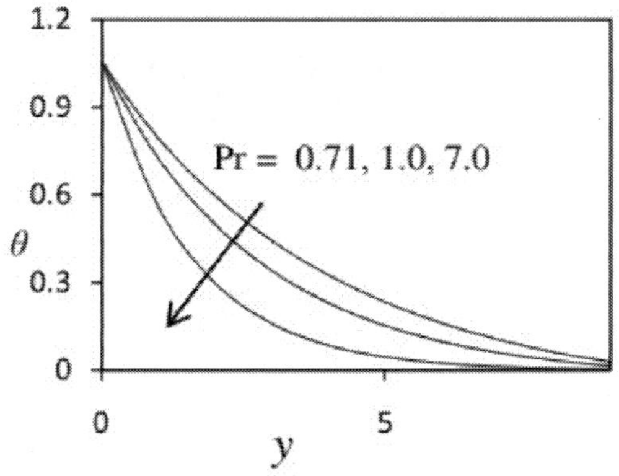

(a) Temperature profiles for different values of Pr

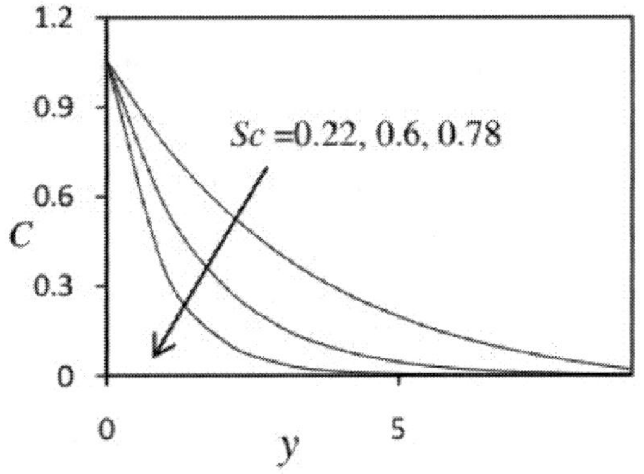

(b) Concentration profiles for different values of Sc

Figure 3: Temperature and Concentration profiles for different values of Pr and Sc.

observed that there is a rise in the velocity due to the enhancement of thermal buoyancy force. Also, as Gr increases, the peak values of the velocity increases rapidly near the porous plate and then decays smoothly to the free stream velocity. The solutal Grashof numberGc defines the ratio of the species buoyancy force to the viscous hydrodynamic force. As expected, the fluid velocity increases and the peak value is more distinctive due to increase in the species buoyancy force. The velocity distribution attains a distinctive maximum value in the vicinity of the plate and then decreases properly to approach the free stream value. It is noticed that the velocity increases with increasing values of the solutal Grashof number. The effect of the Hartmann number M is shown in figure 2(a). It is observed that the velocity of the fluid decreases with the increase of the magnetic field number values. The decrease in the velocity as the Hartmann number M increases is because the presence of a magnetic field in an electrically conducting fluid introduces a force called the Lorentz force, which acts against the flow if the magnetic field is applied in the normal direction, as in the present study. This resistive force slows down the fluid velocity component as shown in figure 2(a). For different values of the permeability parameter K, the velocity on the porous wall are plotted in figure 2(b). It is obvious that the increased values of K tend to increasing of velocity and on the porous wall and so enhance the momentum boundary layer thikness. Figure 3(a) depicts the effect of Prandtl number Pr against y on the temperature field keeping other parameters of the flow field constant. The Prandtl number is a dimensionless number; the ratio of momentum diffusivity (kinematic viscosity) to thermal diffusivity. It is observed that an increase in the prandlt number decreases the temperature of the flow field at all points. In Figure 3(b), we analyze the effect of Schmidt numberSc on concentration profiles of the flow field. Schmidt number is a dimensionless number defined as the ratio of momentum diffusivity (kinematic viscosity) and mass diffusivity. It is observed that a growing Schmidt number decreases the concentration boundary layer thickness of the flow field at all points.

5. Conclusion

In this work, we have studied the effect of heat and mass transfer on magneto hydrodynamics natural convection flow past a moving vertical plate. The governing equations are solved by using finite element method. The results are shown graphically for different values of the parameters considered in the analysis. The following conclusions can be made from the present investigation:

(i) The effect of increasing Grashof number for heat and mass transfer is to accelerate velocity of the flow field at all points.

(ii) A growing Hartmann number retards the velocity of the flow field at all points.

(iii) The velocity of the flow field increases with an increase in permeability parameter.

(iv) The Prandtl number increases the temperature is reduce of the flow field at all points.

(v) The effect of increasing Schmidt number is to reduce the concentration boundary layer thickness of the flow field at all points.

Acknowledgement
The authors are thankful to the University Grant Commission, New Delhi, India for providing financial assistance to carry out this research work under UGC - Major Research Project.

References

[1] Soundalgekar V M and Ganesan P 1981 *International Journal of Engineering Science* **19** 757
[2] Elbashbeshy E M A 1997 *International Journal of Engineering Science* **34** 515
[3] Ganesan P and Palani G 2004 *Proceedings of the Sixth ISHMT/ASME Heat and Mass Transfer Conference and Seventeenth National Heat and Mass Transfer Conference* Kalpakkam. India. 184
[4] Das S S, Tripathy U K and Das J K 2010 *International Journal of Energy and Environment* **1** 467
[5] Palani G and Kim K. Y 2012 *Applied Appl. Math. Mech. Engl. Ed* **33** 605
[6] Sivaiah S and Srinivasa Raju R 2013 *Appl. Math. Mech. Engl. Ed* **34** 559
[7] Sivaiah S, Srinivasa-Raju R and Anand-Rao J 2012 it Journal of Energy Heat and Mass Transfer **34** 12
[8] Anand-Rao J and Sivaiah S 2011 it Appl. Math. Mech. Engl. Ed.**32** 1065
[9] Bathe KJ 1996 '(Prentice Hall-New Jersey)'
[10] Reddy J N 2006 '(McGraw-Hill Book Company New York)' **3**

Journal of Physics: Conference Series **662** (2015) 012014 doi:10.1088/1742-6596/662/1/012014

Effects of chemical reaction on MHD mixed convection stagnation point flow toward a vertical plate in a porous medium with radiation and heat generation

Niranjan Hari[1], S Sivasankaran[1], M Bhuvaneswari[2] and Zailan Siri[1]

[1]Institute of Mathematical Sciences, University of Malaya, Kuala Lumpur 50603, Malaysia.
[2]Department of Mechanical Engineering, University of Malaya, Kuala Lumpur 50603, Malaysia.

E-mail: sd.siva@yahoo.com

Abstract. The aim of the present study is to analyze the effects of chemical reaction on MHD mixed convection with the stagnation point flow towards a vertical plate embedded in a porous medium with radiation and internal heat generation. The governing boundary layer equations are transformed into a set of ordinary differential equations using similarity transformations. Then they are solved by shooting technique with Runge-Kutta fourth order iteration. The obtained numerical results are illustrated graphically and the heat and mass transfer rates are given in tabular form. The velocity and temperature profiles overshoot near the plate on increasing the chemical reaction parameter, Richardson number and magnetic field parameter.

1. Introduction

The study of combined free and forced convection flow of an incompressible viscous fluid with simultaneous heat and mass transfer past a vertical plate under the influence of a magnetic field and chemical reaction has huge applications in water and air pollutions, fibrous insulation, several manufacturing processes of industry such as polymers, aerodynamic extrusion of plastic sheets, etc. A stagnation point is a point in a flow field where the local velocity of the fluid is zero. Boutros et al. [1] explored the steady two-dimensional stagnation point flow of an incompressible viscous fluid over a stretching sheet embedded in a fluid saturated porous medium. Mahapatra and Gupta [2] examined the steady two-dimensional stagnation point flow of an incompressible electrically conducting viscous fluid over a flat deformable sheet. They observed that the wall shear stress increases with increase in magnetic field parameter. Merikin and Pop [3] studied the unsteady free convection boundary layer flow in a porous medium for small and large values of the amplitude of the surface temperature oscillations. Takhar et al. [4] studied magneto mixed convection flow of a viscous incompressible electrically conducting fluid in the vicinity of an axisymmetric stagnation point adjacent to a heated vertical surface. Ferdows et al. [5] numerically investigated the natural convection heat transfer in an inclined semi-infinite plate in a porous medium. They observed that the temperature increases and the velocity of the fluid decreases with increasing the porosity parameter.

The research of MHD incompressible viscous flow has many engineering applications such as a power generator, cooling of reactors, design of heat exchangers and MHD accelerators. Yih [6]numerically analyzed the effect of heat source on steady two dimensional laminar MHD mixed convection flow in a porous medium over a vertical permeable flat plate. Layek et al. [7] analyzed the structure of boundary layer stagnation point flow over a stretching sheet in a porous medium with heat generation and suction/blowing. Bhattacharyya and Layek [8] studied the stagnation-point steady boundary layer flow over a porous shrinking sheet subjected to applied suction and radiation. Afify and Elagazery [9] numerically examined the stagnation point flow, heat and mass transfer of an incompressible, electrically conducting fluid towards a stretching sheet in a porous medium in the presence of chemical reaction. The temperature profile increases with an increase in heat source or sink parameter. They observed that concentration profile increases with an increase in chemical reaction. Chao et al. [10] investigated the chemically reactive stagnation point flow in a catalytic porous bed. Lee et al. [11] investigated heat transfer and fluid flow in a saturated porous medium by using Lie group analysis. They found that the thermal and momentum boundary layer thicknesses increase with increasing radiation parameter. Bhuvaneswari et al. [12] obtained the effects of radiation, internal heat generation on natural convective flow over an inclined surface in a porous medium. When the heat generation parameter increases, the temperature and velocity profiles increase. Bhuvaneswari et al. [13] examined the heat and mass transfer of an incompressible viscous fluid in a semi-infinite inclined surface by using Lie group analysis with chemical reaction. They observed that while increasing the chemical reaction parameter and Schmidt number the thickness of the momentum boundary layer increases. Singh et al. [14] investigated the heat and mass transfer of a mixed convection stagnation point flow through porous media in the presence of the volumetric rate of heat generation/absorption. Mankinde [15] examined the effects of thermal radiation on hydromagnetic mixed convection stagnation point flow with internal heat generation. The present study extends the work of Makinde [15] to include chemical reaction effect on MHD mixed convection stagnation point flow of a vertical plate embedded in a porous medium.

2. Mathematical model

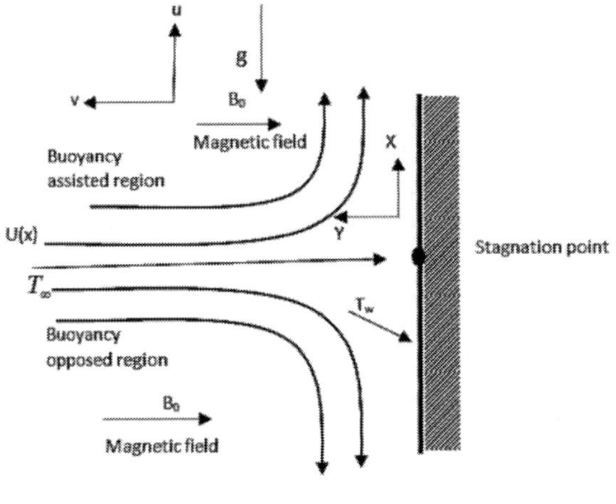

Figure 1. Schematic diagram of the problem.

Consider the two-dimensional steady stagnation-point flow, heat and mass transfer of an incompressible, electrically conducting fluid through a porous medium along a vertical isothermal plate in the presence magnetic field, volumetric heat generation/absorption and first order homogeneous chemical reaction. The magnetic field of constant strength is imposed along the y-axis. The induced magnetic field is neglected. As the fluid hits the wall at the stagnation point, the flow divided into two streams and the viscous flow adheres to the plate within the boundary layer. The velocity distribution in the potential region flow is assumed as $U_\infty = cx$, where c is a positive constant. The following equations represent the flow in a porous medium obeying the Darcy law. Thus, the governing equations of continuity of mass, momentum transfer, energy transfer and species concentration are given by

$$\frac{\partial u}{\partial x} + \frac{\partial v}{\partial y} = 0, \tag{1}$$

$$u\frac{\partial u}{\partial x} + v\frac{\partial u}{\partial y} = \upsilon\frac{\partial^2 u}{\partial y^2} + g\beta(T - T_\infty) + g\beta^*(C - C_\infty)$$

$$-\Big(\frac{\sigma_\varepsilon B_0^2}{\rho} + \frac{\upsilon}{\tilde{K}}\Big)(u - U_\infty) + U_\infty\frac{dU_\infty}{dx}, \tag{2}$$

$$u\frac{\partial T}{\partial x} + v\frac{\partial T}{\partial y} = \alpha\frac{\partial^2 T}{\partial y^2} - \frac{\alpha}{k}\frac{\partial q_r}{\partial y} + Q(T - T_\infty), \tag{3}$$

$$u\frac{\partial C}{\partial x} + v\frac{\partial C}{\partial y} = D\frac{\partial^2 C}{\partial y^2} - \Gamma_0(C - C_\infty). \tag{4}$$

The boundary conditions are

$$u = 0, \quad v = 0, \quad T = T_w, \quad C = C_w \quad \text{at} \quad y = 0,$$

$$u \to U_\infty = cx, \quad T \to T_\infty, \quad C \to C_\infty, \quad \text{as} \quad y \to \infty, \tag{5}$$

where u and v are the components of velocity along x and y directions, respectively, g, $\tilde{K}$, υ, σ_ε, β, β^*, T, c_p, α, k, Q, B_0, D, T_w, and Γ_0 are gravitational acceleration, permeability of the porous medium, kinematic viscosity, electrical conductivity, thermal expansion coefficient, coefficient of concentration expansion, temperature, thermal diffusivity, thermal conductivity, volumetric heat generation/absorption, strength of magnetic field, mass diffusivity, temperature of the plate and chemical reaction rate of solutal concentration. The chemical reaction is taking place at the wall and the substance generated due to chemical reaction is represented as concentration C. The medium is assumed as grey and transparent. It absorbs heat and emits and dispersion of heat are very small. Hence the radiation effect is taken in the medium due to Stefan Boltzmann's law which states that the radiation is proportional to the 4th power of temperature. Heat flux due to radiation is given by,

$$q_r = -\frac{4\sigma^*}{3K'}\frac{\partial T^4}{\partial y}, \tag{6}$$

where σ^* and K' are the Stefan-Boltzmann constant and the mean absorption coefficient, respectively. Using the Rosseland approximation, it is assumed that the differences in temperature within the flow are too small, so that, using Taylor series T^4 may be shown as a linear function of temperature T about the free stream temperature T_∞ i.e., T^4 is approximately $4T_\infty^3 T - 3T_\infty^4$. Then we get

$$q_r = -\frac{16\sigma^* T_\infty^3}{3K'}\frac{\partial T}{\partial y}. \tag{7}$$

Equation (7) is substituted in (3) for temperature. We introduce the following non-dimensional variables:

$$\eta = y\sqrt{\frac{c}{\upsilon}},\ \Psi(x,y) = \sqrt{\upsilon c}\, x f(\eta),\ Cr = \frac{\Gamma_0}{c},\ \theta(\eta) = \frac{T - T_\infty}{T_\omega - T_\infty},$$

$$\phi(\eta) = \frac{C - C_\infty}{C_\omega - C_\infty},\ Gr_T = \frac{g\beta(T_\omega - T_\infty)x^3}{\upsilon^2},\ Gr_c = \frac{g\beta^*(C_\omega - C_\infty)x^3}{\upsilon^2},$$

$$Ri_T = \frac{Gr_T}{Re_x^2},\ Ri_C = \frac{Gr_c}{Re_x^2},\ Re_x = \frac{U_\infty x}{\upsilon},\ Rd = \frac{4\sigma^* T_\infty^3}{kK'},\ K = \frac{\upsilon}{c\tilde{K}},$$

$$Pr = \frac{\upsilon}{\alpha},\ Sc = \frac{\upsilon}{D},\ M = \frac{\sigma_\varepsilon B_0^2}{c\rho},\ S = \frac{Q\upsilon}{\alpha c} = Pr\frac{Q}{c}, \tag{8}$$

where Ψ is the stream function which is defined in the usual form as $u = \frac{\partial\Psi}{\partial y}$ and $v = -\frac{\partial\Psi}{\partial x}$ so that the continuity equation (1) is automatically satisfied. Substituting the expression in (7) together with the variables in (8) into (1)-(5), we obtain the following nonlinear ordinary differential equations:

$$f''' + ff'' - f'^2 + Ri_T\theta + Ri_c\phi - (K + M)(f' - 1) + 1 = 0, \tag{9}$$

$$\left(1 + \frac{4}{3}Rd\right) + Prf\theta' + S\theta = 0, \tag{10}$$

$$\phi'' + Scf\phi' - ScCr\phi = 0, \tag{11}$$

where Cr is a chemical reaction parameter, Gr_c is solutal Grashof number, Gr_T is thermal Grashof number, K is a porous medium permeability parameter, M is magnetic field parameter, Pr is Prandtl number, Rd is a thermal radiation parameter, Ri_c is the solutal Richardson number, Ri_T is the thermal Richardson number, S is an internal heat generation/absorption parameter, Sc is Schmidt number. The corresponding boundary conditions (5) are

$$\begin{aligned} f = 0,\quad f' = 0,\quad \theta = 1,\quad \phi = 1,\quad \text{as}\ \eta = 0,\\ f' = 1,\quad \theta = 0,\quad \phi = 0,\quad \text{as}\ \eta \to \infty. \end{aligned} \tag{12}$$

The set of equations (9)-(11) with the boundary conditions (12) has been solved numerically by applying the shooting technique along with fourth order Runge-Kutta integration scheme.

The skin friction (C_f) is the force along the surface, the Nusselt number (Nu) represents heat transfer through a fluid as a result of convection relative to conduction across the same fluid layer. The Sherwood number (Sh) is the ratio between the convective to diffusive mass transfer. These nondimensional numbers $C_f,\ Nu,\ Sh$ are defined as follows,

$$C_f = \frac{2\tau_\omega}{\rho U_\infty^2},\quad Nu = \frac{xq_\omega}{k(T_\omega - T_\infty)},\quad Sh = \frac{xq_m}{D(C_\omega - C_\infty)}, \tag{13}$$

$$\tau_\omega = \mu \frac{\partial u}{\partial y}\Big|_{y=0}, \quad q_\omega = -k\frac{\partial T}{\partial y}\Big|_{y=0} - \frac{4\sigma^*}{3K'}\frac{\partial T^4}{\partial y}\Big|_{y=0}, \quad q_m = -D\frac{\partial C}{\partial y}\Big|_{y=0}. \tag{14}$$

Substituting (7), (8) and (13) into (14), we obtain the expressions for the skin-friction coefficient, the local Nusselt number and the local Sherwood number as follows.

$$Re_x^{\frac{1}{2}}C_f = f''(0), \; Re_x^{\frac{1}{2}}Nu = -(1+4Rd/3)\theta'(0), \; Re_x^{\frac{1}{2}}Sh = -\phi'(0). \tag{15}$$

The accuracy of our numerical procedure is tested by directly comparing with the previously published work of Singh et al. [14] and Makinde [15] for special cases of the problem under consideration. Table 1 shows that the complete agreement between the comparison results exists. This gives confidence in the numerical results to be reported subsequently.

Table 1. Computation showing the comparison with Singh el al. [14] and Makinde [15] for different values of S when $Ri_T = 1$, $Ri_C = 0.5$, $Pr = 1$, $Sc = 0.5$, $M = 0$, $K = 0$, $Cr = 0$, $Rd = 0$.

	$f''(0)$			$-\theta'(0)$			$-\phi'(0)$		
S	Singh [14]	Makinde [15]	Present	Singh [14]	Makinde [15]	Present	Singh [14]	Makinde [15]	Present
-1	1.8444	1.8444	1.8501	1.3908	1.3908	1.3897	0.4631	0.4631	0.4618
0	1.9995	1.9995	1.9989	0.6392	0.6932	0.6401	0.4789	0.4789	0.4761
1	2.1342	2.1342	2.1299	-0.0730	-0.0730	-0.0727	0.4917	0.4917	0.4902

3. Results and discussions

The results are displayed graphically for different parameters $(K, M, Ri_T, Ri_c, S, Rd, Cr)$, with $Pr = 0.7$ (air) and $Sc = 0.5$. Table 2 illustrates the effect of skin friction, Nusselt number and Sherwood number of different parameters. The skin friction increases with increasing porous medium permeability parameter, magnetic field perimeter and it decreases with increasing the chemical reaction parameter. The local Nusselt number increases with an increasing permeability parameter, magnetic field parameter, thermal radiation parameter and decreases with increasing Richardson number, heat generation parameter. The local Sherwood number increases with an increasing permeability parameter, magnetic field parameter, Richardson number, internal heat generation, chemical reaction parameter.

Figure 2 shows the velocity profiles for different values of K with constant $Ri_T = 1$, $Ri_c = 0.5$, $M = 0.5$, $Rd = 0.2$, $S = 0.2$, $Cr = 0.2$. Figure 2 shows that the fluid velocity increases gradually from the stationary plate surface to its peak value. It is clear that an overshoot in the fluid velocity towards the plate surface is observed. The velocity of the fluid decreases with an increase in the porous medium permeability parameter (K). Figure 3 depicts the influence of different magnetic field parameter values of the velocity field discussed with constant $Ri_T = 1$, $Ri_c = 0.5$, $K = 0.5$, $Rd = 0.2$, $S = 2$, $Cr = 0.2$ values. It is seen from this figure that the velocity decreases initially and then increases with increasing the magnetic field intensity. Boundary layer thickness is very high near the plate.

It is observed from the Figures 4 & 5 that velocity increases as the Richardson number increases. The fluid motion is enhanced by buoyancy forces with increasing values of solutal and thermal (Ri_c, Ri_T) Richardson numbers leading to velocity overshoots near the plate within the boundary layer. The fluid velocity increases gradually from the stationary plate surface to its peak value within the boundary layer region for different values of S with constant Ri_T, Ri_c, K, M, Rd, Cr, see Figure 6. Figures 7 illustrate the temperature profiles for different values of S. We can observe that thermal boundary layer thickness is more than the momentum boundary layer thickness. It is observed from Figure 8 that the temperature increases

Table 2. Different values of f'', $-\theta'$, $-\phi'$ for $Pr = 0.7$ and $Sc = 0.5$.

K	M	Ri_T	Ri_C	S	Rd	Cr	$f''(0)$	$-\theta'(0)$	$-\phi'(0)$
0	0.5	1	0.5	0.2	0.2	0.2	2.282218	0.129818	0.565269
0.5							2.381054	0.131488	0.566446
1.0							2.480144	0.132527	0.567637
2.0							2.658150	0.133415	0.570214
3.0							2.818487	0.134906	0.572237
0.5	0	1	0.5	0.2	0.2	0.2	2.282218	0.129818	0.565269
	2						2.658150	0.133415	0.570214
	4						3.050930	0.139619	0.572251
	6						4.249084	0.157870	0.565790
	8						5.507522	0.135105	0.581117
0.5	0.5	0.1	0.5	0.2	0.2	0.2	1.875287	0.138825	0.546017
		0.5					2.101744	0.135415	0.555178
		1.0					2.381054	0.131488	0.566446
		1.5					2.656359	0.127927	0.577511
		2.0					2.927848	0.124727	0.588375
0.5	0.5	1	0.1	0.2	0.2	0.2	2.203743	0.132965	0.559868
			0.5				2.381054	0.131488	0.566446
			1.0				2.598403	0.129982	0.574389
			1.5				2.811546	0.128839	0.582031
			2.0				3.021085	0.128045	0.589378
0.5	0.5	1	0.5	-1.0	0.2	0.2	2.277775	0.675466	0.561458
				-0.5			2.313538	0.479968	0.563299
				0.0			2.359292	0.242043	0.565472
				0.5			2.418090	-0.054742	0.568029
				1.0			2.493639	-0.432447	0.571028
0.5	0.5	1	0.5	0.2	0	0.2	2.380285	0.11306	0.565899
					0.2		2.381054	0.131488	0.566446
					0.4		2.380925	0.143412	0.566865
					0.8		2.380485	0.177186	0.567461
					1.0		2.381412	0.27606	0.567570
0.5	0.5	1	0.5	0.2	0.2	-1.0	2.425714	0.130183	0.036520
						-0.5	2.404500	0.130797	0.284168
						0	2.387099	0.131309	0.492312
						0.5	2.372845	0.131731	0.669348
						1.0	2.361162	0.132074	0.822065

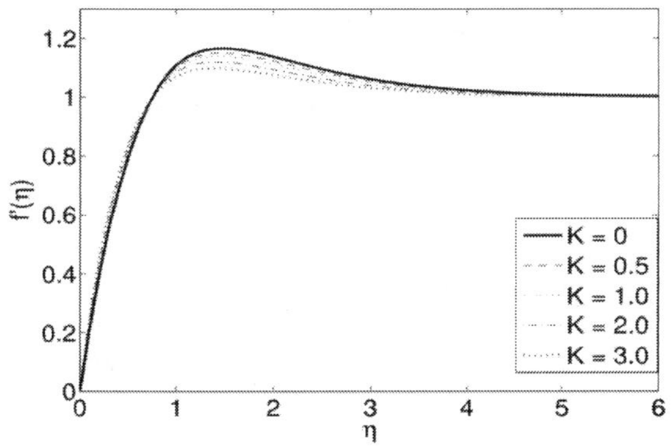

Figure 2. Velocity profiles for different K values with $Ri_T = 1$, $Ri_C = 0.5$, $M = 0.5$, $Rd = 0.2$, $S = 0.2$, $Cr = 0.2$.

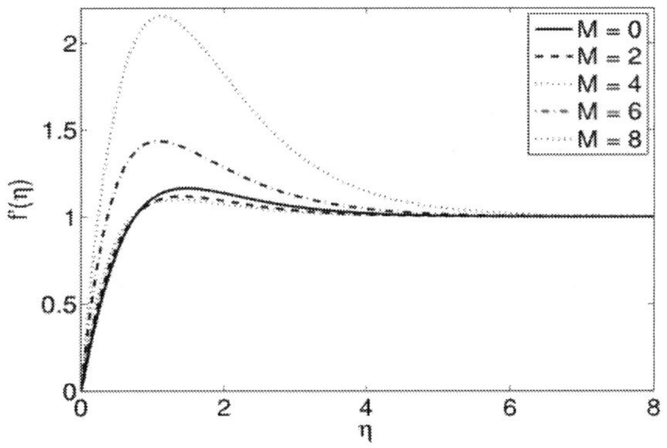

Figure 3. Velocity profiles for different M values with $Ri_T = 1$, $Ri_C = 0.5$, $K = 0.5$, $Rd = 0.2$, $S = 0.2$, $Cr = 0.2$.

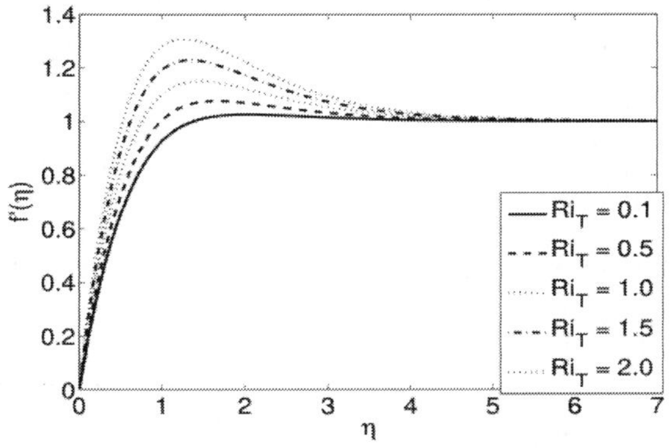

Figure 4. Velocity profiles for different Ri_T values with $Ri_C = 0.5$, $K = 0.5$, $M = 0.5$, $Rd = 0.2$, $S = 0.2$, $Cr = 0.2$.

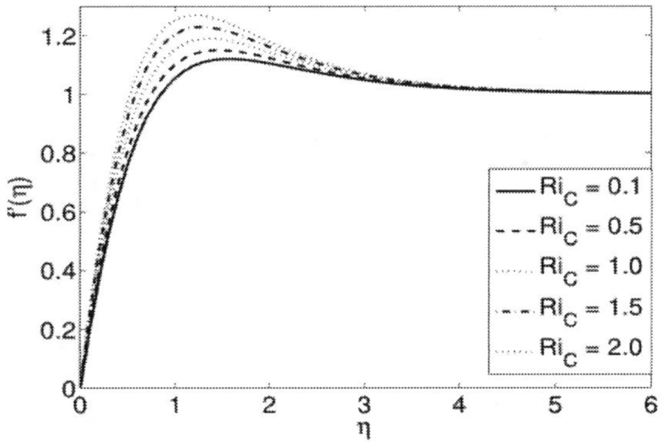

Figure 5. Velocity profiles for different Ri_C values with $Ri_T = 1$, $K = 0.5$, $M = 0.5$, $Rd = 0.2$, $S = 0.2$, $Cr = 0.2$.

with increasing the thermal radiation parameter. Figure 9 illustrates the concentration profiles for different Cr values with constant Ri_T, Ri_c, K, M, S, Rd values. We observe that the concentration gradually decreases exponentially in the boundary layer on increasing Cr values. At the plate surface contains a high chemical species concentration and decay exponentially far away from the plate.

4. Conclusion

In this paper, we studied the MHD mixed convection stagnation point flow, heat and mass transfer toward a vertical plate in a porous medium with radiation, internal heat generation and chemical reaction. The main findings are summarized as follows: The velocity profiles overshoot near the plate surface on increasing the magnetic field, Richardson number and internal heat generation parameter. The velocity decreases on increasing the porous medium permeability parameter. The temperature increases with an increase in the heat generation and thermal radiation parameters. The concentration decreases on increasing the chemical reaction parameter.

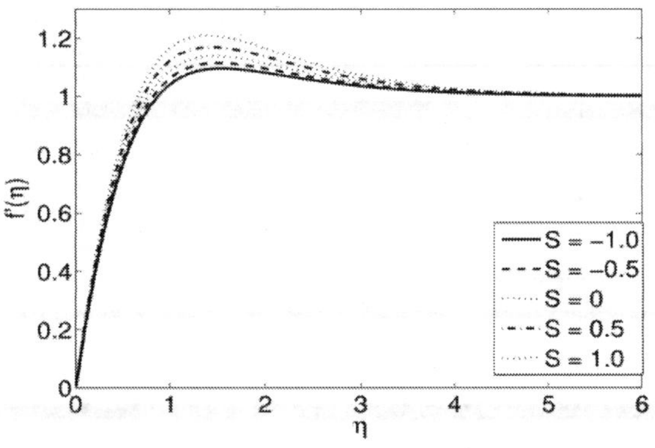

Figure 6. Velocity profiles for different S values with $Ri_T = 1$, $Ri_C = 0.5$, $K = 0.5$, $M = 0.5$, $Rd = 0.2$, $Cr = 0.2$.

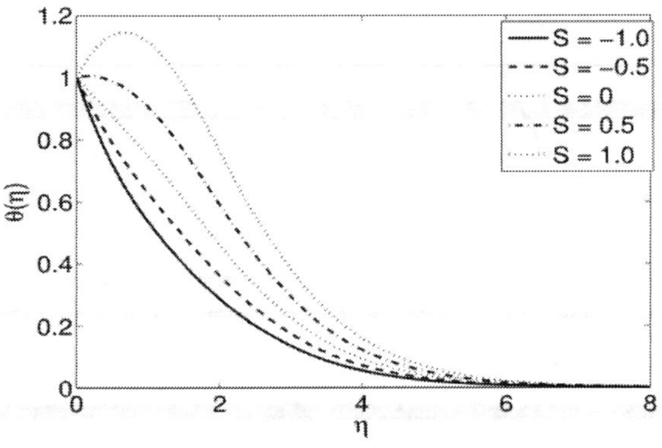

Figure 7. Temperature profiles for different S values with $Ri_T = 1$, $Ri_C = 0.5$, $K = 0.5$, $M = 0.5$, $Rd = 0.2$, $Cr = 0.2$.

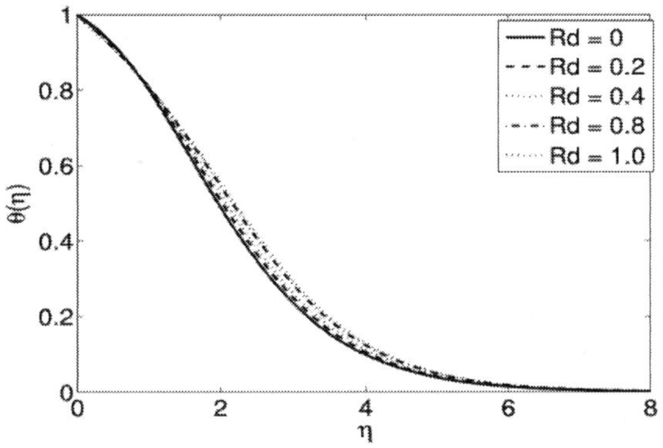

Figure 8. Temperature profiles for different Rd values with $Ri_T = 1$, $Ri_C = 0.5$, $K = 0.5$, $M = 0.5$, $S = 0.2$, $Cr = 0.2$.

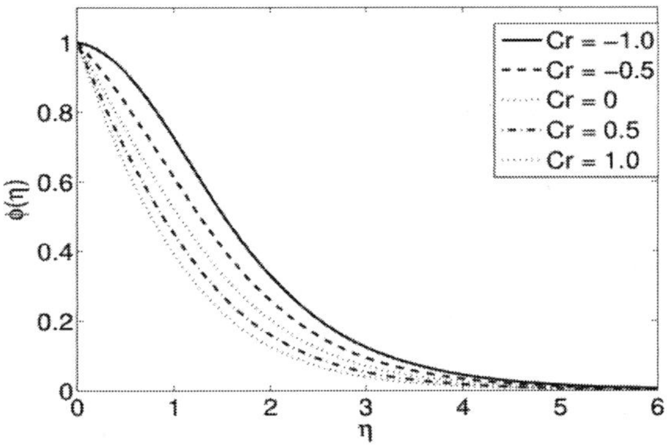

Figure 9. Concentration profiles for different Cr values with $Ri_T = 1$, $Ri_C = 0.5$, $K = 0.5$, $M = 0.5$, $Rd = 0.2$, $S = 0.2$.

Acknowledgments

Authors would like to thank the University of Malaya for the financial support through the research grants, RG216-12AFR and RP011B-13AFR.

References

[1] Boutros Y Z, Malek M B A, Badran N A and Hassan H S 2006 Lie-group method of solution for steady two dimensional boundary layer stagnation point flow towards a heated stretching sheet placed in a porous medium *Meccanica* **41** 681–91

[2] Mahapatra T R and Gupta A S 2000 Magneto hydrodynamic stagnation-point flow towards a stretching sheet *Acta Mechanica* **152** 191–6

[3] Merkin J H and Pop I 2000 Free convection near a stagnation point in a porous medium resulting from an oscillatory wall temperature *Int. J. Heat Mass Transf.* **43** 611–21

[4] Takhar H S, Chamkha A J and Nath G 2003 Unsteady mixed convection on the stagnation point flow adjacent to a vertical plate with a magnetic field *Heat Mass Transf.* **41** 387–98

[5] Ferdows M, Kaino K and Sivasankaran S 2009 Free Convection Flow in an inclined porous surface *J. Porous Media* **12** 997–1003

[6] Yih K A 1998 Heat source/sink effect on MHD mixed convection in stagnation flow on a vertical permeable plate in porous media *Int. Commun. Heat Mass Transf.* **25** 427–42

[7] Layek G C, Mukhopadhyay S and Samad Sk A 2007 Heat and mass transfer analysis for boundary layer stagnation-point flow towards a heated porous stretching sheet with heat absorption/generation and suction/blowing *Int. Commun. Heat Mass Transf.* **34** 347–56

[8] Bhattacharyya K and Layek G C 2010 Effects of suction/blowing on steady boundary layer stagnation-point flow and heat transfer towards a shrinking sheet with thermal radiation *Int. J. Heat Mass Transf.* **54** 302–7

[9] Afify A A and Elagazery Nasser S 2012 Lie group analysis for the effects of chemical reaction on MHD stagnation-point flow of heat and mass transfer towards a heated porous stretching sheet with suction or injection *Nonlinear Analysis: Modelling Control* **17** 1–15

[10] Chao B.H, Wang H and Cheng P 1996 Stagnation point flow of a chemically reactive fluid in a catalytic porous bed *Int. J. Heat Mass Transf.* **39** 3003–19

[11] Lee J, Kandaswamy P, Bhuvaneswari M and Sivasankaran S 2008 Lie group analysis of radiation natural convection heat transfer past an inclined porous surface *J. Mech. Sci. Tech.* **22** 1779–84

[12] Bhuvaneswari M, Sivasankaran S and Kim Y J 2012 Lie group analysis of radiation natural convection flow over an inclined surface in a porous medium with internal heat generation *J. Porous Media* **15** 1155–64

[13] Bhuvaneswari M, Sivasankaran S and Ferdows M 2009 Lie group analysis of natural convection heat and mass transfer in an inclined surface with chemical reaction *Nonlinear Analysis: Hybrid Systems* **3** 536–42

[14] Singh G, Sharma P R and Chamkha A J 2010 Effect of volumetric heat generation/absorption of mixed convection stagnation point flow on an iso-thermal vertical plate in porous media *Int. J. Ind. Math.* **2** 59–71

[15] Makinde O D 2011 Heat and mass transfer by MHD mixed convection stagnation point flow toward a vertical plate embedded in a highly porous medium with radiation and internal heat generation *Meccanica* **47** 1173–84

Resonance type flow due to rectilinear oscillations of a sphere in a micropolar fluid

J V Ramana Murthy[1], G S Bhaskara Rao[2] and T Govinda Rao[2]

[1] Department of Mathematics, N.I.T.W., Warangal - 506004, India.
[2] Department of Basic Sciences and Humanities, Gudlavalleru Engineering College, Gudlavalleru, A.P., India.

E-mail: jvrjosyula@yahoo.co.in

Abstract. The flow of an incompressible micro-polar fluid generated due to rectilinear oscillations of a sphere about a diameter of the sphere is considered. The flow is so slow that oscillations Reynolds number is less than unity and hence nonlinear convective terms in the equations of motion are neglected. A rare but distinct special case in which material constants satisfy a resonance condition is considered. The stream function and drag acting on cylinder are obtained. The effect of physical parameters like micro-polarity and couple stress parameter on the drag due to oscillations is shown through graphs.

1. Introduction

Several Stokes flow problems concerning micro-polar fluids have been studied by researchers over the past a half a century ever since A.C.Eringen introduced the micro-polar fluid theory [1]. S.K.Lakshmana Rao et al in [2–5] studied the oscillatory flows of circular cylinder, sphere, spheroid and elliptic cylinder in incompressible micro-polar fluids, the main thrust of the investigation being the determination of the drag or couple as the case may be on the oscillating body. The drag or couple was expressed in terms of two parameters K and K' and their numerical variation was extensively studied for a spectrum of micro-polarity parameters and oscillation parameters. However in all these problems, a special case, which are branded as oscillatory flows of "Resonance" type that arise when the material parameters of the fluids are constrained in a particular form (to be stated later) have not been investigated. The rare but distinct possibility of resonance flows has been noticed in [4, 5] and the investigation in this case is mathematically more complicated than in the usual non-resonance type flows. This type flows arise whenever oscillations take place in any non-Newtonian fluids. For example this case araises in the papers of [6–8], but these cases were not attempted by the authors. As far as the authors know, these cases were not studied by any researcher. In this paper we propose to investigate this case of resonance type flows in micropolar fluids.

2. Basic equations

The basic equations of an incompressible micro-polar fluid introduced by A.C. Eringen[1] are given by:

$$\frac{\partial \rho}{\partial \tau} + div(\rho \overline{Q}) = 0, \tag{1}$$

$$\rho\Big(\frac{\partial\overline{Q}}{\partial\tau}+\overline{Q}\cdot\nabla\overline{Q}\Big)=-\nabla P+k\ \ \nabla\times\overline{l}-(\mu+k)\nabla\times\nabla\times\overline{Q}, \tag{2}$$

$$\rho J\Big(\frac{\partial\overline{l}}{\partial\tau}+\overline{Q}\cdot\nabla\overline{l}\Big)=-2k\overline{l}+k\ \ \nabla\times\overline{Q}-\gamma\ \ \nabla\times\nabla\times\overline{l}+(\alpha+\beta+\gamma)\nabla(\nabla\cdot\overline{l}), \tag{3}$$

where $\overline{Q},\overline{l}$ fluid velocity and micro-rotation vectors, ρ is density, τ is time, J is gyration coefficient, μ is coefficient of viscosity, k is micro viscosity coefficient and α,β,γ are couple stress viscosity coefficients. For micro-polar fluids, the stress components T_{ij} and couple stress components M_{ij} satisfy the following constitutive equations.

$$T_{ij}=-P\ \delta_{ij}+\frac{1}{2}(2\mu+k)(u_{i,j}+u_{j,i})+k\ e_{ijr}(w_r-l_r), \tag{4}$$

$$M_{ij}=\alpha\ l_{i,i}\ \delta_{i,j}+\beta\ l_{i,j}+\gamma\ l_{j,i}, \tag{5}$$

where $w_r=r^{th}$ component of $\frac{1}{2}(curl\overline{Q})$ and e_{ijr} is permutation tensor$=0$ if any two indices are equal and $=1$ if i,j,r are cyclic and -1 if i,j,r are acyclic.

3. Statement and formulation of the problem

A sphere of radius a is performing rectilinear oscillations with velocity $U_0e^{i\sigma\tau}$ about its diameter in an infinite vat containing incompressible micro-polar fluid. Spherical coordinate system (R,θ,ϕ) with base vectors (e_r,e_θ,e_ϕ) with origin at the centre of the sphere and Z axis along direction of oscillations of the sphere is considered. The flow is axially symmetric, hence the velocity field is independent of toroidal coordinate ϕ and the flow will be in cross sectional plane of the sphere containing the base vectors (e_r,e_θ). The velocity and micro-rotation are assumed in the form:

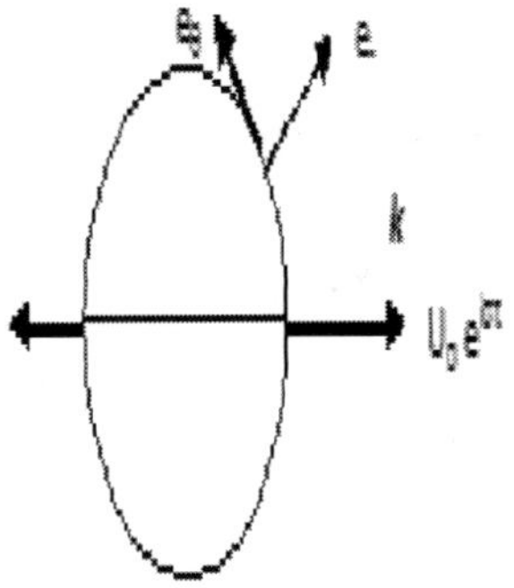

Figure 1. Geometry of the oscillating cylinder.

$$Q=e^{i\sigma\tau}U(R,\theta)e_r+V(R,\theta)e_\theta\quad\text{and}\quad l=e^{i\sigma\tau}C(R,\theta)e_z. \tag{6}$$

The following non-dimensional scheme is introduced. Capitals and LHS terms indicate physical quantities and small letters and RHS terms indicate corresponding non-dimensional quantities.

$$R=ar,\ \ U=U_0u,\ \ V=V_0v,\ \ Q=U_0q,\ \ C=C\frac{U_0}{a},$$
$$l=\nu\frac{U_0}{a},\ \ \Psi=aU_0\psi,\ \ P=p\rho U_0^2,\ \ \tau=U_0\frac{t}{a}. \tag{7}$$

The following are non-dimensional parameters viz, j gyration parameter, ω frequency parameter, s couple stress parameter, c cross viscosity or micro-polarity parameter and Re oscillations Reynolds number for micro-polar fluids.

$$j = \frac{J\rho\sigma a^2}{\gamma}, \quad \omega = \frac{\sigma U_0}{a}, \quad s = \frac{ka^2}{\gamma}, \quad c = \frac{k}{\mu + k}, \quad Re = \frac{\rho U_0 a}{\mu + k}. \tag{8}$$

By the choice of velocity field in (6) and incompressibility condition in (1), we notice that stream function can be introduced as

$$u = \frac{1}{r}\frac{\partial\psi}{\partial\theta} \text{ and } v = -\frac{\partial\psi}{\partial r} \quad i.e. \;\; q = \nabla \times (\psi e_z). \tag{9}$$

Using (6), (7), (8) in (2) and (3) we get

$$\omega \; Re \; q = -Re \cdot \nabla p + c\nabla \times v - \nabla \times \nabla \times q, \tag{10}$$

$$ijv = -2sv + s\nabla \times q - \nabla \times \nabla \times v + \frac{1}{\epsilon}\nabla(\nabla \cdot v). \tag{11}$$

By taking curl to equation (10) pressure p can be eliminated and then using (6) and (9) we get,

$$i \; \omega \; Re \cdot E^2\psi = cE^2C + E^4\psi, \tag{12}$$

$$c(2s + ij)C = E^2C - sE^2\psi, \tag{13}$$

$$E^2 = \frac{\partial^2}{\partial r^2} + \frac{1}{r^2}\frac{\partial^2}{\partial\theta^2} - \frac{\cot\theta}{r^2}\frac{\partial}{\partial\theta}.$$

Taking E^2 operation to (13) and then eliminating E^2C using (12) we get,

$$E^2(E^2 - \lambda_1^2)(E^2 - \lambda_2^2)\psi = 0, \tag{14}$$

$$c(2s + ij)C = -E^4\psi + (i\omega \cdot Re - sc)E^2\psi, \tag{15}$$

where

$$\lambda_1^2 + \lambda_2^2 = (2 - c)s + i(j + \omega Re) \;\text{ and }\; \lambda_1^2\lambda_2^2 = i\omega Re(2s + ij). \tag{16}$$

The equation for C can be re-written as

$$cC = -\frac{i\omega Re}{\lambda_1^2\lambda_2^2}E^2(E^2 - \lambda_1^2 - \lambda_2^2)\psi - E^2\psi. \tag{17}$$

The solution for ψ if $\lambda_1 \neq \lambda_2$ in (14) is given in [3]. The solution for ψ for the case, $\lambda_1 = \lambda_2$ can not be obtained as a limiting case of $\lambda_1 \to \lambda_2$. This case is referred to as "Resonance". This resonance occurs if the material coefficients follow the following relation:

$$\frac{\gamma}{J} = \frac{(2\mu + k)(\mu + k)}{2\mu + 3k} \quad \text{and} \quad \rho\sigma = \frac{(2\mu + k)k + \gamma\rho\sigma}{J(\mu + k)}.$$

In non-dimensional form these conditions are given by

$$(2 - c)s = j - Re \cdot \psi \;\text{ and }\; (2 - c)j = \omega Re(2 + c). \tag{18}$$

In this paper we are interested in the solution for ψ for the case of resonance. In this case the equations for ψ and C are given by

$$E^2(E^2 - \lambda^2)^2\psi = 0 \;\text{ and }\; cC = -\frac{i\omega Re}{\lambda^4}E^2(E^2 - 2\lambda^2)\psi - E^2\psi. \tag{19}$$

The following are the boundary conditions
The sphere is oscillating in the direction of Z-axis. Hence the non-dimensional velocity of the sphere Γ after removing $e^{i\omega t}$ is given by

$$q_\Gamma = k = \cos\theta e_\Gamma - \sin\theta e_\theta,$$

which implies by no-slip condition

$$u = \cos\theta \text{ and } v = -\sin\theta \text{ on } r = 1. \tag{20}$$

By hyper-stick condition given by

$$\nu_\Gamma = \frac{1}{2}(curl q)_\Gamma,$$

reduces to

$$C = 0 \text{ on } r = 1. \tag{21}$$

4. Solution of the problem

Stream function ψ, micro-rotation component C are assumed in the form

$$\psi = f(r)\sin^2\theta \text{ and } C = g(r)\cos^2\theta. \tag{22}$$

Substituting (22) in (19) we get

$$D^2(D^2-\lambda^2)^2 f = 0 \text{ and } cg = -\frac{i\omega Re}{\lambda^4}D^2(D^2-2\lambda^2)f - D^2 f, \tag{23}$$

where

$$D^2 = \frac{d^2}{dr^2} - \frac{2}{r^2}.$$

Substituting (22) in (20) and (21), the conditions on f and g are obtained as:

$$f(1) = \frac{1}{2},\ f'(1) = 1 \text{ and } g(1) = 0. \tag{24}$$

Since the equation (23) for f is linear, f can be taken as

$$f = A_0 f_0 + A_1 f_1 + A_2 f_2,$$

$$with \qquad D^2 f_0 = 0,\ D^2(D^2-\lambda^2)^2 f_1 = 0 \text{ and } D^2(D^2-\lambda^2)^2 f_2 = 0. \tag{25}$$

On solving (25), the solution for f is obtained as

$$f(r) = \frac{A_0}{r} + A_1\sqrt{r}K_{\frac{3}{2}}(\lambda r) + A_2 r^{\frac{3}{2}}K'_{\frac{3}{2}}(\lambda r). \tag{26}$$

The following results are useful to note.

$$D^2 f_1 = \lambda^2 f_1 \text{ and } D^2 f_2 = (-2\lambda f_1 + \lambda^2 f_2). \tag{27}$$

Using (27) in (23), we get

$$cg = A_1(i\omega\cdot Re - \lambda^2)f_1 + A_2(i\omega\cdot Re-\lambda^2)f_2 + 2A_2\lambda^2 f_1. \tag{28}$$

The constants A_0, A_1, A_2 are obtained from the boundary conditions (24) as follows:

$$\begin{bmatrix} 1 & k_{\frac{3}{2}}(\lambda) & k_{\frac{1}{2}}(\lambda) \\ -1 & \frac{1}{2}k_{\frac{3}{2}}(\lambda) + \lambda k'_{\frac{3}{2}}(\lambda) & \frac{3}{2}k_{\frac{1}{2}}(\lambda) + \lambda k'_{\frac{1}{2}}(\lambda) \\ 0 & (i\omega Re - \lambda^2)k_{\frac{3}{2}}(\lambda) & (i\omega Re-\lambda^2)k_{\frac{1}{2}}(\lambda) + 2\lambda k_{\frac{3}{2}}(\lambda) \end{bmatrix} \begin{bmatrix} A_0 \\ A_1 \\ A_2 \end{bmatrix} = \begin{bmatrix} \frac{1}{2} \\ 1 \\ 0 \end{bmatrix}. \tag{29}$$

Hence from (26), (28) and (29), f and g are completely known and hence ψ and C are known.
Pressure: From equation (10) pressure, after cancelling $e^{-i\omega t}$, is obtained as follows.

$$
\begin{aligned}
dp &= \nabla p \cdot dr = \tfrac{\partial p}{\partial r}dr + \tfrac{\partial p}{\partial \theta}d\theta \\
&= \tfrac{1}{Re}\left[\tfrac{1}{r^2 \sin\theta}\tfrac{\partial}{\partial \theta}(-i\omega Re\psi + cC + E^2\psi)dr + \tfrac{1}{\sin\theta}\tfrac{\partial}{\partial r}(-i\omega Re\psi + cC + E^2\psi)d\theta\right],
\end{aligned}
$$

$$
\begin{aligned}
p &= \tfrac{1}{Re}\int(-i\omega Ref + D^2 f + cg)2\cos\theta\tfrac{dr}{r^2} + \tfrac{d}{dr}(-i\omega Ref + D^2 f + cg)\sin\theta d\theta, \\
&= \tfrac{1}{Re}\int(-i\omega Re\cdot A_0 f_0)2\cos\theta\tfrac{dr}{r^2} + \tfrac{d}{dr}(-i\omega Re\cdot A_0 f_0)\sin\theta d\theta,
\end{aligned}
$$

$$
p = \frac{i\omega A_0}{r^2}\cos\theta, \tag{30}
$$

Drag acting on the cylinder per length L:

$$
Drag = D^* = aL\int_0^{2\pi}(T^*_{rr}\cos\theta - T^*_{r\theta}\sin\theta)\mid_{R=a} d\theta,
$$

Multiplying D^* by $\frac{a^2 L}{(\mu+k)}U_0$ the non-dimensional drag D is obtained a

$$
De^{-i\omega t} = aL\int_0^{2\pi}(T_{rr}\cos\theta - T_{r\theta}\sin\theta)\mid_{r=1} d\theta, \tag{31}
$$

$$
T^*_{rr} = -P + (2\mu+k)e_{rr} = -P + (2\mu+k)\frac{\partial U}{\partial R} = -P + \frac{(2\mu+k)U_0}{ar^2}(f' - \frac{2f}{r})2\cos\theta.
$$

Hence on $r = 1, T^*_{rr} = -p$.
Again,

$$
\begin{aligned}
T^*_{r\theta} &= (2\mu+k)e_{r\theta} + ke_{123}(w_3 - \vartheta_3), \\
&= -\tfrac{(2\mu+k)U_0}{2ar}\sin\theta(D^2 f + \tfrac{4f}{r^2} - \tfrac{2f'}{r}) - \tfrac{kU_0}{ar}\sin\theta(\tfrac{1}{2}D^2 f + g).
\end{aligned}
$$

Hence drag in non-dimensional form is given by

$$
\begin{aligned}
D &= \textstyle\int_0^{2\pi}[-Re\cdot p\cdot\cos\theta + ReD^2(D^2 - 2\lambda^2)f\sin\theta]\mid_{r=1} d\theta, \\
&= \textstyle\int_0^{2\pi} -i\omega ReA_0\cos^2\theta + i\omega Re(\tfrac{1}{2} - A_0)\sin^2\theta d\theta, \\
&= \pi i\omega Re(1 - 2A_0.
\end{aligned}
$$

Hence the non-dimensional drag D is given by

$$
D = Real[i\omega Re\cdot\pi(1 - 2A_0)e^{i\omega t}]. \tag{32}
$$

5. Results and discussions

The values of λ are obtained from (16) by solving $x^2 - [(2-c)s + i(j+\omega Re)]x + i\omega Re(2s+ij) = 0$ for x. Then for resonance case, the values of λ are given by

$$
\lambda = \sqrt{x} = \sqrt{\frac{[(2-c)s + i(j+\omega Re)]}{2}}. \tag{33}
$$

This equation involves 5 parameters which are related by two equations in (18). Hence we choose three parameters as independent. Here ω, Re and c are selected independently, with $0 \leq c \leq 1$, $Re \ll 1$ and $\omega \gg 1$ such that $\omega.Re$ is not negligibly small (say>1). For this range of values of Re, the nonlinear convective terms can be neglected but local derivative is retained. After selecting c, Re and ω, the values of s and j are obtained from (18) and then λ is obtained from (33). The values of λ are complex. These values for λ are substituted in (29) and the constants A_0, A_1 and A_2 are obtained. Then the stream function ψ and drag D are obtained from (26) and (32) respectively. Thus obtained ψ will have complex values. To get physical picture, these values are multiplied by $e^{i\omega t}$ and its real part is taken, if the oscillations are in cosine functions. Drag: From figure 2, it is observed that as s the couple stress parameter and j the gyration parameter are increasing, the drag D decreases. The variation drag for different values of j is not observed for small values of c<0.5. From figure 3, we observe that as Re, the Reynolds number and ω, the frequency parameter are increasing, the drag increases for small values of c but as c is nearing unity, the drag decreases. Since $Re.\omega$ appears in the equations as a unit, the behavior of drag with respect to Re and ω is exactly similar. This is clear in figure 3.

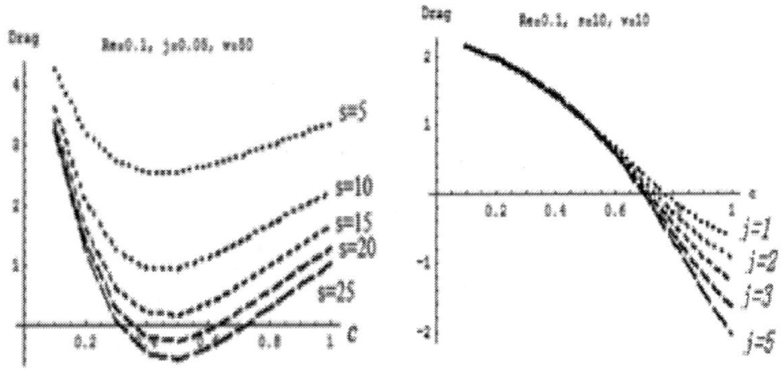

(a) (b)

Figure 2. (a) Drag Vs micro-polarity parameter c at different s. (b) Drag Vs micro-polarity parameter c at different j.

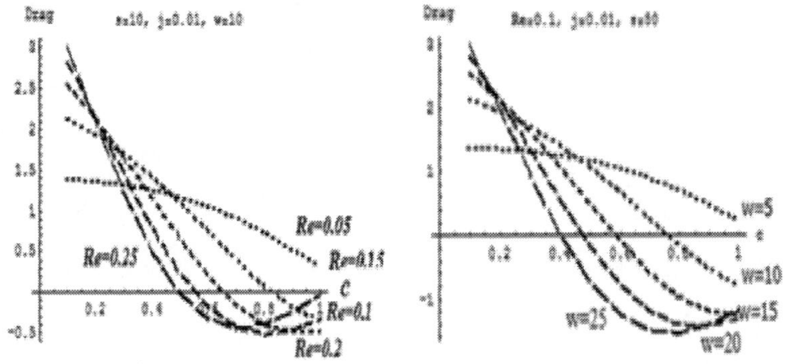

(a) (b)

Figure 3. (a) Drag Vs micro-polarity parameter c at different Re. (b)Drag Vs micro-polarity parameter c at different ω.

Journal of Physics: Conference Series **662** (2015) 012015 doi:10.1088/1742-6596/662/1/012015

5.1. *Stream function*

By looking the stream lines in the six figures given below, it is observed that the entire flow region is symmetric about axis of symmetry and streamlines form closed loops about the equatorial region of the sphere. For small values of cross viscosity(micro-polarity) parameter c, the stream lines are having negative values along the axis of symmetry and positive values near the equatorial region of the sphere. (Black shade represents negative values and white region represents positive values). As c value increases, we can find a drastic change in the pattern of the stream lines. The region along the axis of symmetry becomes white (i.e, positive) and the region about the equatorial line of the sphere becomes black (i.e, negative). This indicates the strong effect of micro-polarity parameter on the rotation of the sphere. It can be seen that as c is having values near to 0.5, the stream lines are forming two closed regions within which the entire flow is contained in. This is an interesting phenomena to note. This may be due to the fact that drag is highest when c is near to 0.1 and decreases as c increases and reaches a minimum value when c is about 0.5 and then the drag increases as c increases again. Since the case $c \rightarrow 0, s \rightarrow \infty$, represents the case of viscous fluids, we can conclude that the drag for viscous fluids is more than the drag of micro-polar fluids. Hence it can be inferred that micro-polar fluids offer less resistance to flow and hence less drag.

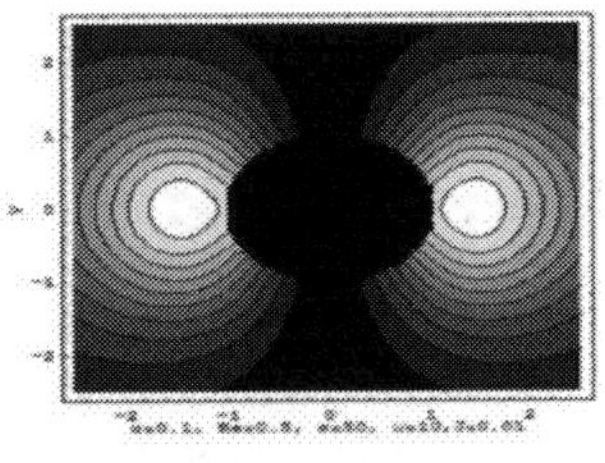

(a) c=0.1

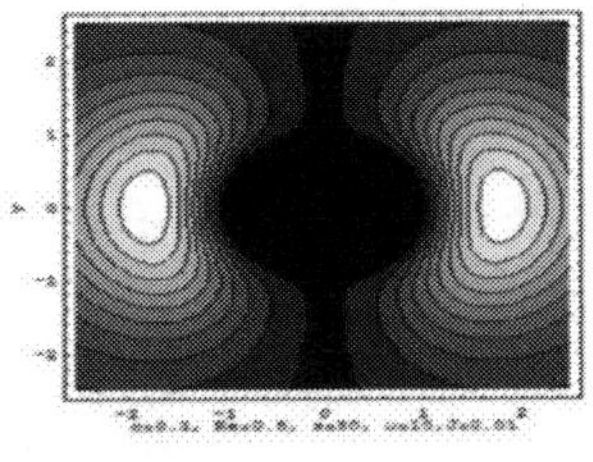

(b) c=0.3

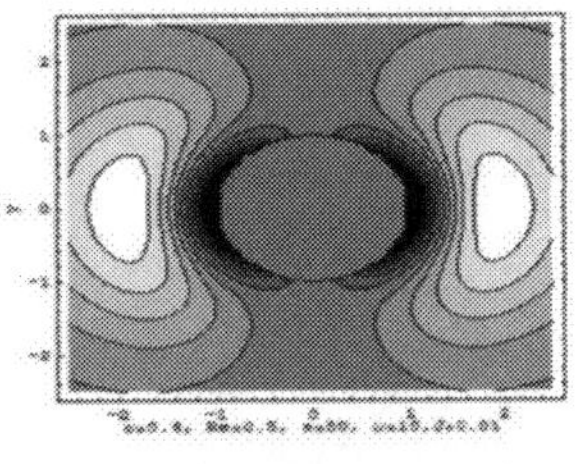

(c) c=0.4

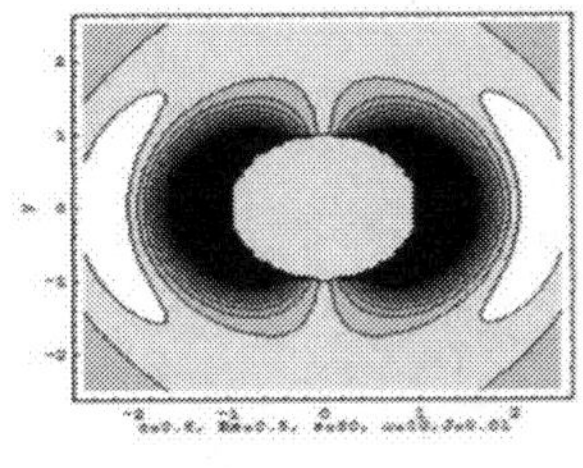

(d) c=0.6

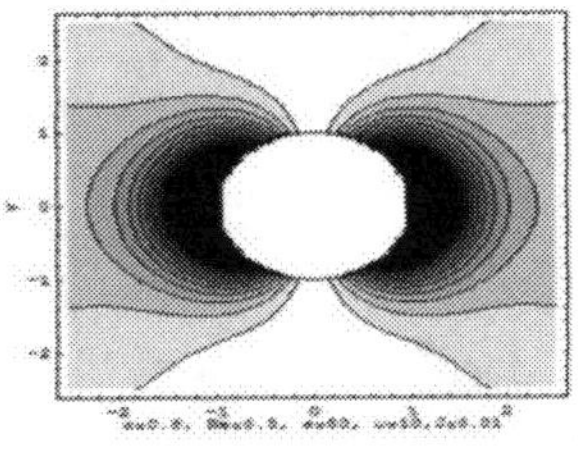

(e) c=0.8

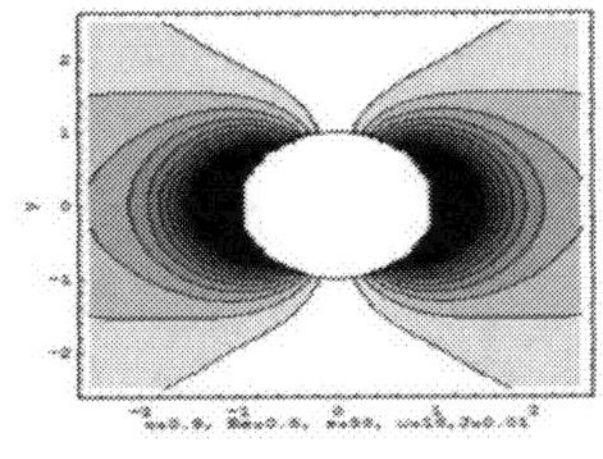

(f) c=0.9

Figure 4. Stream lines at different values of micro-polarity parameter c.

References

[1] Eringen A C 1966 Theory of micropolar fluids *J. Math. Mech.* **16** 1-18
[2] Lakshmana Rao S K and Bhuganga Rao P 1972 Rectilinear oscillations of a circular cylinder about a mean position in an incompressible micro-polar fluid *I.J.E.S* **10** 185-191
[3] Lakshmana Rao S K and Bhuganga Rao P 1971 Oscillations of a sphere in a micro-polar fluid *I.J.E.S* **9** 651 - 672
[4] Lakshmana Rao S K and Iyengar T K V 1983 The rotary oscillations of a spheroid in an incompressible micro-polar fluid *I.J.E.S* **21** 973 - 987
[5] Lakshmana Rao S K, Iyengar T K V and Venkatapathi Raju K 1987 The rectilinear oscillations of an elliptic cylinder in an incompressible micro-polar fluid *I.J.E.S* **25** 531-548
[6] Ramana Murthy J V, Sai K S and Bahali N K 2011 Steady flow of micropolar fluid in a rectangular channel under transverse magnetic field with suction *AIP ADVANCES* **1** 032123
[7] Aparna P and Murthy J V R 2012 Rotary oscillations of a permeable sphere in an incompressible micropolar fluid *Int. J. of Appl. Math and Mech.* **8(16)** 79-91
[8] Nagaraju G and Ramana Murthy J V 2014 Unsteady flow of a micropolar fluid generated by a circular cylinder subject to longitudinal and torsional oscillations *Theoret.Appl.Mech. TEOPM7* **41(1)** 71 - 91

International Conference on Vibration Problems (ICOVP-2015) IOP Publishing
Journal of Physics: Conference Series **662** (2015) 012016 doi:10.1088/1742-6596/662/1/012016

Exact solution for flow over a contaminated fluid sphere for stokes flow

J V Ramana Murthy [1] **and M Phani Kumar** [2]

[1] Department of Mathematics, National Institute of Technology Warangal, India.
[2] Department of Basic Sciences and Humanities, Vignan's Lara Institute of Technology and Science, Vadlamudi, A.P., India.

E-mail: jvrjosyula@yahoo.co.in, phanikumarmeduri@gmail.com

Abstract. Stokes viscous flow past a partially contaminated fluid sphere with no slip condition is considered. No mass transfer for the entire sphere, no slip condition on contaminated part, shear stress continuity on clear part and regularity condition at far away from the body are considered for boundary conditions for evaluation of stream function.

1. Introduction

Study of flow in drops has wide applications in natural and engineering such as nuclear reactors, internal combustion engines, sediment and pollutant transport processes. Stokes [1] solved the creeping flow problem past a sphere neglecting inertial terms in the Navier Stokes equations. Basset [2] calculated drag over a sphere in terms of slip parameter s (Trostel number), $s = \frac{\beta a}{\mu}$ (defined as slip parameter). Creeping flow over a fluid sphere was studied analytically independently by Rybczynski [3] and Hadamard [4]. Happel and Brenner [5] discussed creeping flow past a sphere with no slip boundary condition. Clift et al. [6] and Michaelides [7] in their monographs discussed about viscous flow past a fluid sphere with no slip boundary condition. Sadhal and Johnson [8] derived exact solution of drag force for a fluid sphere in terms of cap angle with stagnation cap over its boundary along the rear of the fluid drop. Stagnation cap is the collection of surfactant at the rear side the fluid sphere. Saboni [9] numerically discussed about the contamination effects on a fluid sphere for Reynolds numbers from 0.1 to 400 and viscosity ratio ranging from 0 to 10 at different stagnation cap angle. Feng et al. [10] has discussed about the viscous flow past a viscous drop with interfacial condition at small but finite Reynolds numbers. Feng derived a formula for drag coefficient. As special cases he derived an expression for drag for solid sphere with interfacial slip condition, solid sphere with no slip boundary condition.

In this paper, stokes viscous flow over a contaminated fluid sphere is considered with the continuous shear stresses over the clear part. Velocity is represented in terms of stream function. The stream lines are drawn for different values of viscosity ratio μ, contamination cap lengths.

2. Formulation of the problem

A fluid sphere of radius a with contamination at rear end, which is held fixed in an uniform viscous flow, is considered. The fluid within the sphere and fluid flowing past the sphere are assumed to be immiscible. The flow is steady, incompressible, axi symmetric, with uniform velocity U_0. A spherical polar coordinate system is considered with origin at the center of the sphere and Z-axis along the direction of uniform flow. μ_i, μ_e, ρ_i, ρ_e are viscosity, density of interior and exterior fluids. The viscosity ratio is taken as $\mu = \frac{\mu_i}{\mu_e}$, $\rho = \frac{\rho_i}{\rho_e}$.

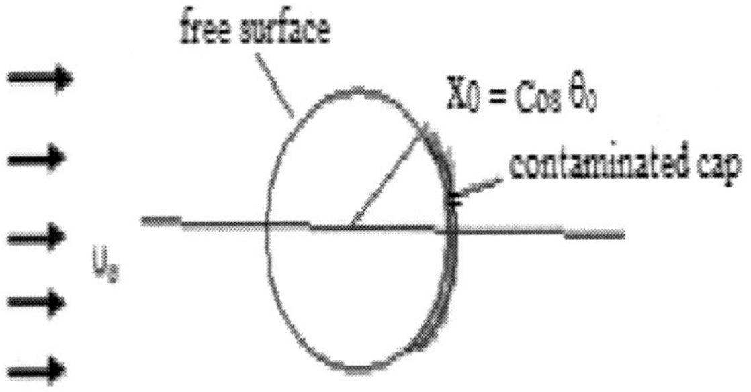

Figure 1. Geometry of the problem.

The viscous fluid is assumed to flow from left to right. In the fluid sphere the clear part (no cap region) is considered for $-1 < x < x_0$ and contaminated part (cap region) is for $x_0 < x < 1$, where $x = \cos\theta$. x_0 is the position of cap or cosine angle of contamination. The velocity components in radial direction U and transverse direction V can be expressed in terms of stream function as

$$U(R,\theta) = \frac{1}{R^2 \sin\theta} \frac{\partial\Psi}{\partial\theta}, \qquad V(R,\theta) = \frac{-1}{R \sin\theta} \frac{\partial\Psi}{\partial R}, \tag{2.1}$$

Any physical (dimensional or non dimensional) quantity for internal flow is represented by fi and external flow by fe. The following non-dimensional scheme is used to obtain equations of motion.

$$R = ar, \quad \bar{Q} = U_0\bar{q}, \quad U = U_0 u, \quad V = V_0 v, \quad \Psi = a^2 U_0 \psi.$$

Reynolds number for external flow is $Re = \frac{2a\rho_e U_0}{\mu_e}$ and for internal flow $Rei = \frac{Re\rho}{\mu}$.
The equation of motion in terms of stream function for the internal flow is

$$E^4\psi i = 0. \tag{2.2}$$

The equation of motion in terms of stream function for the external flow is

$$E^4\psi e = 0, \tag{2.3}$$

where

$$E^2 = \frac{\partial^2}{\partial r^2} + \frac{(1-x^2)}{r^2}\frac{\partial^2}{\partial x^2} = \text{Stokes stream function operator.}$$

Equations (2.2), (2.3) are solved for ψ_e, ψ_i using the following boundary conditions (2.4) - (2.9):

$$at \ \ \theta = 0, \ \ \pi(x=1, x=-1) \ \text{ on } \ \ \text{r=1}, \ \ \psi e = \psi i = 0, \tag{2.4}$$

Tangential velocity is zero along the contaminated part

$$\frac{\partial\psi e(1)}{\partial r} = \frac{\partial\psi i(1)}{\partial r} = 0 \quad \text{for } \ x_0 < x < 1. \tag{2.5}$$

The shear stress is continuous along the clear surface at the interface, which implies

$$\frac{\partial}{\partial r}\Big(\frac{1}{r^2}\frac{\partial \psi e}{\partial r}\Big) = \mu\frac{\partial}{\partial r}\Big(\frac{1}{r^2}\frac{\partial \psi i}{\partial r}\Big) \quad \text{for} \quad -1 < x < x_0, \tag{2.6}$$

Far away from the sphere, regularity condition is taken

$$\lim_{r\to\infty} \psi e = \frac{1}{2}r^2(1-x^2), \tag{2.7}$$

The physical condition, that velocity at the origin is finite, can be taken as

$$\lim_{r\to\infty} \psi i < \infty (\text{it is taken as zero}).$$

3. Solution of the problem

The Stream function ψ for stokes flow is taken in the form

$$\psi = \sum_{n=1}^{\infty} F_n(r)G_n(r).$$

Hence (2.2) or (2.3) takes the form given by

$$E^4\psi = \sum_{n=1}^{\infty} D_n^4 F_n(r)G_n(r), \tag{3.1}$$

where

$$D_n^2 F_n = F_n^{''} - \frac{n(n-1)}{r^2}F_n.$$

For external flow the solution for ψ of (3.1) which satisfies the regularity condition (2.7) at infinity is given by

$$\psi_e(r,x) = (A_2 r^2 + \frac{B_2}{r} + D_2(r))G_2(x) + \sum_{n=3}^{\infty}\Big(\frac{B_n}{r^{n-1}} + \frac{D_n}{r^{n-3}}\Big)G_n(x), \tag{3.2}$$

Using the boundary condition $\psi_e(1,x) = 0$, stream function in (3.2) reduces to

$$\psi_e(r,x) = (r^2 - r + B_2(\frac{1}{r} - r))G_2(x) + \sum_{n=3}^{\infty} B_n\Big(\frac{1}{r^{n-1}} - \frac{1}{r^{n-3}}\Big)G_n(x), \tag{3.3}$$

Similarly for internal flow, the solution for ψ of (3.1) that vanishes at r = 1 (i.e $\psi_i(1,x) = 0$) is given by

$$\psi_i(r,x) = \sum_{n=2}^{\infty} A_n(r^{n+2} - r^n)G_2(x). \tag{3.4}$$

The solution of (3.1) for cap and free regions is assumed to be in the form

$$\psi_e(r,x) = \begin{cases} fe_n(r)\ G_2(x), & \text{for} -1 < x \le x_0 \\ fe_c(r)\ G_2(x), & \text{for} \quad x_0 < x \le 1, \end{cases}$$

$$\psi_e(r,x) = \begin{cases} fi_n(r)\ G_2(x), & \text{for} -1 < x \le x_0 \\ fi_c(r)\ G_2(x), & \text{for} \quad x_0 < x \le 1. \end{cases} \tag{3.5}$$

Again solving (3.1) for the region of free flow ($x_0 < x < 1$), the solutions for the coefficient of $G_2(x)$ which satisfy the continuity of interfacial stress for the external and internal flows are given by

$$fe_n(r) = r^2 - \frac{2+3\mu}{2+2\mu}r + \frac{\mu}{2+2\mu}\frac{1}{r}, fi_n(r) = \frac{r^4 - r^2}{2+2\mu}. \tag{3.6}$$

Similarly for the region of contaminated cap ($-1 < x < x_0$) , the solutions for the coefficient of $G_2(x)$ which satisfy the no-slip condition for the external and internal flows are given by

$$fe_c(r) = r^2 - \frac{3}{2}r + \frac{1}{2r}, fi_c(r) = 0. \tag{3.7}$$

Hence, for external flow, LHS of condition (2.6) takes the form

$$\psi_e''(1) - 2\psi_e'(1) = \begin{cases} (fe_{n,2}''(r) - 2fe_{n,2}'(r))G_2(x), & \text{for } -1 < x \le x_0 \\ fe_c''(r)\ G_2(x), & \text{for } \quad x_0 < x \le 1. \end{cases} \tag{3.8}$$

Substituting (3.6) and (3.7) in (3.8) we get

$$\psi_e''(1) - 2\psi_e'(1) = \begin{cases} (\frac{3\mu}{1+\mu})G_2(x), & \text{for } -1 < x \le x_0 \\ 3G_2(x), & \text{for } \quad x_0 < x \le 1. \end{cases} \tag{3.9}$$

Substituting (3.3) for ψ_e in (3.9) we get

$$6B_2G_2(x) + \sum(4n-2)B_nG_n(x) = \begin{cases} (\frac{3\mu}{1+\mu})G_2(x), & \text{for } -1 < x \le x_0 \\ 3G_2(x), & \text{for } \quad x_0 < x \le 1. \end{cases} \tag{3.10}$$

By using the orthogonal property of Gegenbauer polynomials, the constants in (3.10) are obtained as below

$$(4m-2)B_m = \frac{(2m-1)m(m-1)}{2}\Big[\int_{-1}^{x_0} \frac{3\mu}{1+\mu}\frac{G_2(x)G_m(x)}{1-x^2}dx + \int_{x_0}^{1}\frac{3G_2(x)G_m(x)}{1-x^2}dx\Big]. \tag{3.11}$$

This on simplification gives

$$B_m = \frac{m(m-1)}{4}\Big\{\frac{\mu}{1+\mu}\delta_{2,m} + \frac{3}{1+\mu}\int_{x_0}^{1}\frac{G_2(x)G_m(x)}{1-x^2}dx\Big\}, \tag{3.12}$$

where $\delta_{2,m}$ is kronecker delta which takes value 1 if m=2 and 0 if m$\neq$ 2.
For internal flow, by substituting (3.4) in RHS of (2.6), the condition of continuity of interfacial stress is expressed as:

$$\mu\Big(\frac{\partial^2\psi_i}{\partial r^2} - 2\frac{\partial\psi_i}{\partial r}\Big)\Big|_{r=1} = \mu\sum_{n=2}^{\infty}(4n-2)A_nG_n(x), \tag{3.13}$$

$$\mu\Big(\frac{\partial^2\psi_i}{\partial r^2} - 2\frac{\partial\psi_i}{\partial r}\Big)\Big|_{r=1} = \sum_{n=2}^{\infty}\begin{cases} (\mu fi_{n,2}''(r) - 2fi_{n,2}'(r))G_2(x), & \text{for } -1 < x \le x_0 \\ 0, & \text{for } \quad x_0 < x \le 1, \end{cases}$$

$$\mu\Big(\frac{\partial^2\psi_i}{\partial r^2} - 2\frac{\partial\psi_i}{\partial r}\Big)\Big|_{r=1} = \sum_{n=2}^{\infty}\begin{cases} (\frac{3\mu}{1+\mu})G_2(x), & \text{for } -1 < x \le x_0 \\ 0, & \text{for } \quad x_0 < x \le 1. \end{cases} \tag{3.14}$$

Using the orthogonal property of Gegenbauer polynomials, the constants A_n are obtained as below

$$A_n = \frac{3n(n-1)}{4(1+\mu)}\int_{-1}^{x_0}\frac{G_2(x)G_n(x)}{1-x^2}dx. \tag{3.15}$$

Now substituting the values of the constants A_n and B_n from (3.15) and (3.12) in (3.3) and (3.4), the flow pattern for external and internal flows can be obtained.

Journal of Physics: Conference Series **662** (2015) 012016 doi:10.1088/1742-6596/662/1/012016

4. Results and discussions

The stream function ψ for external flow and for internal flow is taken in the infinite series of Gegenbauer polynomials as in (3.3) and (3.4). This function is to satisfy the discontinuity condition at the junction point of free and contaminated cap regions. Previously this was accomplished by using double series proposed by Collins [11]. By using the present procedure, this infinite series can be obtained easily by taking solutions in the free and cap regions and matching them with the series solution. When the number of terms in the series are limited to twenty (i.e. $N = 20$) the error in the solution is about 10^{-3}. Hence we did not take $N > 20$. The following figure 2, for different values of cap lengths x_0, illustrates variation in the original curve and truncated series solution. The wavy curve represents the stream function represented by infinite series at the boundary and the smooth curve is the stream function at the boundary. We can observe that the difference between these two curves is very small i.e, the series satisfies the boundary condition, within the reasonable tolerance. Both the external and internal stream functions represented by infinite series satisfy the viscous stream function differential equation (3.1). Hence we have obtained exact solution for this contaminated cap problem.

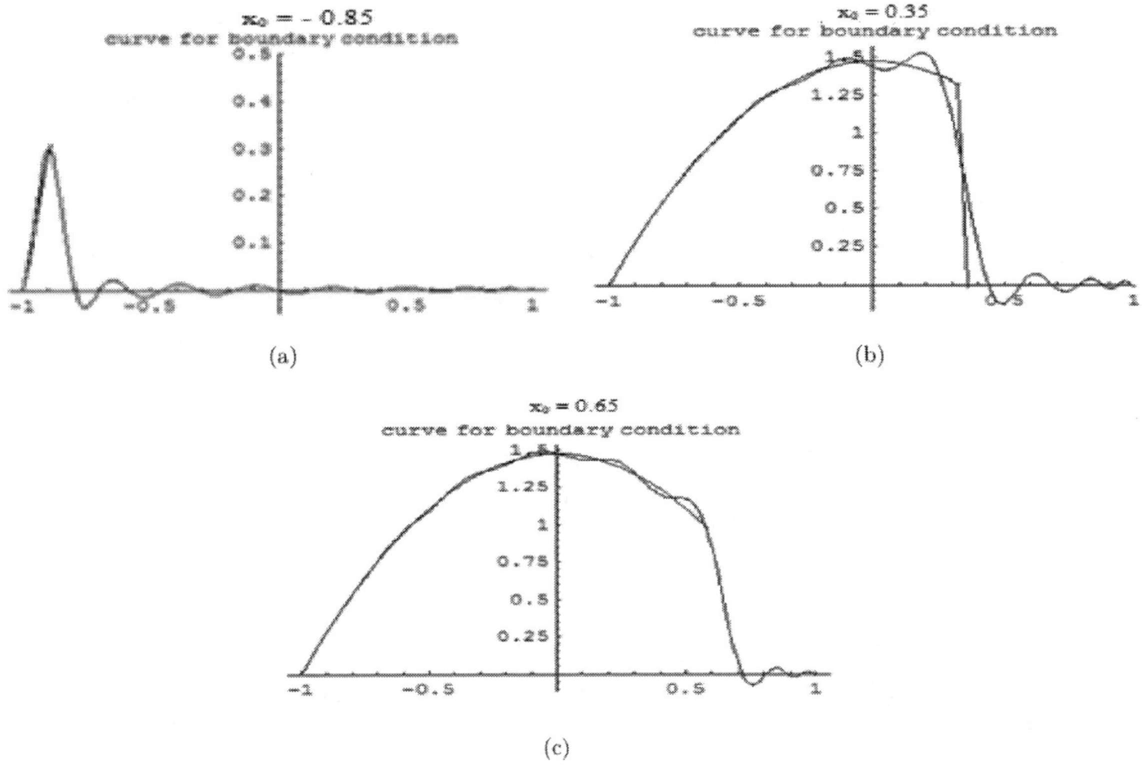

Figure 2. Stress at boundary for various values of cap length x_0.

Now by taking $N = 20$ terms in the stream function, stream line pattern for different values of cap lengths x_0 is obtained. It can be clearly observed from the figure 3 below that when x_0 is near to -1 flow in the internal region of the fluid sphere is very low. As the value of x_0 increases, internal circulation also increases. These flow patterns are similar to the results obtained by Feng and Michaelides [12] for low Reynolds numbers. These results are reasonably in very good agreement with the numerical results of Feng et al. [10]. The difference between the values of stream function at the boundary and the values of stream function obtained by infinite series is very small or insignificant ($\psi = 1 - 0.000001$) for $x_0 < -0.85$ i.e when contaminated cap is almost covering the fluid sphere. When the cap size is decreasing, i.e when x_0 is nearing 1, we can find more circulations within the sphere in the negative direction to the flow direction outside the sphere. The values of the stream function are decreasing and numerically increasing as x_0 is increasing. We can also observe that as Peclet number Pe is increasing, these circulations within the sphere are also increasing. When μ the ratio of viscosities increases, it implies that that the fluid is becoming thick and as $\mu \to \infty$, the sphere will be impermeable and the internal circulations will be absent.

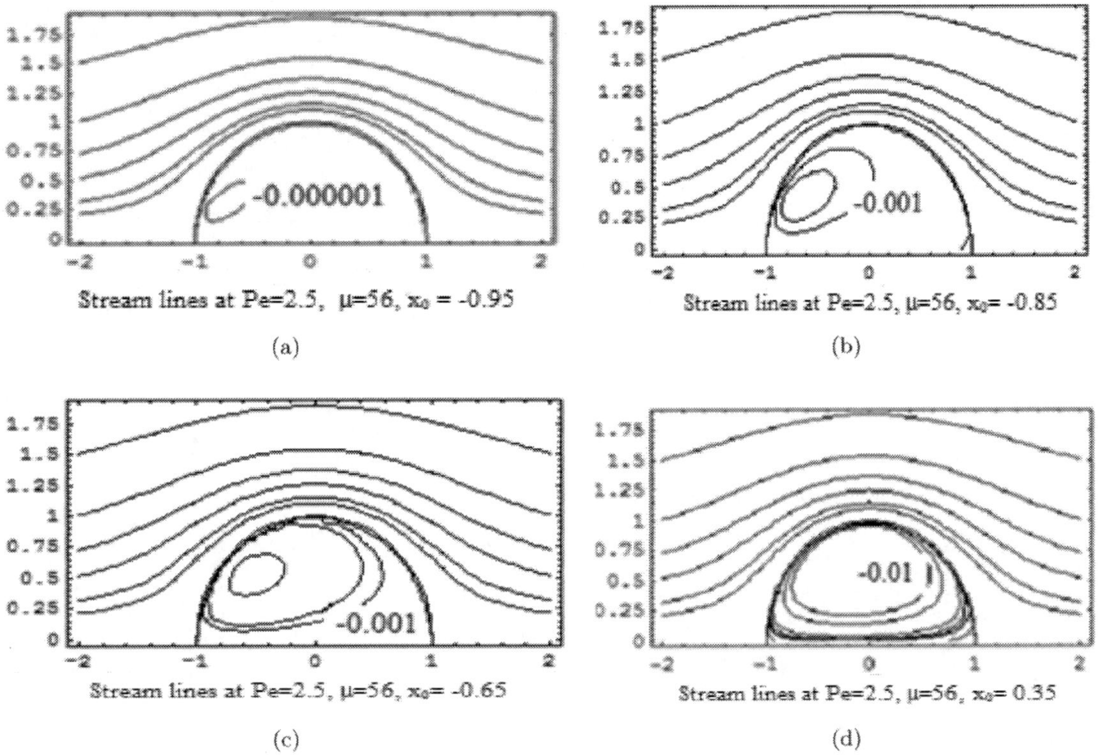

Figure 3. Stream lines for internal and external flows for various cap lengths x_0.

References

[1] Stokes G G 1851 *Cambridge Philos. Trans.* 9:8
[2] Basset A B 1888 *A Treatise on Hydrodynamics* (London: Bell)
[3] Rybczynski W 1911 Uber die fortschreitende bewegung einer flussigen kugel in einem zahenmedium *Bull Acad. Sci. Cracovie, Ser. A* **1** 40-46
[4] Hadamard J S 1911 Mouvement permanent lent dune sphere liquid et visqueuse dans u liquid visqueux *Compt. Rend. Acad. Sci. Paris* **152** 1735-38
[5] Happel John and Brenner Howard 1965 *Low Reynolds number hydrodynamics* (N.J., Prentice Hall Englewood Cliffs)
[6] Clift R, Grace J R and Weber M E 1978 *Bubble drops and particles* (New York: Academic Press)
[7] Michaelides E E 2006 *Particles Bubbles and Drops: Their motion Heat and Mass Transfer* (Singapore: World Scientific)
[8] Sadhal S S and Robert E Johnson 1983 Stokes flow past bubbles and drops partially coated with thin films, Part 1 stagnant cap of surfactant film-exact solution *J. Fluid Mech.* **126** 237-250
[9] Saboni A Alexandrova S Mory M 2010 Flow around a contaminated fluid sphere *International Journal of Multi phase flow* (ijmulflow) **36** 503-512
[10] Feng Z G, Michaelides E E and Shaolin Mao 2012 On the drag force of a viscous sphere with interfacial slip at small but finite Reynolds numbers *Fluid Dyn. Res.* **44** 1-16
[11] Collins W D 1961 On some dual series equations and their applications to electrostatic problems for spherodial caps *Proc. Camb. Phil. Soc.* **57** 367
[12] Feng Z G and Michaelides E E 2000 A numerical study on the transient heat transfer from a sphere at high Reynolds and Peclet numbers *International Journal of Heat and Mass Transfer* **43** 219-229

Finite element analysis of natural convective heat transfer in a porous square cavity filled with nanofluids in the presence of thermal radiation

Balla Chandra Shekar[1] **and Naikoti Kishan**[2]

Department of Mathematics, Osmania University, Hyderabad, Telangana State, India.

E-mails: [1]`shekar.balla@gmail.com`, [2]`kishan_n@rediffmail.com`

Abstract. Free convection heat transfer in a square cavity filled with nanofluid-saturated porous medium with the effects of different nanoparticles in the presence of thermal radiation is investigated in this paper. The top and bottom horizontal walls of cavity are considered adiabatic, while the vertical walls are kept at constant temperatures. The governing partial differential equations are solved by finite element method of Galerkin weighted residual scheme. Numerical results are obtained for different values of the Rayleigh number, radiation parameter and nanofluid volume fraction. The overall investigation of variation of streamlines, isotherms and Nusselt numbers is presented graphically. To examine the accuracy, the present results are compared with the available results.

1. Introduction

Natural convective heat transfer is a significant phenomenon in engineering and industry with widespread applications in diverse fields, such as, geophysics, solar energy, electronic cooling, and nuclear energy. The classical problem of natural convection in square enclosures has many engineering applications such as the cooling systems of electronic components, the building and thermal insulation systems, the built-in-storage solar collectors, the nuclear reactor systems, the food storage industry and the geophysical fluid mechanics.

Low thermal conductivity of conventional heat transfer fluids such as oil, ethylene glycol, and water is a primary restriction in improving the performance and the compactness of many engineering electronic devices. Therefore, there is a strong need to develop advanced heat transfer fluids with substantially higher conductivities. With recently introduced the term of nanofluids by Choi [1], which are the fluids with suspended solid particles of higher thermal conductivity such as metals within it, the need mentioned before has been overcome. The nanofluids have many applications in the industry since materials of nanometer size have unique physical and chemical properties. Convective heat transfer of nanofluids in cavities has been extensively studied by many researchers in recent years. Khanafer et al. [2] investigated the flow and heat transfer performance in a rectangular enclosure using nanofluids for various pertinent parameters and taking into account the solid particle dispersion. Tiwari and Das [3] studied mixed convection in a differentially heated two-sided lid-driven square cavity. One of their conclusions was that nanoparticles were able to change the flow pattern of fluid from natural to forced convection regime. Hence, it is worth it to mention here that all of the above surveyed works dealt with viscous clear fluids (non-porous media) and that most of them reported that it was not always true that increasing the volume fraction of solid particles enhanced the heat transfer rate. It is found that a relatively a limited papers dealing with nanofluids saturated in porous media were published; and most of these papers studied the boundary layer flow.

Nield and Kuznetov [4] examined the influence of nanoparticles natural convection past a vertical plate. Their analytical study was based on Brownian motion and thermophoresis (for nanofluids) and Darcy model (for porous media). Ahmad and Pop [5] studied numerically mixed convection boundary layer flow using three different nanoparticles based on the conventional model of Tiwari and Das [3] which incorporates only the nanofluid volume fraction. They followed Darcy model also to interpret the flow in porous media. Balla and Naikoti [6] studied free convective heat transfer in fluid saturated porous cavity. Cimpean and Pop [7] studied fully developed steady-state mixed convection flow of nanofluids in a inclined porous channel. A lattice Boltzmann model is developed to simulate the convection heat transfer utilizing Al_2O_3-water nanofluids in a square cavity by He et al. [8]. However, the field of nanofluids saturated in porous cavities is found published in very little works as in Sun and Pop [9] and Chamkha and Ismael [10]. Sun and Pop [9] considered a triangular enclosure heated by a wall heater and filled with a porous medium and saturated with three different nanofluids. A two-dimensional numerical study has been performed by Sheikholeslami et al. [11] to investigate free convection inside a square cavity filled with Cu-water nanofluid. Jou and Tzeng [12] numerically investigated the heat transfer performance of nanofluids inside the two-dimensional rectangular enclosures and found that increasing the volume fraction causes a significant increase in average heat transfer rate. A side heated nanofluid-filled square cavity using a non-homogeneous model has been numerically studied by Celli [13]. Thermal radiation effects are extremely important in the flow processes involving high temperature and controlling heat transfer in industry where heat controlling factors determine the quality of the final product. Thermal radiation has a significant role in the overall surface heat transfer when the convection heat transfer coefficient is small. The influence of radiation on mixed convection from vertical surface in a porous medium was studied by Bakier [14]. The effects of radiation on the thermal boundary layer flow have been considerably researched ([15]-[17]). Zahmatkesh [18] has found that the presence of thermal radiation makes temperature distribution nearly uniform in the vertical sections inside the enclosure and causes the streamlines to be nearly parallel with the vertical walls. Radiation effects on transient natural convection flow of nanofluid has been investigated by Balla and Naikoti [19]. The effect of thermal radiation on boundary layer flow and mass transfer over a moving vertical porous surface has been reported by Makinde [20]. Chiu et al. [21] studied numerically mixed convection and heat transfer inside a horizontal rectangular duct in the presence of thermal radiation. It reveals that radiation effect tends to balance the temperature in the flow at the downstream. It is seen that the bulk temperature development was enhanced with the larger Rayleigh number and smaller conduction-to-radiation parameter.

Recently, Sheremet et al. [22] investigated free convection in a nanofluid-saturated porous square cavity using Tiwari and Das [3] nanofluid model. Thus the motivation of the study is to examine the influence of thermal radiation on natural convection flow field and temperature distribution in a square cavity filled with nanofluid saturated porous medium. In the present study the three types of nanoparticles are considered to analyse the flow and heat transfer characteristics. The numerical results for streamlines, isotherms and average Nusselt number are presented graphically.

2. Mathematical formulation

Consider the free convection in a two-dimensional porous square cavity filled with nanofluid based on water and different type of nanoparticles: Cu, Al_2O_3, TiO_2 in the presence of thermal radiation. A schematic geometry of the problem is shown in figure 1. Let u and v be the

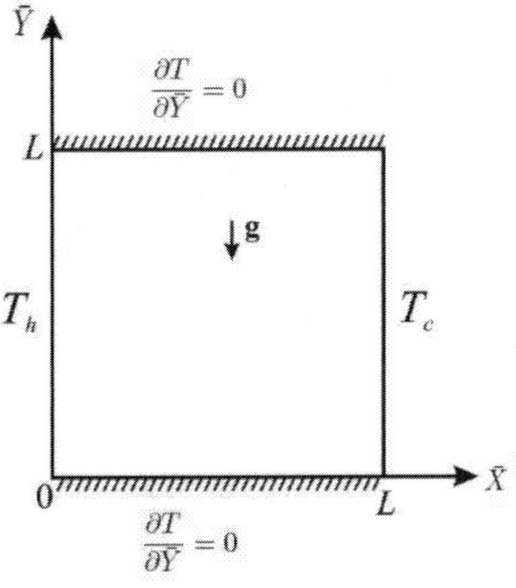

Figure 1. Physical model and coordinate system.

velocity components along $\bar{X}$ and $\bar{Y}$ directions respectively and L is the height of the cavity. It is assumed that the top and bottom walls of the cavity are adiabatic. Also it is assumed that the left wall is the source of heat and maintained at T_h, while the right walls is kept at T_c, where $T_h > T_c$. It is further assumed that the base fluid and the nanoparticles are in thermal equilibrium, the nanofluid is newtonian and incompressible and the flow is laminar. Using the Darcy-Boussinesq approximation $\rho = \rho_0 (1 - \beta(T - T_0))$, the basic equations are

$$u = -\frac{K}{\mu_{nf}} \left(\frac{\partial P}{\partial \bar{X}} \right), \tag{1}$$

$$v = -\frac{K}{\mu_{nf}} \left(\frac{\partial P}{\partial \bar{Y}} + \rho_{nf} g \right), \tag{2}$$

$$u\frac{\partial T}{\partial \bar{X}} + v\frac{\partial T}{\partial \bar{Y}} = \alpha_{nf} \left(\frac{\partial^2 T}{\partial \bar{X}^2} + \frac{\partial^2 T}{\partial \bar{Y}^2} \right) - \frac{1}{(\rho C_P)_{nf}} \frac{\partial q_r}{\partial \bar{Y}}. \tag{3}$$

On eliminating the pressure from equations (1) and (2) and using the stream function $\bar{\psi}$ which is defined as $u = \frac{\partial \bar{\psi}}{\partial \bar{Y}}$ and $v = -\frac{\partial \bar{\psi}}{\partial \bar{X}}$ we obtain,

$$\frac{\partial^2 \bar{\psi}}{\partial \bar{X}^2} + \frac{\partial^2 \bar{\psi}}{\partial \bar{Y}^2} = -\frac{gK \left[\phi \rho_p \beta_p + (1 - \phi) \rho_f \beta_f \right]}{\mu_{nf}} \frac{\partial T}{\partial \bar{X}}, \tag{4}$$

where K is the permeability of the porous medium, g is the acceleration due to gravity, p is the pressure, ρ is the density, β is the thermal expansion coefficient, μ is the dynamic viscosity, k is the thermal conductivity and C_p is the specific heat at a constant pressure. The physical properties of the nanofluid are given by

$$\mu_{nf} = \frac{\mu_f}{(1 - \phi)^{2.5}}, \quad (\rho C_p)_{nf} = (1 - \phi)(\rho C_p)_f + \phi(\rho C_p)_p,$$

$$\frac{k_{nf}}{k_f} = \frac{k_p + 2k_f - 2\phi(k_f - k_p)}{k_p + 2k_f + \phi(k_f - k_p)}, \quad (\rho\beta)_{nf} = (1-\phi)(\rho\beta)_f + \phi(\rho\beta)_p.$$

The radiative heat flux term $\frac{\partial q_r}{\partial \bar{Y}}$ in the energy equation is simplified by invoking Rosseland approximation Brewester [27] as follows:

$$q_r = -\frac{4\sigma^*}{3k^*}\frac{\partial T^4}{\partial \bar{Y}}. \tag{5}$$

Equation (5) can be linearized by expanding T^4 into the Taylor series about T_∞. Neglecting higher order terms, we obtain

$$T^4 \simeq 4T_\infty^3 T - 3T_\infty^4. \tag{6}$$

Using equations (5) and (6), We obtain,

$$\frac{\partial q_r}{\partial \bar{Y}} = -\frac{16\sigma^* T_\infty^3}{3k^*}\frac{\partial^2 T}{\partial \bar{Y}^2}. \tag{7}$$

Introducing the following dimensionless variables

$$x = \frac{\bar{X}}{L}, \quad y = \frac{\bar{Y}}{L}, \quad \psi = \frac{\bar{\psi}}{\alpha_f}, \quad \theta = \frac{T - T_c}{T_h - T_c}.$$

The equations (4) and (3) transform to the following form

$$\frac{1}{(1-\phi)^{2.5}}\left(\frac{\partial^2\psi}{\partial x^2} + \frac{\partial^2\psi}{\partial y^2}\right) = -Ra\left[\phi\left(\frac{\rho_p}{\rho_f}\right)\left(\frac{\beta_p}{\beta_f}\right) + (1-\phi)\right]\frac{\partial\theta}{\partial x}, \tag{8}$$

$$\frac{\partial\psi}{\partial y}\frac{\partial\theta}{\partial x} - \frac{\partial\psi}{\partial x}\frac{\partial\theta}{\partial y} = \frac{\alpha_{nf}}{\alpha_f}\left[\frac{\partial^2\theta}{\partial x^2} + \left(1 + \frac{4}{3}R_d\right)\frac{\partial^2\theta}{\partial y^2}\right]. \tag{9}$$

where Ra is the Rayleigh number and R_d is the radiation parameter

$$Ra = \frac{gK\beta_f(T_h - T_c)L}{\nu_f\alpha_f}, \quad R_d = \frac{4\sigma^* T_\infty^3}{K_{nf}k^*}.$$

The corresponding boundary conditions are

$$\begin{array}{l} \psi = 0, \ \ \theta = \frac{1}{2} \ \text{ at } \ x = 0, \\ \psi = 0, \ \ \theta = -\frac{1}{2} \ \text{ at } \ x = 1, \\ \psi = 0, \ \ \frac{\partial\theta}{\partial y} = 0 \ \text{ at } \ y = 0, 1. \end{array} \tag{10}$$

The physical quantity of interest is the local Nusselt number which is defined as

$$Nu = -\frac{k_{nf}}{k_f}\left(1 + \frac{4}{3}R_d\right)\left(\frac{\partial\theta}{\partial x}\right)_{x=0}.$$

Table 1. Thermo physical properties of base fluid and nanoparticles (Oztop and Abu-Nada [23]).

Physical properties	water	Cu	Al_2O_3	TiO_3
$C_p\ (J/KgK)$	4179	385	765	686.2
$\rho\ (Kg/m^3)$	997.1	8933	3970	4250
$k\ (W/mK)$	0.613	400	40	8.9538
$\alpha \times 10^{-7}\ (m^2/s)$	1.47	1163.1	131.7	30.7
$\beta \times 10^{-5}\ (1/K)$	21	1.67	0.85	0.9

3. Method of solution

The dimensionless partial differential equations (8) and (9) are solved by weighted residual Galerkin finite element method [26]. In weighted residual approach the unknowns are replaced by approximate trial solution which in the context of a discretized domain are given by polynomial relationships to obtain the residuals. These residuals are then multiplied by weight functions and their integrals over an element domain are set to be zero. Let the approximate solutions of ψ, θ are $\psi = \sum_{i=1}^{3} \psi_i \xi_i$; $\theta = \sum_{i=1}^{3} \theta_i \xi_i$; where ξ_i are the linear interpolation functions for a triangular element. The Galerkin finite element model for a typical element Ω_e is given by

$$\begin{bmatrix} [K^{11}] & [K^{12}] \\ [K^{21}] & [K^{22}] \end{bmatrix} \begin{bmatrix} \{\psi\} \\ \{\theta\} \end{bmatrix} = \begin{bmatrix} \{F^1\} \\ \{F^2\} \end{bmatrix},$$

where,

$$K^{11} = \frac{1}{(1-\phi)^{2.5}} \iint_{\Omega_e} \left(-\frac{\partial \xi_j}{\partial x}\frac{\partial \xi_i}{\partial x} - \frac{\partial \xi_j}{\partial y}\frac{\partial \xi_i}{\partial y} \right) dxdy \, ,$$

$$K^{12} = Ra \left[\phi \left(\frac{\rho_p}{\rho_f} \right) \left(\frac{\beta_p}{\beta_f} \right) + (1-\phi) \right] \iint_{\Omega_e} \left(\xi_j \frac{\partial \xi_i}{\partial x} \right) dxdy \, , F^1 = 0 \, , K^{21} = 0 \, , F^2 = 0 \, ,$$

$$K^{22} = \iint_{\Omega_e} \left(\xi_j \frac{\partial \bar{\Psi}}{\partial y}\frac{\partial \xi_i}{\partial x} - \xi_j \frac{\partial \bar{\Psi}}{\partial x}\frac{\partial \xi_i}{\partial y} + \frac{\alpha_{nf}}{\alpha_f}\frac{\partial \xi_j}{\partial x}\frac{\partial \xi_i}{\partial x} + \frac{\alpha_{nf}}{\alpha_f}\left(1 + \frac{4}{3}R_d\right)\frac{\partial \xi_j}{\partial y}\frac{\partial \xi_i}{\partial y} \right) dxdy \, ,$$

where $\bar{\Psi} = \sum_{j=1}^{3} \bar{\Psi}_j \xi_j$. The whole domain is divided into 800 triangular elements of equal size. Each element is three nodded, therefore whole domain contains 441 nodes. Each element matrix is of order 6 × 6, since at each node two functions are to be evaluated. Hence after assembly of the elemental equations, we obtain a system of 882 non-linear coupled equations. To linearize the system of equations the function $\bar{\Psi}$ is incorporated, which is assumed to be known. After imposing the boundary conditions a matrix of system of linear equations of order 722 × 722 is remained. This system has been solved by using Gauss seidel iteration method.

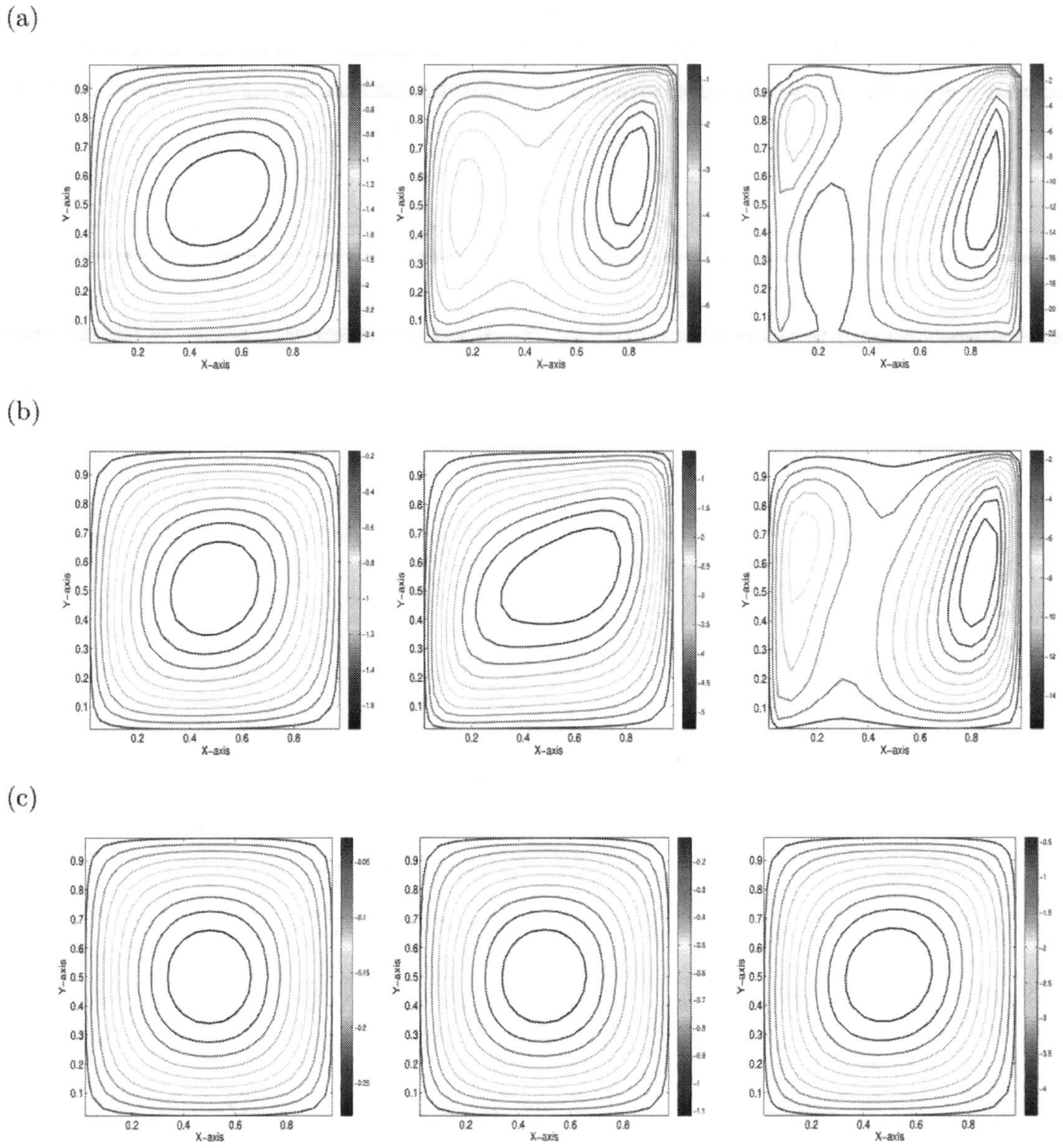

Figure 2. Stream lines for (a) $\phi = 0.02$ (b) $\phi = 0.2$ (c) $\phi = 0.6$ and Ra=50,200,800.

4. Results and discussions

The numerical results are presented by the streamlines, isotherms and local Nusselt number. In the present study the following values have been taken for key parameters: Rayleigh number (Ra=50,200,800), thermal radiation parameter (R_d=0,0.5,1) and the volume fraction of nanofluid (ϕ=0.02,0.2,0.6). The effect of nanoparticle volume fraction ϕ on the flow and heat transfer characteristics has been considered for three types of nanofluids namely Cu, Al_2O_3 and TiO_3. To check the validity of the present method and code, a comparison with selective data from the published work was carried out. The present results are displayed in table 2 for some comparative data with those reported by Manole and Lage [24], Baytas and Pop [25] and Sheremet et al. [22].

Table 2. Comparison of the Nusselt numbers at the hot wall for ϕ=0.

	Ra=10	Ra=100	Ra=1000	Ra=10000
Manole and Lage [24]		3.118	13.637	48.117
Baytas and Pop [25]	1.079	3.16	14.06	48.33
Sheremet et al. [22]	1.079	3.115	13.667	48.823
Present results	1.08	3.129	13.7353	48.5451

The mechanism of forming such streamlines is as follows: the fluid near the left hot wall (interface) is heated, and then it moves to the right cold wall (due to differentially heating) and moves up due to convection (Buoyancy force). When it becomes closer to the upper adiabatic wall, it turns to the right and falls down to the lower adiabatic wall. Hence, a single vortex of clockwise rotation (negative sign) is formed within the porous domain. Figure 2 displays the streamlines for different values of the Rayleigh number Ra=50,200,800 and for copper nanoparticles with different values of volume fractions ϕ=0.02,0.2,0.6. From the figure it can be observed that when the Rayleigh number Ra increases, the flow convection is strengthened, the flow cell is extended along the vertical axis and the boundary layers become more significant. Depending on the Ra and ϕ single or multiple cell convection was found. Further it can be noticed that for large value of Rayleigh number (Ra=800), the convection mode is pronounced, the flow region in the cavity split into two cells. From the figure it is also revealed that the strength of the streamlines decreases with the increase of nanoparticle volume fraction ϕ. It is interestingly noticed that the influence of ϕ is predominating for the higher Ra values and it is meager for lower values of Ra. In case of $\phi = 0.6$, for different values of Ra the convection is dominant, the circulation cell is very stronger and it is in the centre of the cavity.

Figure 3 illustrates the isotherms for different values of Ra=50,200,800 for copper nanoparticles with volume fractions ϕ=0.02,0.2,0.6. For low Rayleigh number Ra=50, the conduction raising is dominant which defines weak heat transfer. It can be seen from the figure that the isotherms are vertically stratified. As the Rayleigh number Ra increases the convection is dominant and the isotherms become denser along the vertical walls. Thus Rayleigh number helps in minimizing the conduction heat transfer and increasing the convection heat transfer. The isotherms become parallel to hot wall with the increase in the nanopartical volume fraction ϕ, indicating weak heat transfer. However, the effect of Rayleigh number on heat transfer increases with the lower values of ϕ.

Figures 4 and 5 illustrate the effect of Rayleigh number Ra on streamlines and isotherms respectively, for three types of nanoparticles Cu, Al_2O_3, TiO_3 with nanoparticle volume fraction ϕ=0.2. From figure 4 it is observed that for Ra=50 a vortex with clock wise rotation is formed. The vortex is elongated in horizontal direction and further split into two vortices as Ra is increased to 800 in case of Cu. The flow pattern is nearly same as that of Cu, but with the change in the flow intensity for the nanofluids Al_2O_3, TiO_3. It is also observed that the circulation strength is higher for Cu and it is lower for Al_2O_3 regardless of the Rayleigh number. From figure 5 it is observed that, for the nanofluid Cu, the isotherms change almost from vertical lines to horizontal lines in the central region of the cavity as Ra increases from 50 to 800. For higher values of Ra the isotherms are closer near the vertical walls. The increase in the Rayleigh number Ra enhances the rate of heat transfer. Similar pattern of isotherms is observed for the nanofluids Al_2O_3, TiO_3 except the values of isotherms. It is also observed that Cu has higher isotherm value than the other nanofluids.

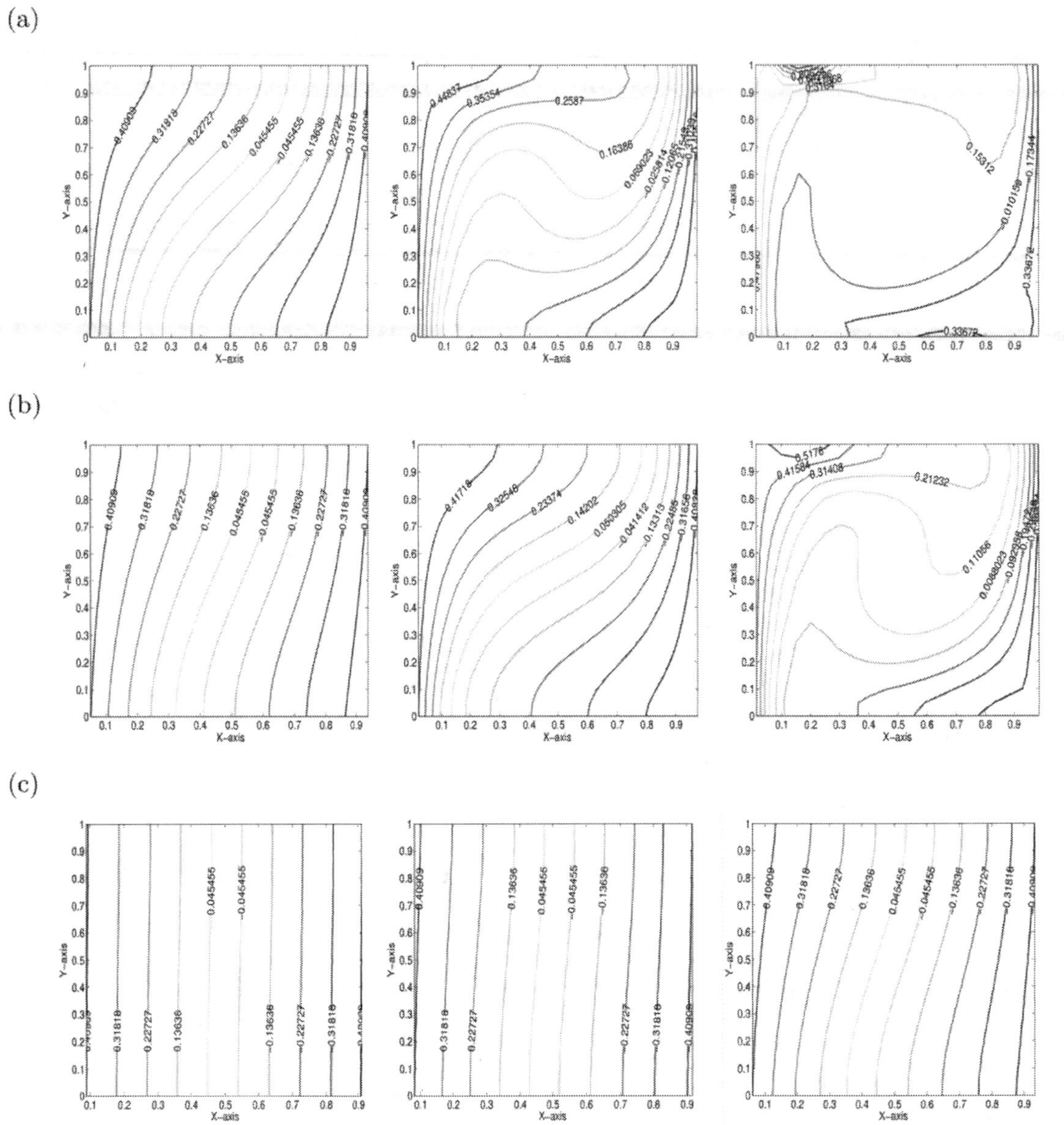

Figure 3. Isotherms for (a) $\phi = 0.02$ (b) $\phi = 0.2$ (c) $\phi = 0.6$ and Ra=50,200,800.

The effect of nanoparticle volume fraction ϕ on the behaviour of streamlines is presented for the radiation parameter R_d=0,0.5,1 in figure 6. Due to heating of the left wall the fluid rises towards the cold wall. From the figure 6(a) it is observed that the streamlines in the cavity form in to two cells for ϕ=0.02. The effect of radiation parameter is to strengthen the flow intensity. For ϕ=0.2, one flow cell is formed and elongated horizontally. As R_d increases, the circulation strength enhances. A centrally circulated cell is appeared in case of ϕ=0.6 and its strength increases with the increase in the radiation parameter. Figure 7 depicts the effect of radiation parameter R_d on isotherms for different volume fractions of nanofluid ϕ=0.02,0.2,0.6. The thermal radiation significantly alters the wall temperature and generates thermal instabilities while reducing the overall heat transfer rates. The effect of R_d is to reduce the heat transfer for

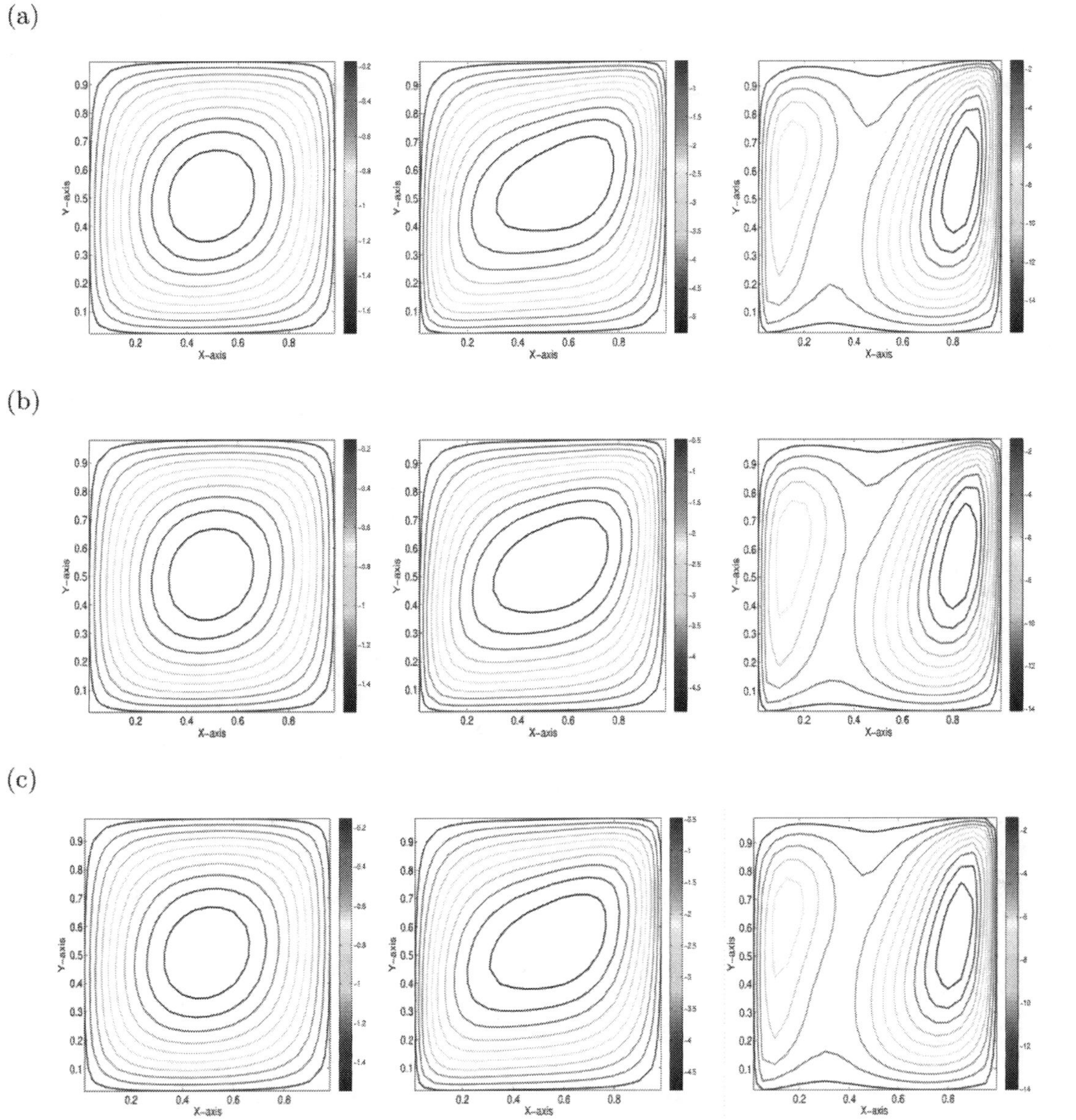

Figure 4. Stream lines for (a) Cu (b) Al_2O_3 (c) TiO_3 and Ra=50,200,800.

lower values of ϕ. As the value of ϕ increases the effect of radiation parameter is less.

Figures 8-11 present the distribution of local Nusselt number Nu along the hot vertical wall ($x = 0$) for different values of ϕ, Ra, R_d and for the three nanofluids Cu, Al_2O_3, TiO_3. Figure 8 shows a linear variation of the local Nusselt number with the nanoparticle volume fraction. Moreover, an increase in ϕ leads to decrease the local Nusselt number near the bottom wall and the opposite phenomena is noticed at the top wall. Figure 9 depicts that as the Rayleigh number Ra increases, the local Nusselt number also increases near bottom wall $y = 0$, and the variation is less at the top wall. From the figure 10, it can be observed that the influence of radiation parameter on Nusselt number leads to increase the local Nusselt number. Figure 11 reveals that the local Nusselt number value decreases when the nanofluid changes in the order

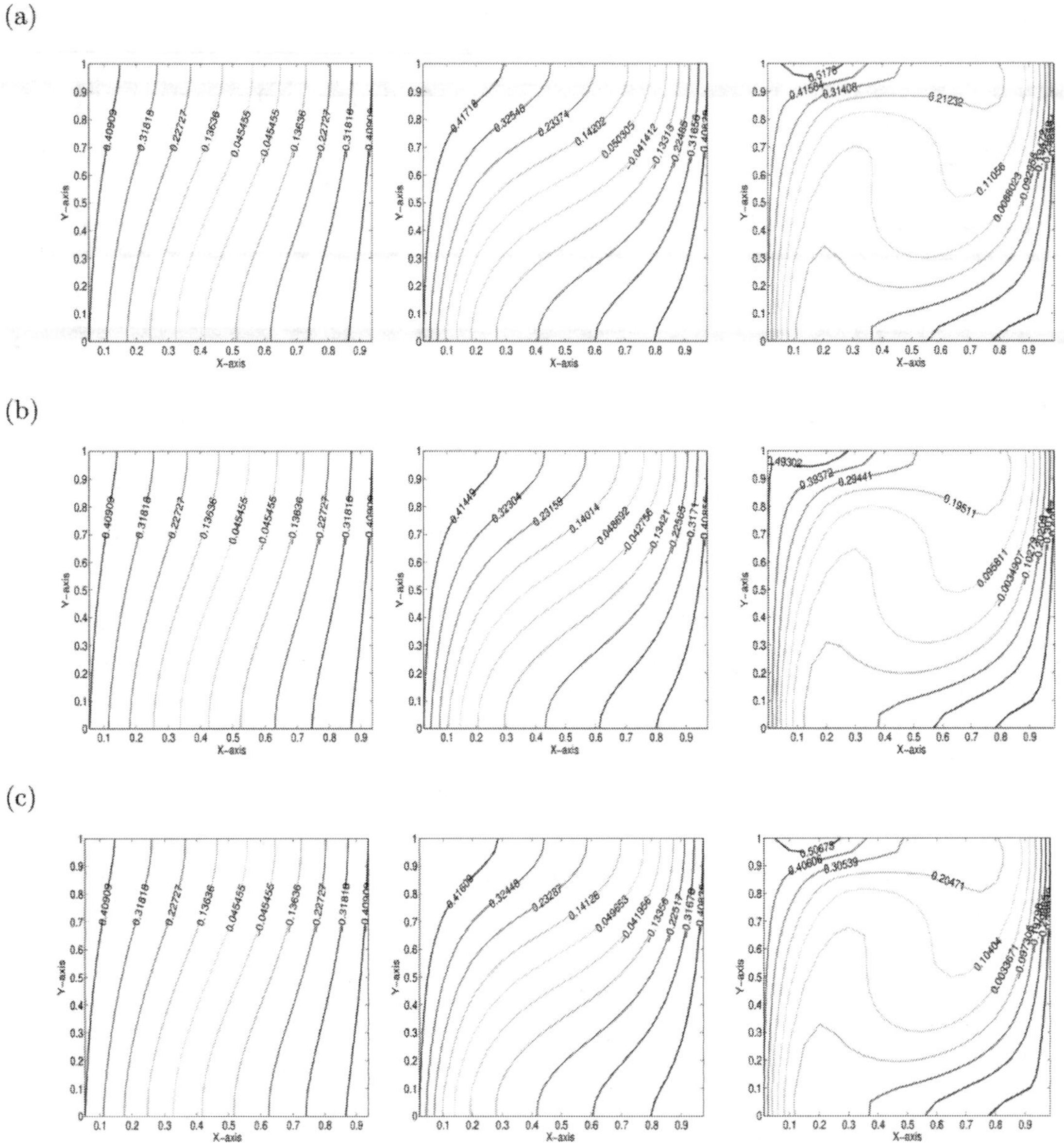

Figure 5. Isotherms for (a) Cu (b) Al_2O_3 (c) TiO_3 and Ra=50,200,800.

as Cu, TiO_3, Al_2O_3.

5. Conclusion

Natural convection flow and heat transfer in a porous square cavity filled with nanofluid under the influence of thermal radiation is studied. The flow and temperature fields as well as heat transfer rate are analysed for various values of flow parameters such as Rayleigh number Ra, nanoparticle volume fraction ϕ and radiation parameter R_d. The computational results lead to the following conclusions:

- The flow pattern takes single or double cell form depending on Ra and ϕ. The flow circulation is pronounced with higher values of Ra and ϕ. The heat transfer increases

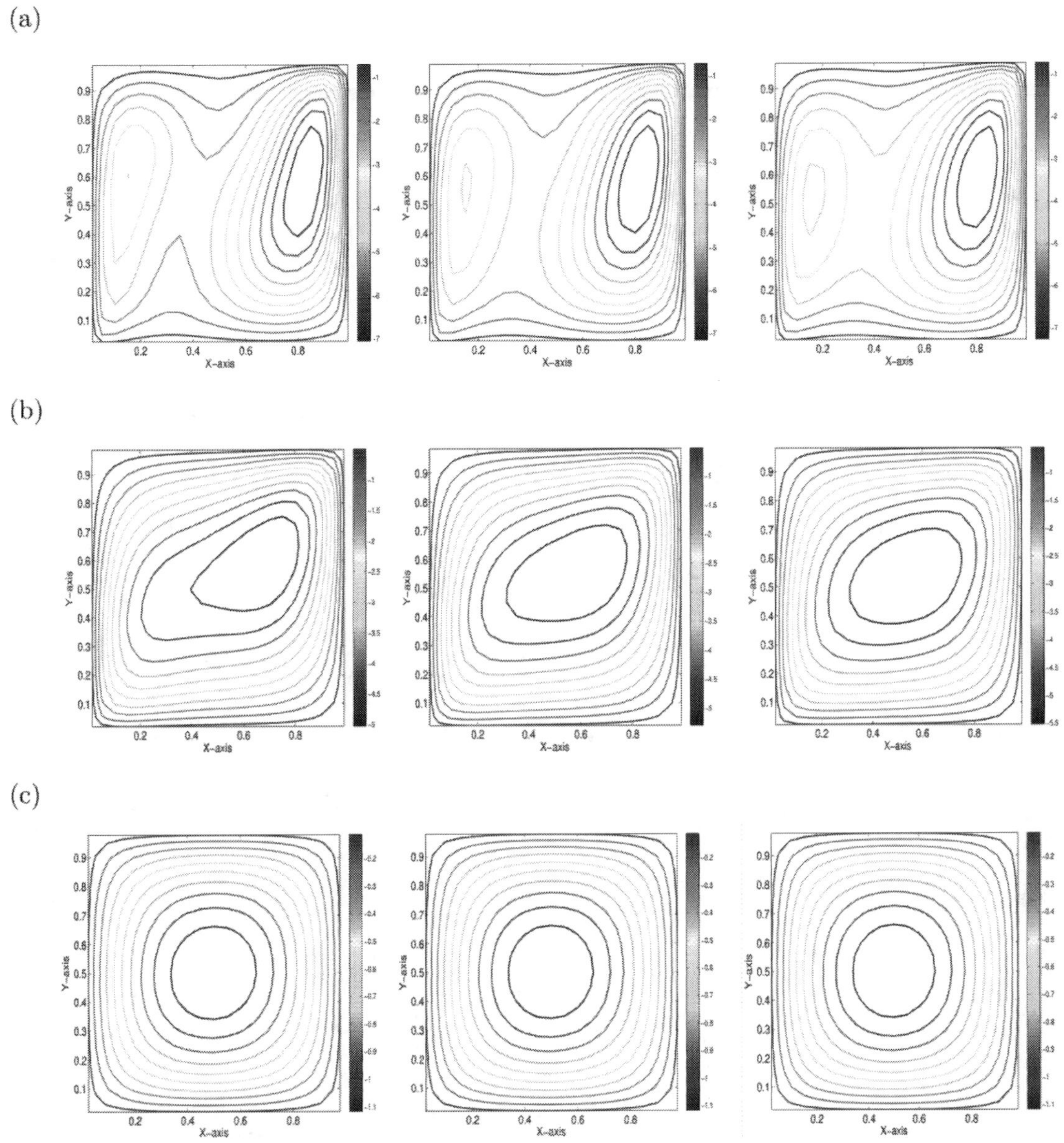

Figure 6. Streamlines for (a) $\phi = 0.02$ (b) $\phi = 0.2$ (c) $\phi = 0.6$ and R_d=0,0.5,1.

with the increase in Ra, while decreases with the increase in ϕ.

- The strength of the flow increases with the increase in the radiation parameter R_d. The effect of R_d leads to reduce the isotherm values.
- The heat transfer and flow intensity are higher for the nanoparticles Cu while they are lower for Al_2O_3.
- The Nusselt number is increase with the increase in radiation parameter R_d and Rayleigh number Ra, while it decreases with the increase in nanoparticle volume fraction ϕ. Further, it decreases when the nanofluid changes in the order as Cu, TiO_3 and Al_2O_3.

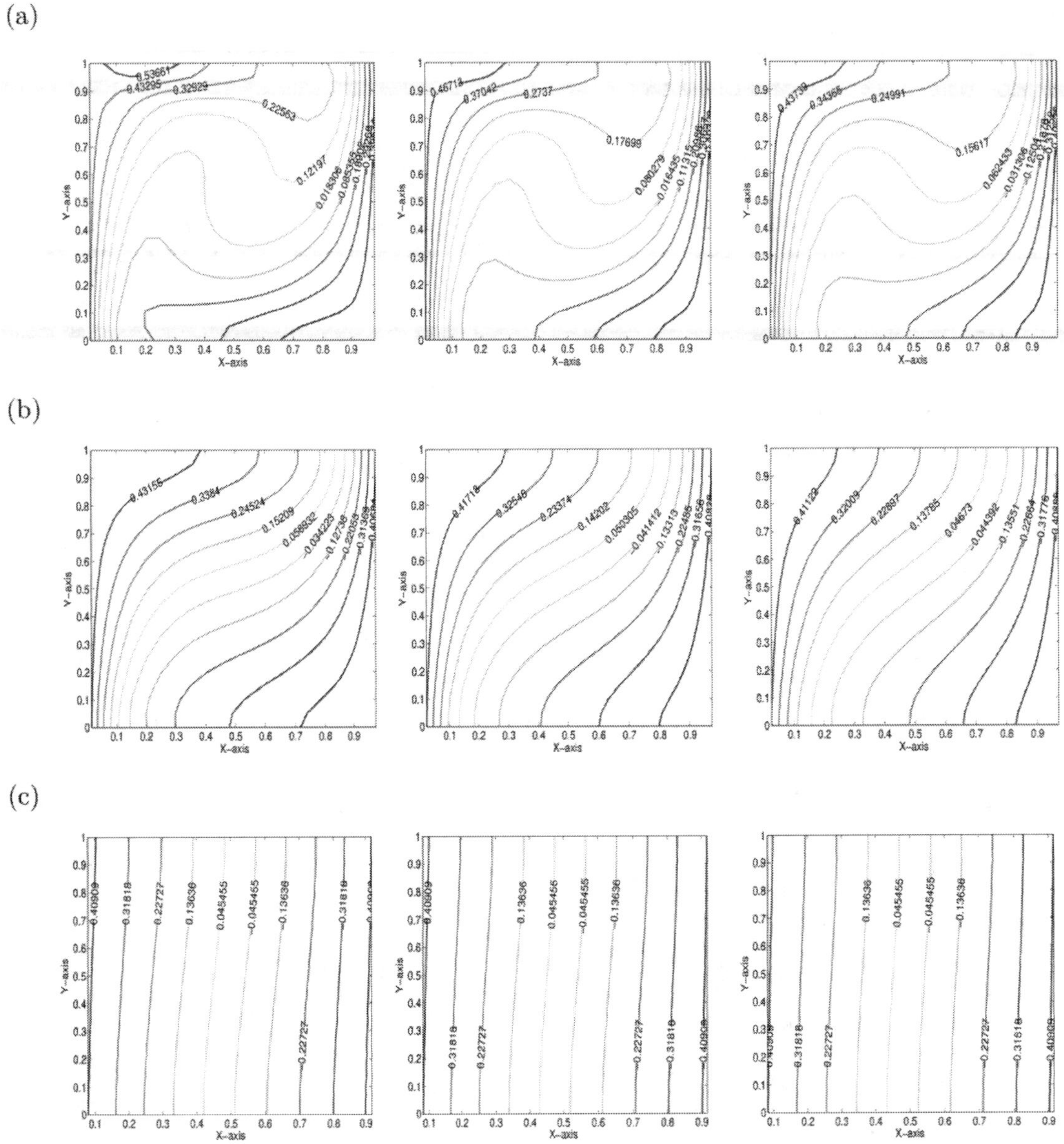

Figure 7. Isotherms for (a) $\phi = 0.02$ (b) $\phi = 0.2$ (c) $\phi = 0.6$ and R_d=0,0.5,1.

References

[1] Choi S U S 1995 Enhancing thermal conductivity of fluids with nanoparticles *Proceedings of the 1995 ASME International Mechanical Engineering Congress and Exposition* vol 66 (San Francisco, USA: ASME, FED 231/MD) 99-105

[2] Khanafer K, Vafai K and Lightstone M 2003 Buoyancy-driven heat transfer enhancement in a two-dimensional enclosure utilizing nanofluids *Int. J. Heat Transf.* **46** 3639-53

[3] Tiwari R K and Das M K 2007 Heat transfer augmentation in a two-sided lid-driven differentially heated square cavity utilizing nanofluids *Int. J. Heat Mass Transfer.* **50** 2002-18

[4] Nield D A and Kuznetsov A V 2009 The ChengMinkowycz problem for natural convective boundary-layer flow in a porous medium saturated by a nanofluid *Int. J. Heat Mass Transfer.* **52** 5792-95

[5] Ahmad S and Pop I 2010 Mixed convection boundary layer flow from a vertical flat plate embedded in a porous medium filled with nanofluids *Int. Commun. Heat Mass Transfer.* **37** 987-91

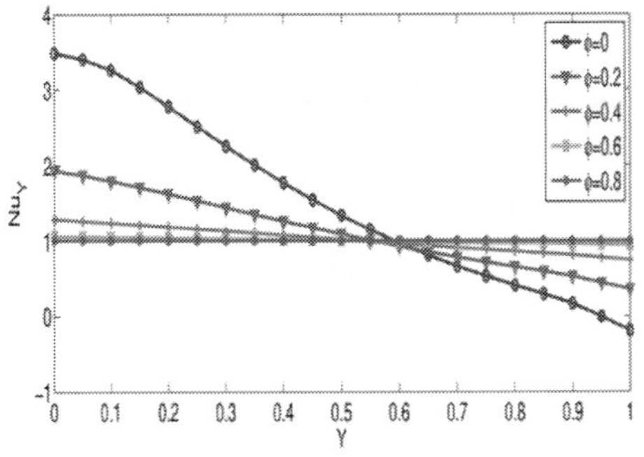

Figure 8. Effect of nanoparticle volume fraction ϕ on Nusselt number.

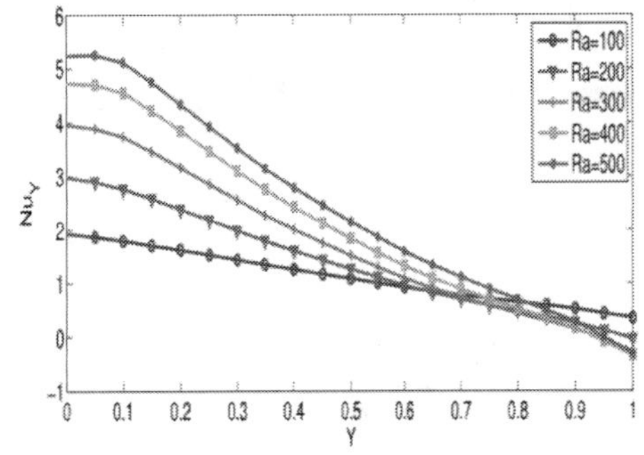

Figure 9. Effect of Rayleigh Ra on Nusselt number.

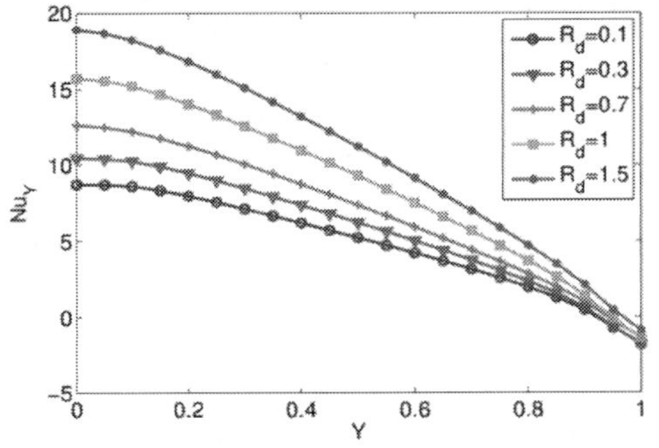

Figure 10. Effect of radiation parameter R_d on Nusselt number.

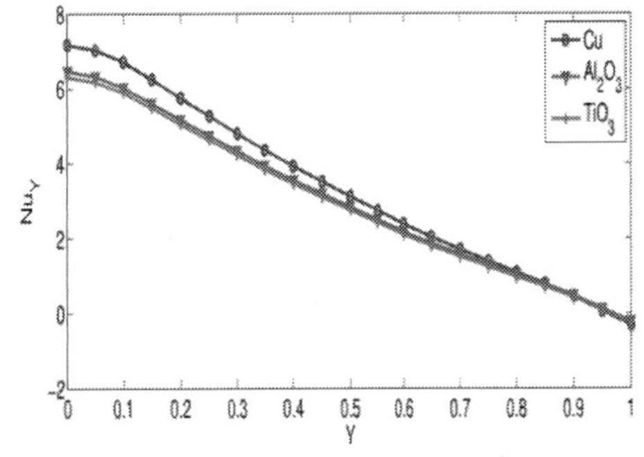

Figure 11. Effect of nanofluid on Nusselt number.

[6] Balla C S and Naikoti K 2015 Soret and Dufour effects on free convective heat and solute transfer in fluid saturated inclined porous cavity *Engineering Science and Technology, an International Journal* **18(4)** 543-54

[7] Cimpean D S and Pop I 2012 Fully developed mixed convection flow of a nanofluid through an inclined channel filled with a porous medium *Int. J. Heat Mass Transfer.* **55** 907-14

[8] He Y, Qi C, Hu Y, Qin B, Li F and Ding Y 2011 Lattice Boltzmann simulation of alumina-water nanofluid in a square cavity *Nanoscale Research Letters* **6** 184

[9] Sun Q and Pop I 2011 Free convection in a triangle cavity filled with a porous medium saturated with nanofluids with flush mounted heater on the wall *Int. J. Thermal Sci.* **50** 2141-53

[10] Chamkha A J and Ismael M A 2013 Conjugate heat transfer in a porous cavity filled with nanofluids and heated by a triangular thick wall *Int. J. Therm. Sci.* **67** 135-51

[11] Sheikholeslami M, Gorji-Bandpy M, Seyyedi S M, Ganji D D, Houman B R and Soheil S 2013 Application of LBM in simulation of natural convection in a nanofluid filled square cavity with curve boundaries *Powder Technology* **247** 87-94

[12] Jou R Y and Tzeng S C 2006 Numerical research of nature convective heat transfer enhancement filled with nanofluids in rectangular enclosures *Int. Commun. Heat Mass Transfer.* **33** 727-36

[13] Celli M 2013 Non-homogeneous model for a side heated square cavity filled with a nanofluid *Int. J. Heat Fluid Flow.* **44** 327-55

[14] Bakier A Y 2001 Thermal radiation effect on mixed convection from vertical surface in saturated porous media *Int. Commun. Heat Mass Transf.* **28** 119-26

[15] Cortell R 2008 Effects of viscous dissipation and radiation on the thermal boundary layer over a nonlinearly stretching sheet *Phys. Lett.* A **372** 631-36

[16] Govardhan K, Kishan N and Balaswamy B 2012 Effect of viscous dissipation and radiation on MHD gas flow and heat and mass transfer over a stretching surface with a uniform free stream *Journal of Engineering*

Physics and Thermophysics **85** 909-16

[17] Naikoti K and Vadithya M 2014 Thermal radiation effects on magneto hydro dynamic flow and heat transfer in a channel with porous walls of different permeability *Thermal Science* **18(suppl.2)** 563-72

[18] Zahmatkesh I 2007 Influence of thermal radiation on free convection inside a porous enclosure *Emir. J. Eng. Res.* **12(2)** 47-52

[19] Balla C S and Naikoti K 2014 Finite element analysis of magnetohydrodynamic transient free convection flow of nanofluid over a vertical cone with thermal radiation *Proceedings of Institution of Mechanical Engineers Part N: Journal of Nanoengineering and Nanosystems* DOI: 10.1177/1740349914552879.

[20] Makinde O D 2005 Free-convection flow with thermal radiation and mass transfer past a moving vertical porous plate *Int. Commun. Heat Mass Transfer.* **32** 1411-19

[21] Chiu H C, Jang J H and Yan W M 2007 Mixed Convection Heat Transfer in Horizontal Rectangular Ducts with Radiation Effects *Int. J. of Heat Mass Transfer.* **50** 2874-82

[22] Sheremet M A, Grosan T and Pop I 2015 Free convection in a square cavity filled with a porous medium saturated by nanofluid using Tiwari and Das' nanofluid model *Transp. Porous Med.* **106** 595-610

[23] Oztop H F and Abu-Nada E 2008 Numerical study of natural convection in partially heated rectangular enclosures filled with nanofluids *International Journal of Heat and Fluid Flow* **29** 1326-36

[24] Manole D M and Lage J L 1992 Numerical benchmark results for natural convection in a porous medium cavity *Heat and Mass Transfer Porous Media* **105** 44-59

[25] Baytas A C and Pop I 1999 Free convection in oblique enclosures filled with a porous medium *Int. J. Heat Mass Transf.* **42** 1047-57

[26] Reddy J N 1984 *An introduction to the finite element method* (New York: Mc Graw-Hill)

[27] Brewester M Q 1992 *Thermal Radiative Transfer and Properties* (New York: John Wiley and Sons)

Analysis of noise emitted from diesel engines

S Narayan

Mechanical Engineering Department,University of Roma Tre, Scala B10, Int A,Via Berengaro, Rome, Italy-8400162.

E-mail: rarekv@gmail.com

Abstract. In this work combustion noise produced in diesel engines has been investigated. In order to reduce the exhaust emissions various injection parameters need to be studied and optimized. The noise has been investigated by mean of data obtained from cylinder pressure measurements using piezo electric transducers and microphones on a dual cylinder diesel engine test rig. The engine was run under various operating conditions varying various injection parameters to investigate the effects of noise emissions under various testing conditions.

1. Introduction

To meet emissions norms new technologies are being developed all over the world. Many clean engine technologies are being used in diesel engines including EGR, Electronic fuel injection systems etc[1]. However there is a trade off between the exhaust emissions reduction and noise generated from diesel engines [2]. Modern EGR methods allow reductions of particulate matter as well as NOx at a time [3]. Variation in injection processes cause effectiveness of combustion processes which can reduce noise. Pilot injection also causes reduction in emissions from engine. This work is focusses on optimizing parameters of injection without change in structure to reduce the noise levels emitted from engine. Diesel engine combustion process begins when the atomized fuel enters from the injection nozzle orifice into combustion chamber where it gets mixed up with compressed air forming air-fuel mixture. Hence there is ignition delay between actual start of combustion process and fuel injected inside diesel engine [4]. A typical heat release curve of a diesel engine is shown here in figure 1. There is rapid change in pressure of engine cylinder during the phase of pre- mixed combustion which causes oscillation of whole engine structure [5]. The amplitude of these vibrations depends upon injection delay period as well as on rate of injection of fuel. Cylinder pressure can be transformed from time domain to frequency domain to correlate it with engine noise.

2. Background

Due to high efficiency, diesel engines have been a favourite choice for heavy duty applications including trucks . However they suffer from drawbacks of high noise, weight and vibrations. These engines are of two types: 1. Direct Injection(D.I.) Engines and 2. In direct injection engines. In the D.I. engines , the fuel is directly injected inside combustion chamber and due to lesser time for mixing, a heterogamous mixture consisting of both rich and lean parts is formed in the chamber.

Modern diesel injection systems use multiple injection process to control emissions like soot and Nitrous oxide (NOx) formation. These generally use three phases of injection namely pre-

Published under licence by IOP Publishing Ltd

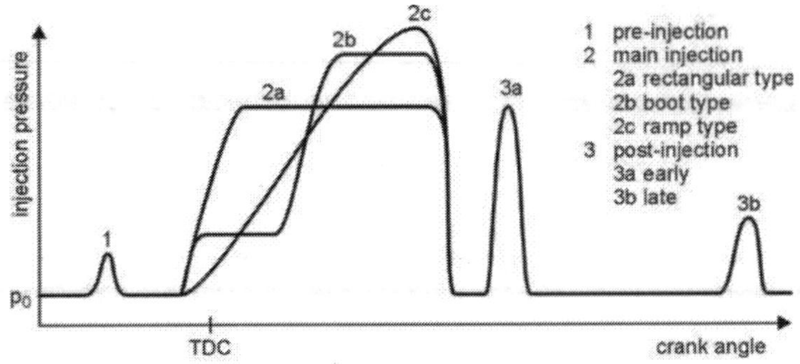

Figure 1. Multiple injection process in diesel engines.

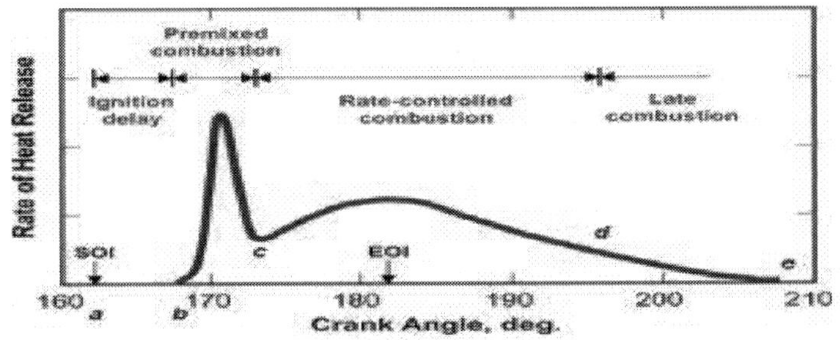

Figure 2. Ideal heat release curve in engine.

injection period, main -injection period and post injection period as seen from figure 1. There is delay period between the start of ignition process and fuel injected inside diesel engine. More this ignition delay, more is the temperature during combustion and hence better condition for NOx formation. To shorten the delay period, small amount of fuel is pre-injection before main injection during the phase pre-mixed combustion phase. The torque and power produced in engine depends upon main injection period. It is advantageous to vary injected fuel mass with time to reduce the specific fuel consumption. This method is known as rate shaping as depicted in figure 1. Rate shaping may be rectangular, step or boot in shape. Post-injection of fuel is done to reduce soot emissions and in some cases may be useful for exhaust gas recirculation treatment of gases [6]. It has been reported that post injection reduces soot by about 70% without increasing the fuel consumption [7].

3. Experimental test rig

Experiments were conducted on a dual cylinder Lombardini LDW442CRS common rail direct injection test rig having specifications as presented in Table 1. This engine test rig has a piezo electric type Kistler 6056A make pressure transducer for in cylinder pressure measurements and an optical crank angle encoder for detection of TDC position as well as engine speed. The given system can do maximum of 2 injections per cycle. The injection strategy for the engine is shown in figure 2. Dual injection strategy was used to take emissions and consideration's discussed in previous section. The engine was run at 2000 RPM and the signals were obtained at both motored and 100% load condition. The data obtained is shown in Table 2.

Table 1. Engine specifications.

Type	Direct Injection
Bore	60.6mm
Stroke	68mm
Displacement	440cc
Compression ration	20:1
Maximum Power	8.5kW@4400RPM
Maximum Torgue	25Nm2000 RPM

Table 2. Testing cases.

Case	Prail	Qpre	Qmain	SOIpre	SOImain
B3	700	1	14.1	13.2°	6°
BASE	720	1	14.6	16.2°	6°
B1	700	2	14.1	17.1°	6°
B2	700	1	14.1	20.1°	6°

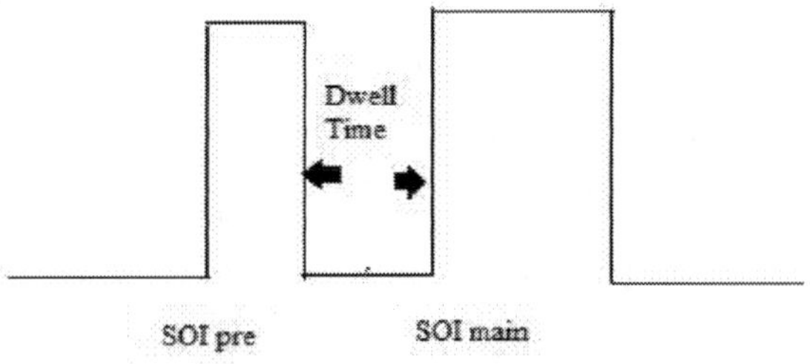

Figure 3. Injection process.

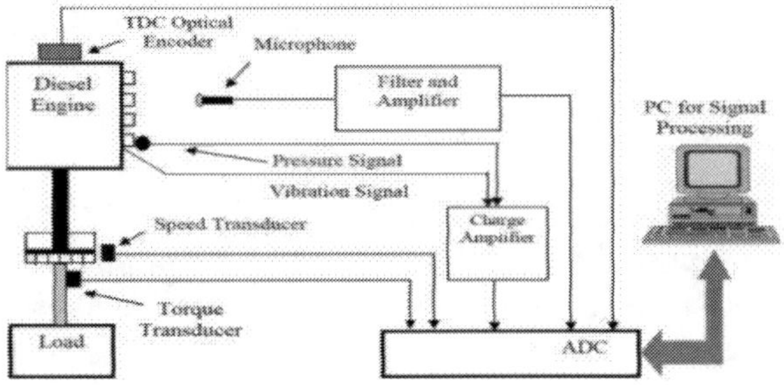

Figure 4. Engine rig showing microphone.

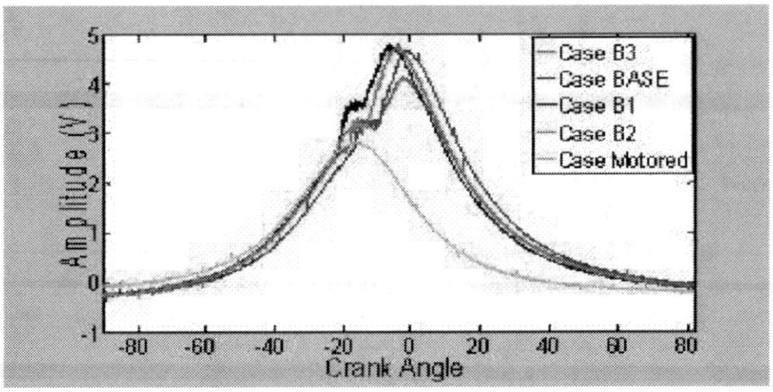

Figure 5. In cylinder pressure.

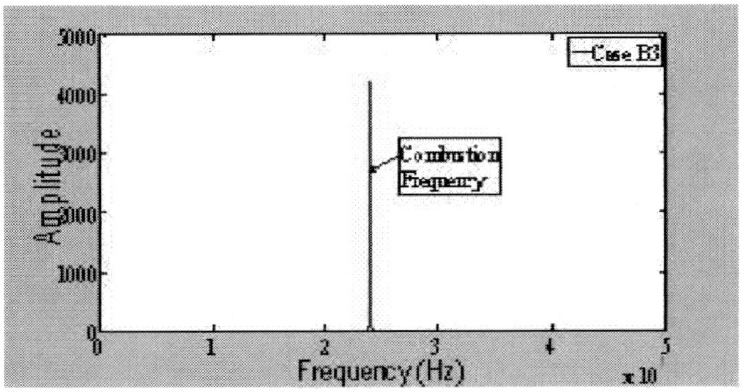

Figure 6. Acoustic emission spectrum(Case B3).

4. Results and discussions

As seen from figure 5, an increase in pre-injection timings causes increase in the injection delay which in turn leads to more fuel being injected inside the combustion chamber, hence more cylinder pressure rise. The frequency of combustion oscillations can be evaluated from relationship[8].

$$F_c = \frac{Z_n}{120}, \tag{1}$$

where, n -Engine RPM and Z -number of cylinders.

Acoustic emissions spectrums for the test conditions have been plotted in figures 6-9 using FFT transformations. An increase in pre injection timing causes increase in pressure and energy increase in acoustic spectrum. This is due to increase in energy released. In order to take into account the energy content, rate of heat release equation is considered next for the test conditions. For combustion engines ideal rate of heat release equation is given by relationship[9].

$$ROHR = \left(\frac{du}{d\theta} * V * \frac{1}{r-1} + \frac{r}{r-1} * P * \frac{du}{d\theta}\right), \tag{2}$$

here p , v represent the specific heat ratio, pressure, volume corresponding to crank angle. Figures 10 and 11 show the traces of ROHR and Cumulative ROHR for the test conditions. As evident from these plots high ignition delay causes impulsive combustion which increases the

Journal of Physics: Conference Series **662** (2015) 012018 doi:10.1088/1742-6596/662/1/012018

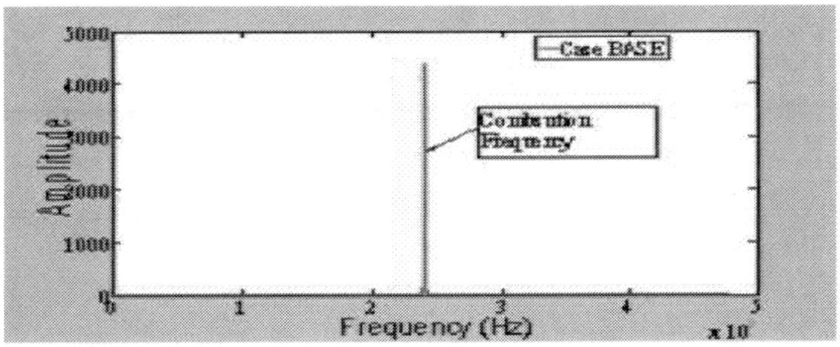

Figure 7. Acoustic emission spectrum(BASE).

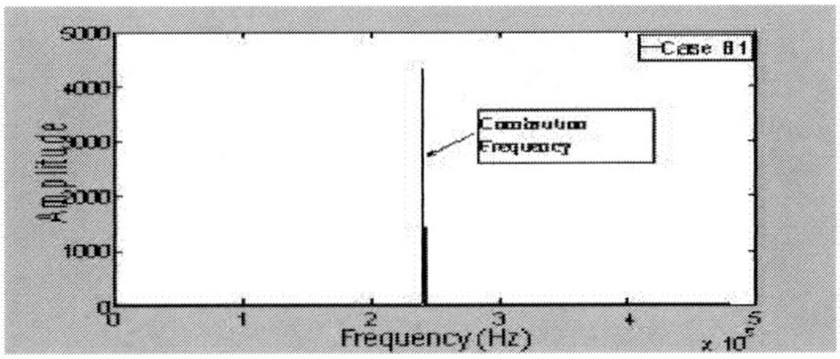

Figure 8. Acoustic emission spectrum(Case B1).

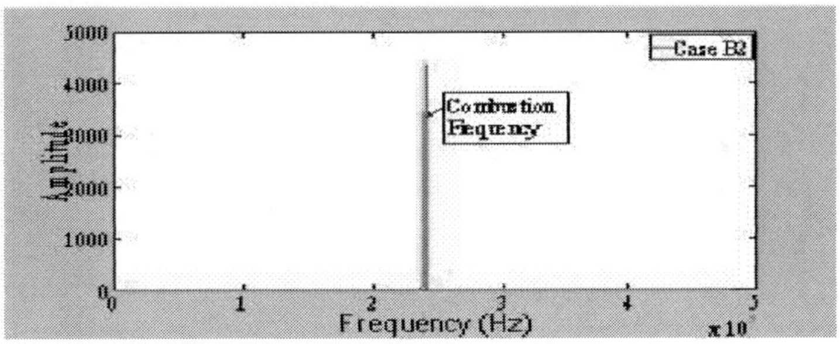

Figure 9. Acoustic emission spectrum(Case B2).

amplitude of audio spectrum from engine. The vibrations from the structure cause oscillations of engine rig which have been investigated in figures 12-15.

Table 3. Difference in injection parameters.

Case	Parameter
C	BASE
D	Q_{pre} increased by 1 mm^3 / stroke
A	Injection angle retarded by 3°
B	Injection Pressure Increased by 20 Bars

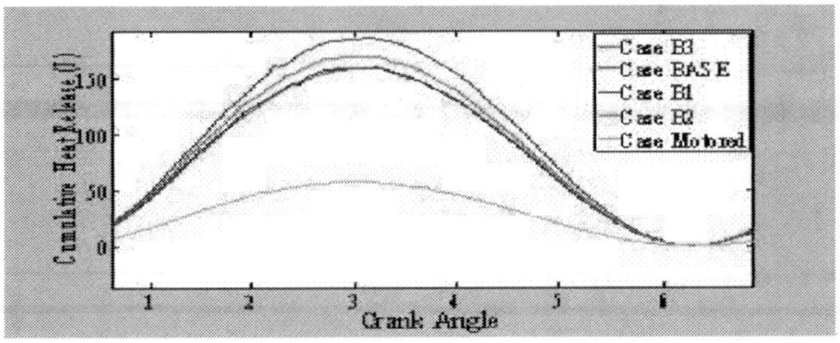

Figure 10. Rate of heat release(J/Degree).

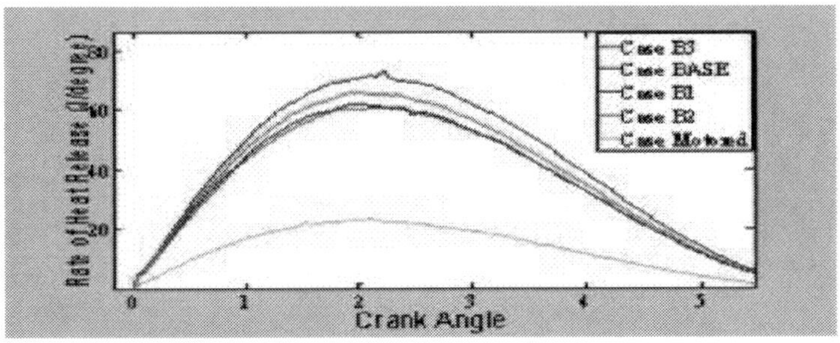

Figure 11. Cumulative heat release(J).

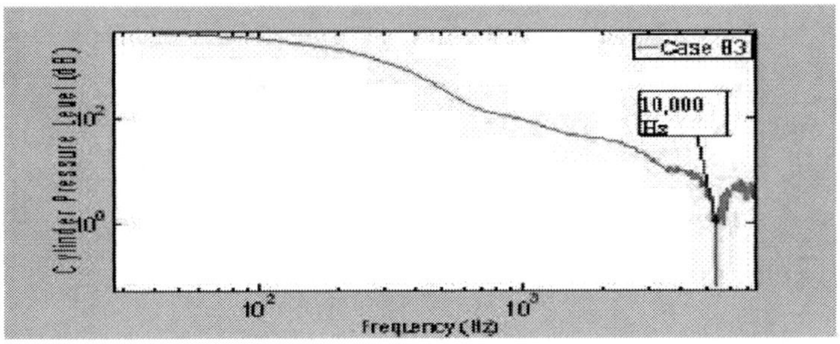

Figure 12. Cylinder pressure spectrum(Case B3).

In general spectrum plot of in cylinder pressure can be divided into three regions[6]:
a) Region of low frequency-in this region the maximum pressure depends upon integration of cylinder pressure curve.
b) Second one is that of medium frequency range in which cylinder pressure falls linearly in logarithmic scale.
c) Third region of high frequency zone is the beginning of combustion phenomenon where resonance of engine structure takes places which depends upon gas temperature and geometry of cylinder.
Due to changing temperatures of gas, resonance is unsteady process. As seen from these curves, the region 3 begins around a frequency of 3000Hz. In order to find the range corresponding to region 1, change in the fuel injection parameters were again varied for the given test conditions [10]. The parameters can be seen in Table 3.

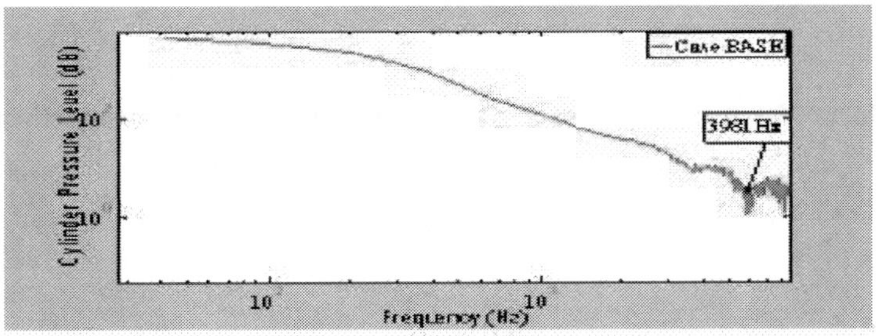

Figure 13. Cylinder pressure spectrum (Case BASE).

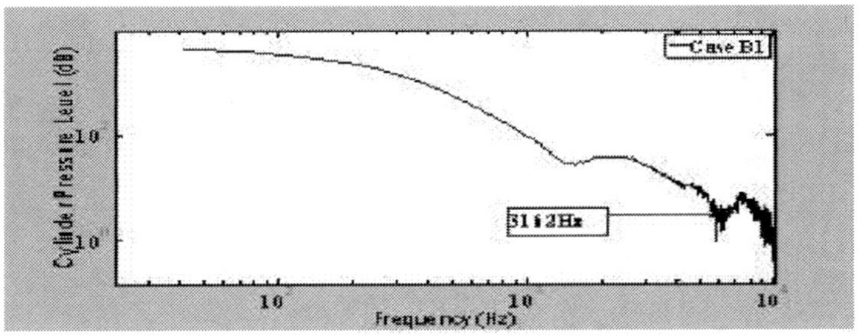

Figure 14. Cylinder pressure spectrum(Case B1).

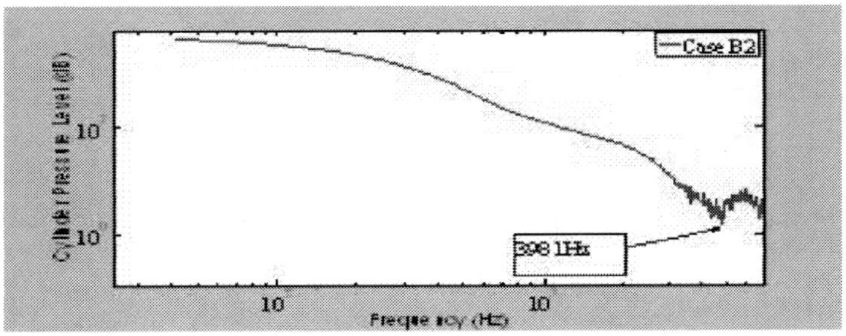

Figure 15. Cylinder pressure spectrum(Case B2).

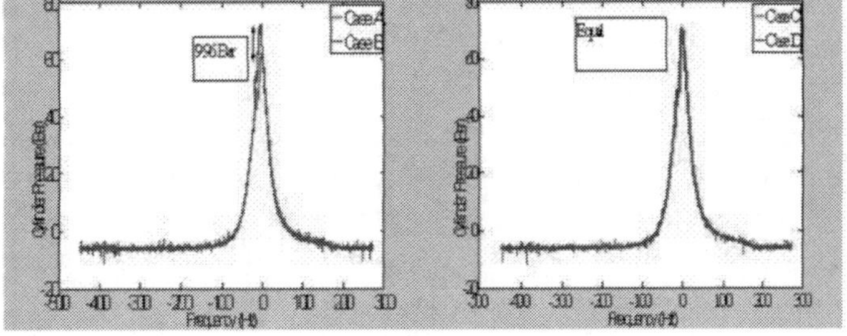

Figure 16. Cylinder pressure after change of parametres.

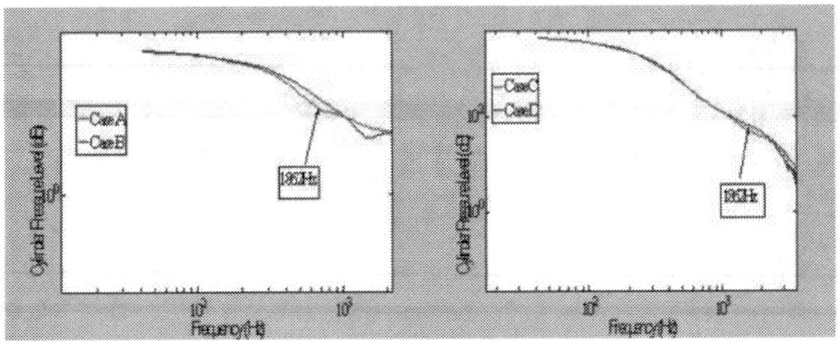

Figure 17. Cylinder pressure ppectrum after change of parametres.

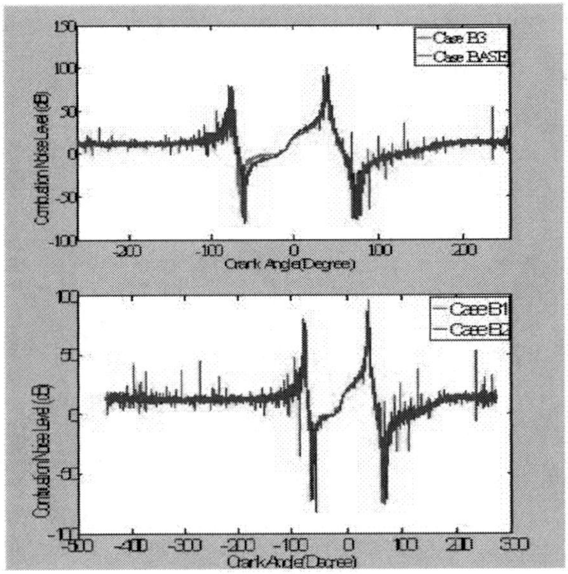

Figure 18. Combustion noise levels for given test conditions.

As seen from figure 15 and 16, differences in maximum pressures for conditions A, B and C, D is 9.96 bar and none respectively. Region 1 of pressure spectrum falls below 1800Hz band. In order to further evaluate the combustion noise, the engine was fired under motored condition. It has been assumed that total air borne noise (ON) is sum of combustion noise(CN) and motored engine noise (MN)as reported in [11]. i.e.

$$ON=MN+CN,$$

where MN-engine noise under motored condition, CN-Combustion noise as elaborated in works [12]. Peak values of both combustion as well as mechanical noise occurs at near TDC position, hence it is difficult to separate both of these components. Bearing these conditions in mind, combustion noise for given test conditions was evaluated and results are plotted as seen in figure 17. As seen from these plots, there are two peaks corresponding to double injection events during engine cycle.

5. Conclusion

It is evident from plots that increase in pre-injection causes an increase in combustion noise levels. In order to reduce emissions from engine combustion control should be optimized. This work presented here allows comparison of noise emitted from engines by varying various testing parameters. The combustion noise was seen to be function of rate of heat release and load on engine. Based on both parameters a loop system can be developed for control of the centre of combustion. The correlation thus obtained can be used to stop or start the pre injection system for optimizing combustion noise keeping pollution emissions at minimum levels.

Nomenclature

SOI-Start of injection
TDC-Top dead centre
QMain -Amount of fuel injected per stroke in pilot(pre)injection
Prail -Common rail injection pressure
QPre-Amount of fuel injected per stroke in pilot(pre)injection
EGR-Exhaust gas recirculation
ROHR-Rate of heat release
FFT-fast Fourier transformations

References

[1] Spontl Moro and Ravagllioli 2012 Combustion noise real time evaluation and processing for combustion control *Proc., 4th ASME* I.C. Engines Division Torino Italy
[2] Mendez S and Thirouard B 2009 Using multiple injection strategies in diesel combustion:Potential to improve emissions, noise and fuel economy trade off in low CR engines'*SAE International Journal of fuel lubricfation* **1** 662-674 doi-10.4271/2008-01-1329
[3] Yun Hanho and Reitz 2005 Combustion optimization in low temperature diesel combustion regime *International Journal of Engine Research*
[4] Heywood John B (Internal Combustion Engines Fundamentals)
[5] Schaberg W and Priede T 1990 Effects of a rapid pressure rise on engine vibration and noise *SAE International paper* 900013 doi:10,4271/900013
[6] Carsten Baumgarten 2006 Mixture formation in internal combustion engines *Springer-Verlag Berlin Heidelberg New York*
[7] Fessler H Langride S Eckhardt T and Gstrein W 2003 Prospects for the diesel engine with stricter emission laws *9th Symp., The Working Process of the Internal Combustion Engine* Institute for Internal Combustion Engines and Thermodynamics Graz University of Technology 1-26
[8] Gang Sheng 2012 Vehicle noise, Vibration, and Sound quality *SAE international* **3**
[9] Andersson and Mcckelvey 2012 The Torque ratio concept for combustion monitoring of internal combustion engines *Elsevier Journal of Control Engineering Practice* 561-568
[10] Jung Jin and Kwangmim Won 2013 An advanced method for developing combustion noise through the analysis of diesel combustion Comprehensive Combustion *SAE International Journal of Engines* 2013-01-1901
[11] Arndt and Brandl 2001 Comprehensive combustion noise optimization *SAE International paper* 2001-10-1510 doi:10,4271/2001-014-1510
[12] Saad and El Sabai 1999 Combustion noise predication inside diesel engine *SAE International paper* 1999-01-1774 doi:10,4271/1999-01-1774

Large amplitude free vibrations of Timoshenko beams at higher modes using coupled displacement field method

K.Krishna Bhaskar
Department of Mechanical Engineering, University College of Engineering, JNTUK, Kakinada, Andhra Pradesh, India.
E-mail: krisbhaskar22@yahoo.co.in

K.Meera Saheb
Department of Mechanical Engineering, University College of Engineering, JNTUK, Kakinada, Andhra Pradesh, India.
E-mail: meera.aec@gmail.com

Abstract. A simple but accurate continuum solution for the shear flexible beam problem using the energy method involves in assuming suitable single term admissible functions for the lateral displacement and total rotation. This leads to two non-linear temporal differential equations in terms of the lateral displacement and the total rotation and are difficult, if not impossible, to solve to obtain the large amplitude fundamental frequencies of beams as a function of the amplitude and slenderness ratios of the vibrating beam. This situation can be avoided if one uses the concept of coupled displacement field where in the fields for lateral displacement and the total rotation are coupled through the static equilibrium equation. In this paper the lateral displacement field is assumed and the field for the total rotation is evaluated through the coupling equation. This approach leads to only one undetermined coefficient which can easily be used in the principle of conservation of total energy of the vibrating beam at a given time, neglecting damping. Finally, through a number of algebraic manipulations, one gets a nonlinear equation of Duffing type which can be solved using any standard method. To demonstrate the simplicity of the method discussed above the problem of large amplitude free vibrations of a uniform shear flexible hinged beam at higher modes with ends immovable to move axially has been solved. The numerical results obtained from the present formulation are in very good agreement with those obtained through finite element and other continuum methods for the fundamental mode, thus demonstrating the efficacy of the proposed method. Also some interesting observations are made with variation of frequency Vs amplitude at different modes.

1. Introduction

The classic work of Woinowsky Krieger [1] on the large amplitude vibrations of slender beams inspired many a researcher to develop simple continuum formulations [2-4] and finite element formulations [5, 6] with some simplifying assumptions. However, large amplitude vibrations of shear flexible beams have not been received much attention till recently except in the work of Rao and Raju [7] who have used the simplified finite element formulation. Introduction of the effects of shear deformation and rotary inertia in the formulation yields coupled nonlinear differential equations. Further the oscillations being non-harmonic with two independent variables, namely, the lateral deflection and total rotation, solution of such complicated equations by assuming both the spatial and temporal distribution is rather difficult. On the other hand, if the two variables appearing in the formulation are coupled to make the final equation a single non-linear equation, then the solution becomes much simpler. The reason being, in this system of single non-linear equation, assuming suitable admissible functions in space can eliminate the space variable and the resulting temporal equation can be solved using any standard method. Another interesting part of the present work is that unlike researchers arrived at finite element analogues based on the continuum theories, the authors are making an attempt to arrive at a continuum analogue of the recent coupled finite element for mutation [8, 9]. The purpose of the present paper is to demonstrate the efficacy of the proposed method by applying to the large amplitude free vibrations of a uniform, shear flexible hinged-hinged and clamped-clamped beams in conjunction with the coupled displacement theory and the principle of conservation of total energy. The results are obtained in closed form expressions of the non-linear to linear radian frequency ratios for simply supported and clamped beams at various modes for various slenderness and maximum amplitude ratios, and are obtained using the harmonic balance method [10]. Comparison of the present results with those existing in literature for the first mode for shear flexible beams shows that the present results obtained from the concept of coupled displacement field analysis are accurate.

2. Shear deformable beam theory for static equilibrium used for coupled displacement field

The kinematics of a shear flexible beam theory (Timoshenko beam theory) can be written as

$$\bar{u}(x,z) = z\theta, \tag{1}$$

$$\bar{w}(x,z) = w(x,z), \tag{2}$$

where $\bar{u}$ and $\bar{w}$ are the axial and transverse displacements at a generic point of the beam, z is the distance of the generic point from the neutral axis, w represents the transverse displacement and θ represents the total rotation anywhere on the beam axis and x, z are the independent spatial variables. The axial strain and shear strain are given by

$$\varepsilon_x = z\frac{\partial \theta}{\partial x}, \tag{3}$$

$$\gamma_{xz} = \frac{\partial w}{\partial x} + \theta, \tag{4}$$

Now, the expressions for the strain energy U and potential energy of the externally applied loads W are given by

$$U = \frac{EI}{2}\int_0^L \left(\frac{\partial w}{\partial x}\right)^2 dx + \frac{kGA}{2}\int_0^L \left(\frac{\partial w}{\partial x} + \theta\right)^2 dx, \tag{5}$$

$$W = \int_a^b p(x)w(x)dx, \tag{6}$$

where EI is the flexural rigidity, GA is shear rigidity, k is the shear coefficient (taken as 5/6 in the present study), p(x) is the static load per unit length on the beam and L is its length. Applying the principle of minimization of total potential energy, as

$$\delta(U - W) = 0, \tag{7}$$

the following static equilibrium equations can be obtained [11]

$$EI\frac{\partial^2\theta}{\partial x^2} - kGA\left(\frac{\partial w}{\partial x} + \theta\right) = 0, \tag{8}$$

$$kGA\left(\frac{d^2w}{dx^2} + \frac{d\theta}{dx}\right) + p = 0. \tag{9}$$

Equations (8) and (9) are coupled equations and can be solved for obtaining the solution for static analysis of shear deformable beams. A close observation of equation (8) shows that it is dependent on the load term p and equation (9) is independent of the load terms p. Hence equation (9) is used to couple the bending rotation θ and the transverse displacement w, so that the two independent variable problem become a single variable problem and the resulting large amplitude free vibration problem becomes much simpler.

3. Coupled displacement field theory

In this theory, we assume an admissible function θ for hinged-hinged beam which satisfies all the essential boundary conditions and symmetric condition, in the beam domain, and evaluate the lateral displacement w distribution using equation (9), θ distribution along the length of the beam is assumed as

$$\theta = a\frac{n\pi}{L}\cos\frac{n\pi x}{L}, \tag{10}$$

where a is the central lateral displacement of the beam and n is the mode number Equation (9) can be rewritten as

$$\frac{\partial w}{\partial x} = -\theta + \frac{EI}{kGA}\theta'', \tag{11}$$

$$\theta'' = -a\left(\frac{n\pi}{L}\right)^2\cos\frac{n\pi x}{L}, \tag{12}$$

Substituting the function θ and θ'' in equation (11), we obtain the displacement field for w, after integration as

$$w = \lambda a \sin\frac{n\pi x}{L}, \tag{13}$$

$$\lambda = \frac{\frac{-n\pi}{L}}{\left(1 + \left(\frac{n\pi x}{L}\right)^2 \frac{EI}{kGA}\right)}. \tag{14}$$

It may be noted here that because of the coupled field theory w, the transverse displacement distribution, containing the same undetermined coefficient a as for θ distribution, w satisfies all the essential boundary and symmetric conditions and given by

$$w(0) = w(L) = \left(\frac{\mathrm{d}w}{\mathrm{d}x}\right)_{x=L/2} = 0. \tag{15}$$

4. Large amplitude vibrations

Large amplitude vibrations can be studied, once the coupled displacement field for the derived lateral displacement w, for an assumed θ distribution is evaluated using the principle of conservation of total energy at any instant of time, which states that

$$U + T + W = constant, \tag{16}$$

The expression for U, T and W are given by

$$U = \frac{EI}{2}\int_0^L (\frac{\partial w}{\partial x})^2 dx + \frac{kGA}{2}\int_0^L (\frac{\partial w}{\partial x} + \theta)^2 dx, \tag{17}$$

$$T = \frac{\rho A}{2}\int_0^L \overset{\bullet}{w}^2 dx + \frac{\rho I}{2}\int_0^L \overset{\bullet}{\theta}^2 dx, \tag{18}$$

$$W = \frac{T_a}{2}\int_0^L \frac{I}{2}\left(\frac{\partial w}{\partial x}\right)^2 dx, \tag{19}$$

where U is the strain energy, T is the kinetic energy, W is the work done by the tension developed because of large amplitudes (deformations), ρ is the mass density, A is the area of cross - section, I is the moment of inertia, T_a is the tension developed in the beam because of large deformations, and ($^\bullet$) denotes differentiation with respect to time. T_a is obtained following Woinowsky Krieger [1] as

$$T_a = \frac{EI}{2Lr^2}\int_0^L \left(\frac{\partial w}{\partial x}\right)^2 dx, \tag{20}$$

where r is the radius of gyration. It is to be noted here that w in Equation (20) does not contain shear flexible terms, and T_a is evaluated in terms of $(q = a/r)$. Substituting the expressions for θ and w (obtained from the coupled field), the expression for U, T and W are obtained as

$$U = \frac{EI\ a^2}{4L}(n\pi)^2\left[1 + \frac{(n\pi)^2 EI}{L^2 kGA}\right], \tag{21}$$

$$T = \frac{\rho A L^3}{4(n\pi)^2}\overset{\bullet}{a}^2\left\{\left[1 + \frac{(n\pi)^2 EI}{L^2 kGA}\right]^2 + \frac{I}{A}\frac{(n\pi)^2}{L^2}\right\}, \tag{22}$$

$$W = \frac{EI}{32r^2}\left(\frac{n\pi}{L}\right)^2 L\ a^4\left[1 + \frac{(n\pi)^2 EI}{L^2 kGA}\right]. \tag{23}$$

Substituting the expressions for U, T and W in equation (16) and simplifying, noting that $I = Ar^2$, we get

$$\overset{\bullet}{q}^2 + \alpha_1 q^2 + \alpha_2 q^4 = constant \tag{24}$$

,

$$\text{where } \alpha_1 = \frac{EI(\frac{n\pi}{L})^4\left[1 + \frac{(n\pi)^2 E}{kG\beta^2}\right]}{\rho A\left\{\left[1 + \frac{(n\pi)^2 E}{kG\beta^2}\right]^2 + \frac{(n\pi)^2}{\beta^2}\right\}}, \tag{25}$$

$$\alpha_2 = \frac{EI(\frac{n\pi}{L})^4\left[1 + \frac{(n\pi)^2 E}{kG\beta^2}\right]^2}{\rho A\left\{\left[1 + \frac{(n\pi)^2 E}{kG\beta^2}\right]^2 + \frac{(n\pi)^2}{\beta^2}\right\}}. \tag{26}$$

In equations (25) and (26) E and G can be eliminated by the standard relation,

$$G = \frac{E}{2(1+v)}, \tag{27}$$

where ν is the Poisson ratio, and $\beta = L/r$, the slenderness ratio of the beam. Using the Harmonic balance method (HBM) [10] as discussed in the next section can solve equation (23).

5. Harmonic balance method

The direct numerical integration method (DNI) [3], proposed by the first author equation (24) can be solved to the desired degree of accuracy. However, for an elegant closed form solution one can advantageously use the HBM which is briefly discussed in this section. Differentiating equation (24), we get

$$\ddot{q} + \alpha_1 q + 2\alpha_2 q^3 = 0. \tag{28}$$

This is the famous Duffing equation and is solved by assuming

$$q = q_m \sin \omega_{NL} t. \tag{29}$$

Where ω_{NL} is the non-linear radian frequency and q_m is the maximum amplitude ratio a_m/r. From equations (29) and (27) and after simplification, and noting that, by neglecting the higher harmonics of ω_{NL} and using the approximate relation $\sin^3 \omega_{NL} t \approx \frac{3}{4} sin \omega_{NL} t$ we get

$$\omega^2{}_{NL} = \alpha_1 + \alpha_2 q^2{}_m. \tag{30}$$

From equation (29), If $q_m = 0$, i.e, for linear vibrations, linear radian frequency,

$$\omega_L = \omega_{NL} \ , \omega^2{}_L = \alpha_1. \tag{31}$$

From equations (25), (26) and (30), after simplification we obtained as,

$$\left(\frac{\omega_{NL}}{\omega_L}\right)^2 = 1 + \frac{3}{16}\left[1 + \frac{(n\pi)^2 E}{kG\,\beta^2}\right]\left(\frac{a_m}{r}\right)^2. \tag{32}$$

Equation (33) is an elegant form to calculate the ratios of $\omega_{NL}^2/\omega_L{}^2$for various values of amplitude and slenderness ratios of the beam. In terms of the Poisson ratio, equation (33) can be written as

$$\left(\frac{\omega_{NL}}{\omega_L}\right)^2 = 1 + \frac{3}{16}\left[1 + \frac{3.12(n\pi)^2}{\beta^2}\right]\left(\frac{a_m}{r}\right)^2, \tag{33}$$

For very large β i.e. for slender beams, where shear deformation can be neglected, equation (34) becomes

$$\left(\frac{\omega_{NL}}{\omega_L}\right)^2 = 1 + \frac{3}{16}\left(\frac{a_m}{r}\right)^2, \tag{34}$$

which is a standard result [4]. In case of a clamped-clamped beam, a single term trigonometric admissible function for θ that satisfies the required boundary conditions is taken as

$$\theta = a\frac{2n\pi}{L}\cos\frac{2n\pi x}{L}, \tag{35}$$

and by following the similar procedure as discussed for hinged-hinged beam the ratios of the nonlinear to the linear radian frequencies can be obtained for the clamped-clamped beam, as

$$\left(\frac{\omega_{NL}}{\omega_L}\right)^2 = 1 + \frac{3}{64}\left[1 + \frac{12.48(n\pi)^2}{\beta^2}\right]\left(\frac{a_m}{r}\right)^2. \tag{36}$$

The relations for calculating the fundamental frequency parameter for both the cases was given in Table 1.

Table 1. Fundamental frequency parameter for hinged-hinged and clamped-clamped beams

	Fundamental frequency $(\lambda) = \frac{\rho A \omega_L^2 L^4}{EI}$
Hinged-hinged beam	$\frac{(n\pi)^4\left[1+\frac{3.12(n\pi)^2}{\beta^2}\right]}{\left\{\left[1+\frac{3.12(n\pi)^2}{\beta^2}\right]^2+\frac{(n\pi)^2}{\beta^2}\right\}}$
Clamped-clamped beam	$\frac{16(n\pi)^4\left[1+\frac{12.48(n\pi)^2}{\beta^2}\right]}{3\left\{\left[1+\frac{12.48(n\pi)^2}{\beta^2}\right]^2+\frac{(2n\pi)^2}{3\beta^2}\right\}}$

6. Numerical results and discussion

Using the formulation described above, the large amplitude behavior of a uniform shear flexible hinged-hinged and clamped-clamped beams are obtained in terms of ω_{NL}/ω_L (ratio of non-linear radian frequency to the linear radian frequency) in terms of various a_m/r (maximum amplitude ratios) and L/r (slenderness ratios). As a demonstration of the proposed formulation, this beam is considered with axially immovable ends (Fig.1) for hinged-hinged beam. Table 1 shows expressions for linear frequency parameters for hinged-hinged and clamped-clamped beams. Numerical results in the form of linear frequency parameters for various slenderness ratios are given in Table 2 and Table 3 at different modes with Poisson ratio of 0.3 for both hinged-hinged and clamped-clamped beams. From these tables linear frequencies are increasing with mode number. The present results in terms of ω_{NL}/ω_L are presented in Table 4 and Table 5 respectively for hinged-hinged and clamped-clamped beams for fundamental mode. For the sake of comparison and validation of the proposed method, the same results obtained by the finite element method (FEM) [7] and by using the direct numerical integration method (DNI) [3] are also included in Table 4 and Table 5 for hinged-hinged case for Poisson ratio 0.3 at mode 1. Similar results are presented in Table 6 and Table 7 for hinged-hinged and clamped-clamped beams at various modes. The present results for the extreme cases of L/r = 10 and a_m/r =5 match very well with the results of FEM and DNI. Further, the present results match excellently with those of Ref. [4] for slender beams, showing the efficacy of the proposed method and the accuracy of results over a range of L/r = 10 to L/r = 100. Then both shear deformable beam and slender beams can be solved using the present formulation. Further, no shear locking phenomenon [12] exists in the present formulation.

7. Conclusions

A coupled displacement field formulation presented in this paper, to study the variation of large amplitude vibrations of a uniform shear flexible hinged-hinged and clamped-clamped beams at higher modes with axially immovable ends. This classical problem is chosen as a demonstration problem, to show the efficacy of the formulation. The increase of frequency ratio is considerable at higher modes and this is more considerable in hinged-hinged beams than clamped-clamped beams. The results obtained using the present formulation are found to be in good agreement with those results obtained by FEM and DNI for a wide range of L/r ratios, including the slender beams, and a_m/r ratios. The preliminary studies on the shear flexible beams with other boundary conditions, with axially immovable ends, give promising results, which indicate the generality of the present formulation. Further the present method proposed does not exhibit shear locking phenomenon commonly encountered in the formulations of shear flexible beams.

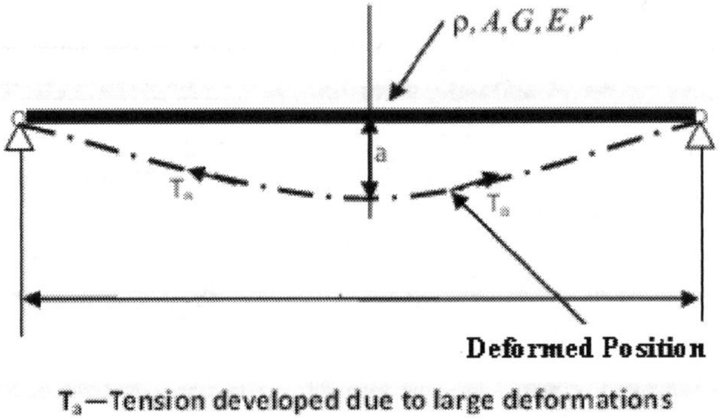

Figure 1. A shear flexible hinged-hinged beam undergoing large deformations

Table 2. Values of non-dimensional linear frequency parameter ($\sqrt{\lambda}$) with shear for hinged-hinged beam

β	Mode					
	1	DNI [13]	2	3	4	5
10	8.3971	–	25.4525	44.398	63.4843	82.367
25	9.5742	9.567	35.3385	71.5827	113.7518	159.0779
50	9.7984	9.79	38.2968	83.1529	141.3539	209.9281
100	9.8575	9.849	39.1934	87.3235	153.1874	235.4498

Table 3. Values of non-dimensional linear frequency parameter ($\sqrt{\lambda}$) with shear for clamped-clamped beam

β	Mode					
	1	DNI [13]	2	3	4	5
10	15.0685	–	37.1896	58.7957	79.9787	100.9363
25	20.6962	20.61	67.3388	121.7498	177.3149	232.4349
50	22.2157	22.17	82.7847	168.9496	269.3554	376.7119
100	22.6574	22.63	88.8629	193.8516	331.1387	493.8748

Table 4. ω_{NL}/ω_L for a uniform shear flexible hinged-hinged beam at mode 1

a_m/r	β									
	10	25			50			100		
	Present Study	Present Study	FEM [7]	DNI [3]	Present Study	FEM [7]	DNI [3]	Present Study	FEM [7]	DNI [3]
0.0	1	1	1	1	1	1	1	1	1	1
0.2	1.0049	1.0039	-	1.0039	1.0037	-	1.0038	1.0037	-	1.0038
0.4	1.0194	1.0156	1.0156	1.0156	1.0150	1.0150	1.0151	1.0149	1.0149	1.0149
0.6	1.0432	1.0348	-	1.0347	1.0336	-	1.0335	1.0333	-	1.0332
0.8	1.0756	1.0610	1.0605	1.0608	1.0589	1.0585	1.0588	1.0584	1.0585	1.0582
1.0	1.1159	1.0939	1.0925	1.0933	1.0907	1.0897	1.0902	1.0899	1.089	1.8940
2.0	1.4075	1.3367	1.3213	1.3313	1.3263	1.3142	1.3212	1.3237	1.3125	1.3186
3.0	1.7910	1.6645	1.6146	1.6501	1.6456	1.6052	1.6318	1.6409	1.6257	1.6272
4.0	2.2191	2.0366	-	2.0116	2.0092	-	1.9850	2.0023	-	1.9783
5.0	2.6706	2.4328	-	2.3968	2.3969	-	2.3619	2.3878	-	2.3531

Table 5. ω_{NL}/ω_L for a uniform shear flexible clamped-clamped beam at mode 1

a_m/r	β									
	10	25			50			100		
	Present Study	Present Study	FEM [7]	DNI [3]	Present Study	FEM [7]	DNI [3]	Present Study	FEM [7]	DNI [3]
0	1	1	1	1	1	1	1	1	1	1
0.2	1.0021	1.0011	–	1.0011	1.001	–	1.001	1.0009	–	1.0009
0.4	1.0083	1.0045	1.039	1.0045	1.0039	1.0039	1.0039	1.0038	1.0035	1.0037
0.6	1.0187	1.0101	–	1.01	1.0088	–	1.0088	1.0085	–	1.0084
0.8	1.0329	1.0178	1.0153	1.0178	1.0156	1.0142	1.0156	1.0151	1.014	1.0149
1	1.051	1.0277	1.0238	1.0276	1.0243	1.0228	1.0242	1.0235	1.0218	1.0231
2	1.1911	1.1066	1.0895	1.1058	1.094	1.08	1.0933	1.0908	1.0834	1.0892
3	1.3935	1.2268	1.1855	1.2239	1.2011	1.1771	1.1987	1.1946	1.1759	1.1902
4	1.6354	1.3777	–	1.3711	1.3368	–	1.3313	1.3264	–	1.3178
5	1.9017	1.5502	–	1.5389	1.4932	–	1.4836	1.4786	–	1.4647

Table 6. ω_{NL}/ω_L for a uniform shear flexible hinged-hinged beam for modes 2,3,4 and 5

a_m/r	n=2				n=3			
	β							
	10	25	50	100	10	25	50	100
0	1	1	1	1	1	1	1	1
0.2	1.0083	1.0045	1.0039	1.0038	1.0141	1.0054	1.0042	1.0038
0.6	1.0727	1.0396	1.0348	1.0336	1.1201	1.0476	1.0368	1.0341
0.8	1.126	1.0694	1.0611	1.059	1.2053	1.0832	1.0646	1.0599
1	1.1911	1.1066	1.094	1.0908	1.3067	1.1273	1.0992	1.0921
2	1.6354	1.3777	1.3368	1.3264	1.9571	1.4432	1.354	1.3307
3	2.1835	1.7379	1.6645	1.6457	2.7144	1.8537	1.6955	1.6536
4	2.7746	2.1428	2.0366	2.0092	3.5101	2.309	2.0815	2.0207
5	3.3861	2.5714	2.4328	2.3969	4.3231	2.7871	2.4915	2.412

a_m/r	n=4				n=5			
	β							
	10	25	50	100	10	25	50	100
0	1	1	1	1	1	1	1	1
0.2	1.022	1.0067	1.0045	1.0039	1.0321	1.0083	1.0049	1.004
0.6	1.1834	1.0587	1.0396	1.0348	1.26	1.0727	1.0432	1.0357
0.8	1.3083	1.1021	1.0694	1.0611	1.4299	1.126	1.0756	1.0627
1	1.4533	1.1556	1.1066	1.094	1.6224	1.1911	1.1159	1.0963
2	2.3341	1.5303	1.3777	1.3368	2.7438	1.6354	1.4075	1.3445
3	3.3179	2.0047	1.7379	1.6645	3.9609	2.1835	1.791	1.6785
4	4.335	2.5233	2.1428	2.0366	5.2071	2.7746	2.2191	2.057
5	5.3667	3.0636	2.5714	2.4328	6.4655	3.3861	2.6706	2.4594

Table 7. ω_{NL}/ω_L for a uniform shear flexible clamped-clamped beam for modes 2,3,4 and 5

a_m/r	n=2				n=3			
	β							
	10	25	50	100	10	25	50	100
0	1	1	1	1	1	1	1	1
0.2	1.0055	1.0017	1.0011	1.0010	1.0113	1.0026	1.0014	1.0010
0.4	1.0220	1.0067	1.0045	1.0039	1.0444	1.0104	1.0054	1.0042
0.6	1.0488	1.0150	1.0101	1.0088	1.0973	1.0231	1.0121	1.0093
0.8	1.0853	1.0265	1.0178	1.0156	1.1674	1.0408	1.0214	1.0165
1	1.1305	1.0411	1.0277	1.0243	1.2518	1.0631	1.0333	1.0257
2	1.4533	1.1556	1.1066	1.0940	1.8077	1.2330	1.1273	1.0992
3	1.8714	1.3247	1.2268	1.2011	2.4703	1.4733	1.2685	1.2119
4	2.3341	1.5303	1.3777	1.3368	3.1735	1.7554	1.4432	1.3540
5	2.8196	1.7597	1.5502	1.4932	3.8953	2.0621	1.6407	1.5172

a_m/r	n=4				n=5			
	β							
	10	25	50	100	10	25	50	100
0	1	1	1	1	1	1	1	1
0.2	1.0192	1.0039	1.0017	1.0011	1.0294	1.0055	1.0021	1.0012
0.4	1.0749	1.0155	1.0067	1.0045	1.1129	1.0220	1.0083	1.0049
0.6	1.1618	1.0345	1.0150	1.0101	1.2397	1.0488	1.0187	1.0110
0.8	1.2735	1.0605	1.0265	1.0178	1.3980	1.0853	1.0329	1.0194
1	1.4041	1.0931	1.0411	1.0277	1.5784	1.1305	1.0510	1.0302
2	2.2104	1.3339	1.1556	1.1066	2.6393	1.4533	1.1911	1.1159
3	3.1213	1.6593	1.3247	1.2268	3.7978	1.8714	1.3935	1.2457
4	4.0673	2.0290	1.5303	1.3777	4.9863	2.3341	1.6354	1.4075
5	5.0285	2.4228	1.7597	1.5502	6.1876	2.8196	1.9017	1.5915

Acknowledgments
The authors are highly thankful to authorities of University College of Engineering(A), JNTUK-Kakinada for extending the necessary support for publishing the paper.

References

[1] Woinowsky and Krieger S 1950 The effect of an axial force on the vibration of hinged bars *Journal of Applied Mechanics* **17** 35-36
[2] Srinivasan A V 1965 Large amplitude free oscillations of beams and plates *AIAA J* **3** 1951-1953
[3] Rao G V and Raju K K 2002 A direct numerical integration method to study the large amplitude vibration of slender beams with immovable ends *Journal of Institute of Engineers (I)* **83** 42-44
[4] Rao G V and Raju K K 2003 Large amplitude free vibrations of beams an energy approach *ZAMM* **83** 493-498
[5] Chuh Mei 1972 Non-linear vibrations of beams by matrix displacement method *AIAA J* **10** 355-357
[6] Rao G V, Raju I S and Raju K K 1976 Finite element formulation for the large amplitude free vibrations of beams and orthotropic circular plates *Computers and Structures* **6** 169-172
[7] Rao G V, Raju I S and Raju K K 1976 Non-linear vibrations of beams considering shear deformation and rotary inertia *AIAA J* **14** 685-687
[8] Singh G and Rao G V 2000 Shear flexible finite elements retrospect and prospects *Journal of Institution of Engineers (I)* **81** 12-19
[9] Raveendranath P, Singh G and Rao G V 2001 A three noded shear flexible curved beam element based on coupled displacement field interpolation *International Journal of Num. Meth. Engg* **51** 85-102
[10] Azrar L and white R G 1999 A semi - analytical approach to the nonlinear dynamic response problem of s-s and c-c beams at large vibration amplitudes. Part1: General theory and application to the single mode approach to free and forced vibration analysis *Journal of Sound and Vibration* **224** 183-207
[11] Reddy J N 1993 *An introduction to the finite element method* (McGraw-Hill Inc, New York)
[12] Prathap G and Bhashyam G R 1982 Reduced integration and the shear flexible beam element *International Journal of Num. Meth. Engg* **18** 195-210
[13] Rao G V, Meera sahib K and RangaJanardhana G 2006 Concept of coupled displacement field for large amplitude free vibrations of shear flexible beams *Journal of Vibration and Acoustics* **128**

Effect of Link Flexibility on tip position of a single link robotic arm

E Madhusudan Raju[1], L Siva Rama Krishna[1],YSharath Chandra Mouli[1] and VNageswara Rao[1]

[1]Department of Mechanical Engineering, University College of Engineering, Osmania University, Hyderabad, Telengana, India 500007

E-mail: ettaboina@gmail.com, lsrkou@gmail.com, ysharathchandramouli@gmail.com and vnageswararao.ou@gmail.com

Abstract. The flexible robots are widely used in space applications due to their quick response, lower energy consumption, lower overall mass and operation at high speed compared to conventional industrial rigid link robots. These robots are inherently flexible, so that the kinematics of flexible robots can't be solved with rigid body assumptions. The flexibility in links and joints affects end-point positioning accuracy of the robot. It is important to model the link kinematics with precision which in turn simplifies modelling of dynamics of flexible robots. The main objective of this paper is to evaluate the effect of link flexibility on a tip position of a single link robotic arm for a given motion. The joint is assumed to be rigid and only link flexibility is considered. The kinematics of flexible link problem is evaluated by Assumed Modes Method (AMM) using MAT LAB Programming. To evaluate the effect of link flexibility (with and without payload) of robotic arm, the normalized tip deviation is found for flexible link with respect to a rigid link. Finally, the limiting inertia for payload mass is found if the allowable tip deviation is 5%.

1. Introduction

The Conventional rigid-link robots are widely used in industrial automation. However, to obtain high accuracy in the end-point position control of these robots, the weight to payload ratio of the robot must be high and the operation speed is normally quite slow. At the same time, large power supply requirements and thus considerable energy consumption is inevitable to operate these heavy-weight robots. These drawbacks greatly limit the applications of rigid-link robots. Flexible link robots have potential advantages, such as faster operation, low energy consumption, and higher load-carrying capacity for the amount of energy expended by using light-weight flexible link manipulators over rigid manipulators. The Light weighted flexible link robots, being inherently flexible, can no longer be modelled with rigid body assumptions. During the motion of such robots or due to external disturbances, the links vibrate. These vibrations must be damped out or controlled to achieve required tasks. There are two commonly used methods to discretize an infinite number of degrees of freedom systems into a finite dimensional system, Namely, Assumed Mode Method and Finite Element Method. The finite element method may require a fewer number of operations but the resulting stiffness of the link is over-estimated which intern results in unstable closed-loop responses of the original manipulator. Whereas assumed mode method is accurate with some complex mathematical formulations and consumes

Published under licence by IOP Publishing Ltd

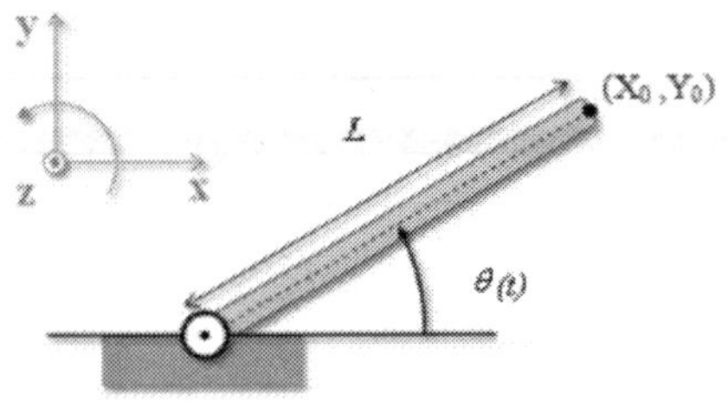

Figure 1. Single rigid link robot arm.

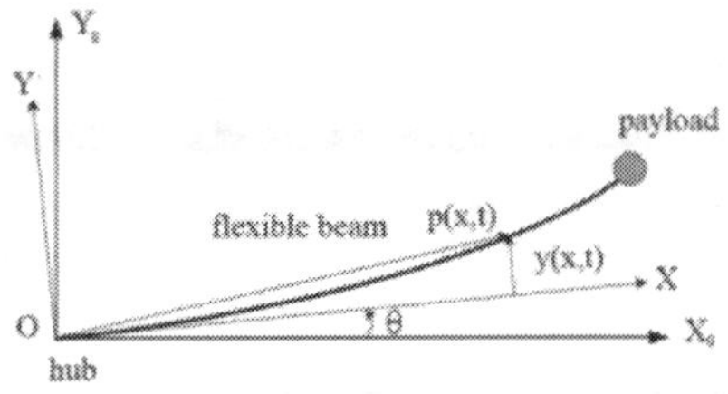

Figure 2. Single flexible link robotic arm.

less time compared to Finite Element Method. The present paper deals with estimating the deviations in the tip position of a single link robotic arm by Assumed Modes Method (AMM) using MATLAB Programming for specified joint motion (Constant Angular Velocity).

2. Literature review

Book et.al.[1] in their paper discussed about the dynamics and control of flexible manipulators.Cetinkunt, and Book [2] in their work discussed the implementation of symbolic and dynamic simulation manipulator with compliant joints and links with recursive Lagrangian-Assumed Modes formulation.The FE modelling impact on One-Link flexible Robotic Arm is illustrated in the work of Chapnik et.al. [3]. Vincentefeliu, et,al. [4] have addressed special class of single-link flexible arms, i.e. flexible mass less structures having masses concentrated at certain points of the beam.

3. Problem definition

The main objective of this paper is to evaluate the effect of link flexibility on tip position of a single link robotic arm with and without payload for a specified joint motion. Firstly tip positions are found for rigid link and using Assumed Modes Method the flexible links tip positions were found. As a case study when a joint is subjected to constant angular velocity i.e. θ (t)=0.5t radians for a simulation time of 12 sec with step size of 0.1 sec, the tip deviations with respect to rigid link were found and compared.The Assumptions made for modelling a flexible link with Assumed mode method (AMM).

- The flexible link with uniform density and flexural rigidity is considered.
- Euler-Bernoulli beam is assumed i.e. vibration of beam considered only in transverse direction.
- The deflection of the flexible link is small compared to the length of the link.
- The payload attached to the free tip of the flexible robot is a concentrated mass.

4. Research methodology

Considering a single rigid link robotic arm as shown in the Figure 1, the tip position of the link can be easily expressed by Equation 1.

$$(X_0, Y_0) = (L\cos(\theta t), L\sin(\theta t)) \tag{1}$$

Considering the flexible link robotic arm as shown in the Figure 2, the flexible link undergoes deformation in motion due to the flexibility of the link. One can observe that a point on this link has a deviation y(x,t)from the un-deformed position.Therefore the motion of the point, related to y (x, t), is not completely determined by joint angle θ and it can also be concluded that an infinite number of θ's needed to describe the motion of the entire link. This paper attempts to compute the effect of flexibility on tip position of a single robotic arm at a given time instant for a given joint motion with and without payload using Assumed Mode Method (AMM) and compute the effect of variation of payload on the tip position. Table 1 illustrates the input parameters for modeling the link. The tip position is computed using Equation 1 in MATLAB.

Table 1. Input parameters.

Link input parameters	Link
Length of the link(L)	=0.5m
Width of the Link(W)	=0.04m
Depth of the cross section of link (d)	= 0.002m
Cross Section Area (A)	= w x d m^2
Moment of inertia(I)	= (w x d^3)/12 m^4
Youngs Modulus of the material (E)	= 7x10^4 N/mm^2
Density (\rho)	= 2700 kg/m^3
Payload mass (M\tiny{p})	= 1 kg
Payload Inertia (J\tiny{p})	= 1 to 5 kgm^2
Joint Motion (\theta(t))	=0.5t radians

4.1. Calculation of Tip position with constant angular velocity

4.1.1. Case I: Motion of rigid link without payload at link tip. When the link is subjected to a joint motion, the tip of the single rigid link starts at X_0=0.5m, at t=0 sec and reaches X_0=0.48 m, in 12 sec. The resultant figures in 3 to 5 provide the positions of tip of the link in x and y axis and variation of link length as the joint is moved with constant angular velocity.

4.1.2. Case II: Motion of rigid link with payload at link tip . When the link tip is provided with payload mass and subjected to above motion, as the link is assumed to be rigid there will not be any change in tip positions of the link as shown in figures 3 to 5.

4.1.3. Motion of a Flexible Link The position of link tip is not independent of payload for joint motion of flexible link as in rigid link. This results in complex kinematics for flexible link. One of the widely used methods to solve flexible links is Assumed Mode Method (AMM) Consider a flexible link robot with body attached reference frame XOY and fixed reference frame X_0OY_0 as shown in figure 2.The different notations being used in subsequent sections are enumerated below:

Journal of Physics: Conference Series **662** (2015) 012020 doi:10.1088/1742-6596/662/1/012020

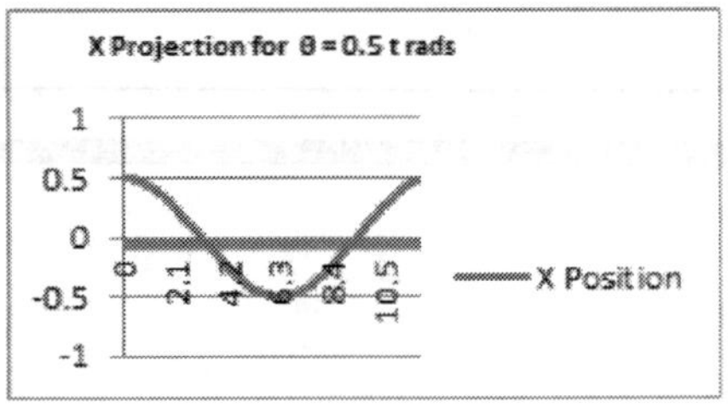

Figure 3. Tip position along X-axis of rigid link.

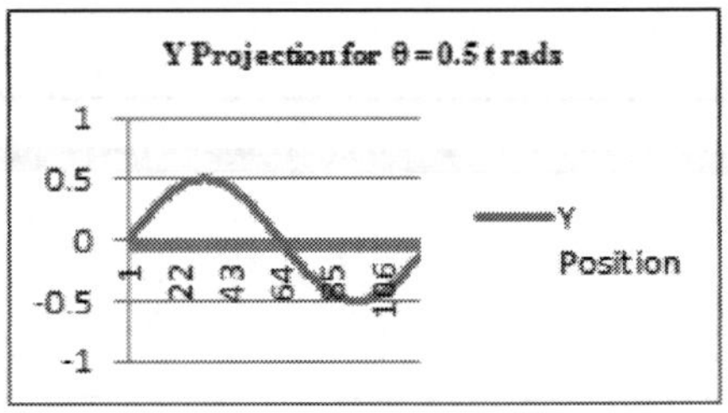

Figure 4. Tip position along Y-axis of rigid link.

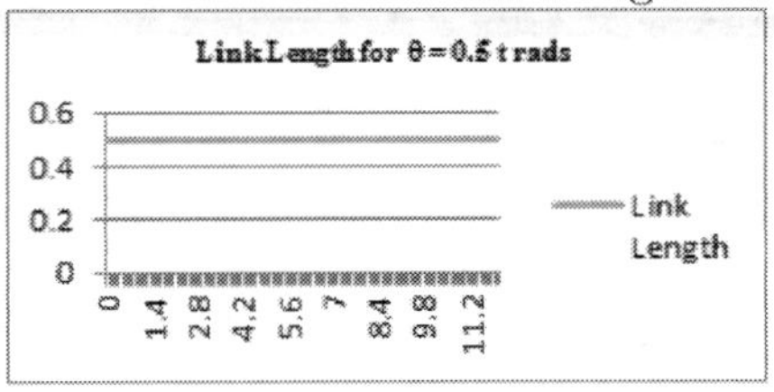

Figure 5. Link Length variation of rigid link

L :Length of the flexible beam.
E :Young's modulus of the flexible beam.
I :Area moment of inertia of flexible beam.
ρ :Density of the flexible beam.
M_L :Mass of the tip payload.
J_L :Inertia of the tip payload.
x :Projection of any point of a deformed link in X axis w.r.t. body attached frame.
y :Projection of any point of a deformed link in Y axis w.r.t. body attached frame.
X_0 :Projection of any point of a deformed link in X axis w.r.t. fixed reference frame.
y_0 :Projection of any point of a deformed link in Y axis w.r.t. fixed reference frame.
$P_0(x,t)$:Position vector of any point on the robot with respect to the fixed reference frame.
$P(x,t)$:Position vector of any point with respect to body attached frame.

When the link undergoes angular motion the tip of the link vibrates which dies down with time. Therefore the position of any point and the tip of the link can be defined as

$$P_0(x,t) = A_R P(x,t) \tag{2}$$

where A_R the rotation matrix.

$$A_R = \begin{bmatrix} \cos(\theta) & -\sin(\theta) \\ \sin(\theta) & \cos(\theta) \end{bmatrix}$$

The position vector of any point on link with body attached frame of reference is given by

$$P(x,t) = \begin{bmatrix} x \\ y \end{bmatrix} = \begin{bmatrix} x \\ y(x,t) \end{bmatrix} \tag{3}$$

The position vector along the length of the link depends on the lateral deformation 'y' of the link at that section at a given time. Value of y can be found by Assumed Mode Method (AMM). The problem of flexible link can be solved assuming it as a Euler- Bernoulli's cantilever beam with a payload 'M_P' at the tip of the beam undergoing free vibration, the governing equation to represent the vibration of link can be written as follows

$$EI\frac{\partial^4 y(x,t)}{\partial x^4} + \rho A\frac{\partial^2 y(x,t)}{\partial t^2} = 0 \tag{4}$$

Boundary Conditions: Since, the equation of motion Eq.4 involves a second order derivative with time and a fourth order derivative with x, two initial conditions and four boundary conditions are needed for finding a unique solution for y(x, t) and they are given in following Equations 5 to 10

$$y(x, t=0) = y_i \tag{5}$$

$$\frac{\partial y(x,t=0)}{\partial t} = \frac{\partial y(x,t)}{\partial t}\bigg|_{t=0} = y_i \tag{6}$$

$$y(x,t)\bigg|_{x=0} = 0 \tag{7}$$

$$\frac{\partial y(x,t)}{\partial x}\bigg|_{x=0} = 0 \tag{8}$$

$$EI\frac{\partial^2 y(x,t)}{\partial x^2}\bigg|_{x=L} = -J_L\frac{\partial^2}{\partial t^2}\left(\frac{\partial y(x,t)}{\partial x}\bigg|_{x=L}\right) \tag{9}$$

$$EI\frac{\partial^3 y(x,t)}{\partial x^3}\bigg|_{x=L} = -M_L\frac{\partial^2}{\partial t^2}\left(y(x,t)|_{x=L}\right) \tag{10}$$

The solution of equation 4 can be expressed as follows

$$y(x,t) = \sum_{j=1}^{n} C_{1,j}\sin(\omega_j t)\left\{(\cos(\beta_j x) - \cosh(\beta_j x)) - \alpha(\sin(\beta_j x) - \sinh(\beta_j x))\right\} \tag{11}$$

where,

$$\alpha = \frac{-\beta_j^3\cos(\beta_j L) - \beta_j^3\cosh(\beta_j L) + \frac{M_L}{\rho}\beta_j^4\sin(\beta_j L) - \frac{M_L}{\rho}\beta_j^4\sinh(\beta_j L)}{-\beta_j^3\sin(\beta_j L) - \beta_j^3\sinh(\beta_j L) - \frac{M_L}{\rho}\beta_j^4\cos(\beta_j L) + \frac{M_L}{\rho}\beta_j^4\cosh(\beta_j L)} \tag{12}$$

$$C_{1,j} = \left[\frac{1}{\int_0^L \phi_i^2(x)dx + \frac{M_L}{\rho}\phi_i^2(L)}\right]^{1/2} \tag{13}$$

On substituting the Equation 12 and 13 in Equation 11 and taking x=L, the transverse deflection of a tip with respect to body fixed frame of the link can be obtained. Using Equation 4.10, deviations in the tip position of flexible link are calculated as a normalized deviation of the tip with respect to link length.

$$\text{Deviation in tip position\%} = \left(\frac{|Deviation|}{\text{Link Length}}\right)100\% \tag{14}$$

4.1.4. Case III: Motion of a Flexible Link with no payload at the link tip It is assumed that link tip is free without any payload, the link is subjected to specified joint motion as follows and the tip position of the link is found and resultants graphs are shown in Figures 6 to 8.

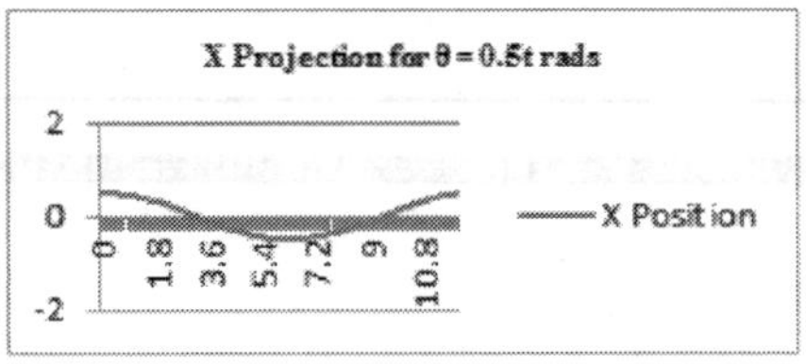

Figure 6. Tip position along X-axis..

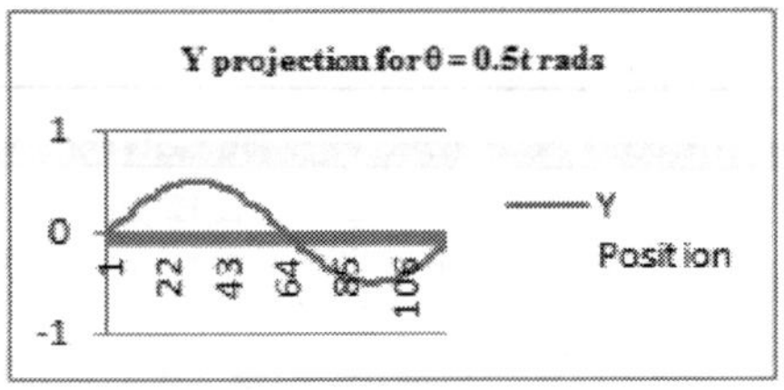

Figure 7. Tip position along Y-axis..

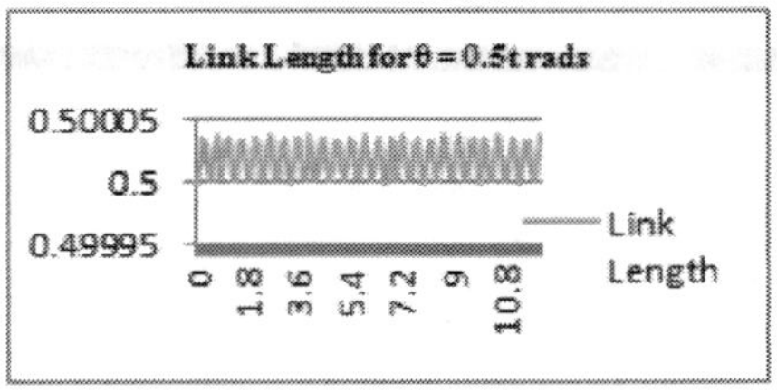

Figure 8. Link Length variations without tip mass

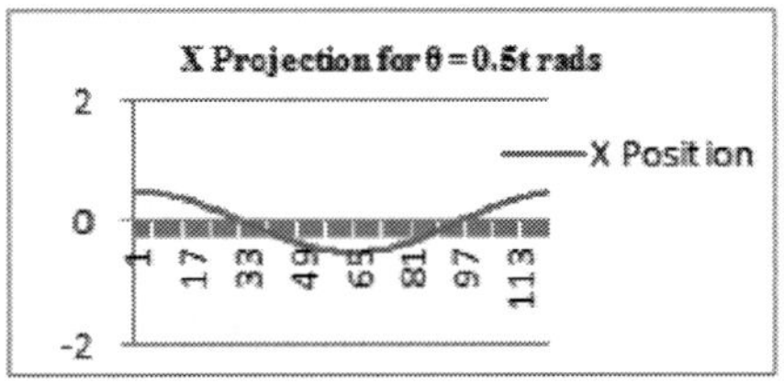

Figure 9. Tip position along X-axis..

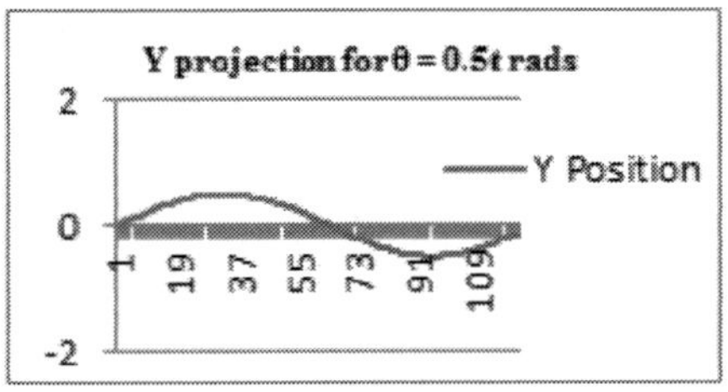

Figure 10. Tip position along Y-axis..

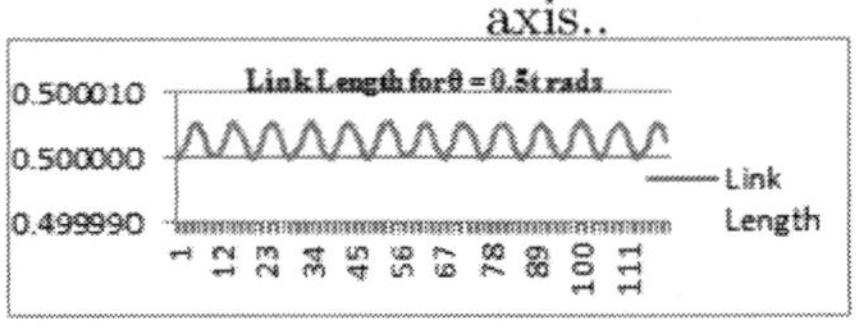

Figure 11. Link Length variations for constant angular velocity

4.1.5. Case IV: Motion of a Flexible Link with payload at the link tip It is assumed that link tip is with a payload M_L=1kg with inertia J_L=1kgm^2 at the tip, the link is subjected to specified joint motion and the tip position of the link is shown in figures 9 to 11.The tip position of the flexible link is found by varying the payload inertia from J_L=1 to 5 kgm^2 with keeping payload mass constant.

5. Results and Discussions

The Figures 1, 2 and 3 gives the percentage deviation in the link length, X Position and Y Position of the flexible link for θ = 0.5t radians with payload no mass and with payload of varying inertia.Form the above figures it can be inferred that the flexible link without the payload has negligible tip deviation. Whereas, with payload of M_L=1kg and maximum payload inertia of J_L=4 kgm^2, the tip deviation is found to be less than 5%.

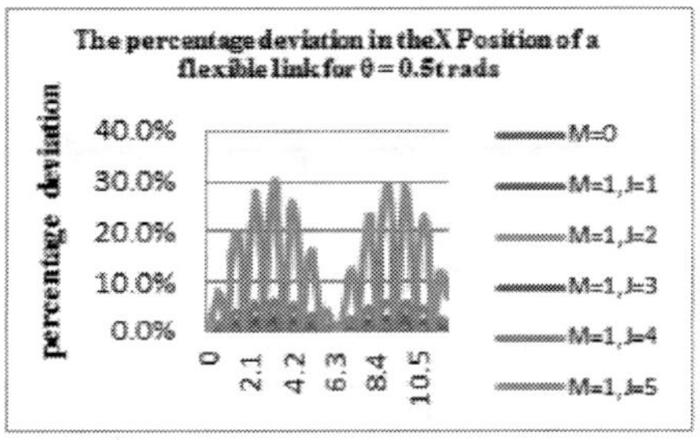

Figure 12. The Percentage deviation in X Position.

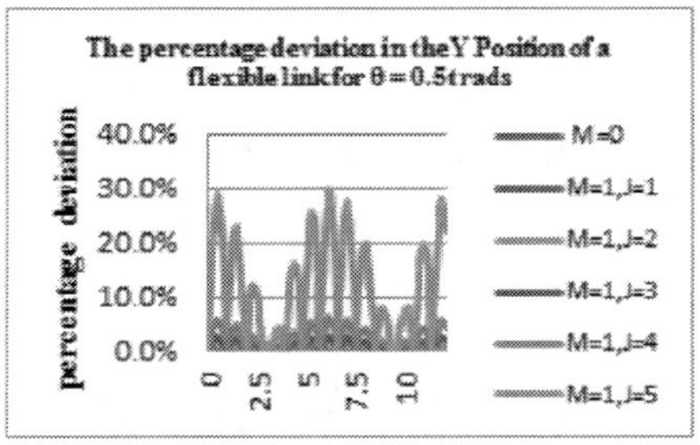

Figure 13. The percentage deviation in the Y Position.

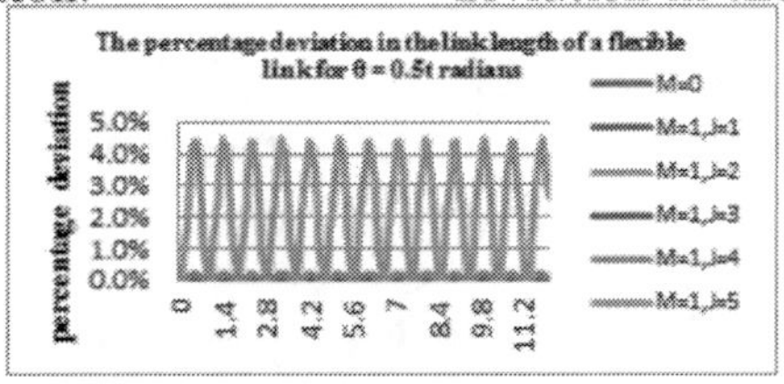

Figure 14. The percentage deviation flexible link length for = 0.5t radians

6. Conclusions

The effective tip positions and link lengths for a flexible link with and without tip payload mass for different payload inertias are compared with that of rigid link for constant angular velocity as joint motion. Following are the conclusions drawn from above work

- The link length variation has no effect of flexibility in the case of tip carrying no payload and for the case of link carrying tip payload mass, the deviation in link length is observed to be less than 0.2% till its inertia becomes 4 kgm^2 and a drastic change in link length variation is observed from 5 kgm^2 limiting the allowable payload inertia to 4 kgm^2.

- The maximum percentage deviation in X and Y-position of tip with no payload mass is observed to be less than 1% and that for the case of link carrying tip payload mass, they are less than 5% until its inertia becomes 4 kgm^2 and a drastic change in link length variation is observed from 5 kgm^2 limiting the allowable payload inertia as 4 kgm^2.

References

[1] Book W.J, Maizzo-Neto O and Whiteny D.E 1975 Feedback control of two beam, two joint systems with distributed flexibility *ASME J of Dynamics systems Measurements and Control* 97(4), 424-431

[2] Cetinkunt, S. and Book, W.J. 1989 Symbolic Modeling and Dynamic Simulation of Robotic Manipulators with Compliant Links and Joints *Robotics and Computer-Integrated Manufacturing* 5 (4), 301-310

[3] Chapnik B.V, Hepplerand G.R and Aplevich J.D 1991 Modeling Impact on a One-Link Flexible Robot Arm *IEEE Transaction on Robotics and Automation* 7(4), 479 - 488

[4] VincenteFeliu, Kuldip S Rattan and Benjamin Brown H 1990 Adaptive Control of a Single-Link Flexible Manipulator *IEEE Control systems Magazine* 0272-1708 0200-0029

Thermoelastic steam turbine rotor control based on neural network

Romuald Rzadkowski
Department of Dynamics of Machines, Institute of Fluid Flow Machinery, Polish Academy of Sciences, Gdansk ,Poland.

E-mail: z3@imp.gda.pl

Krzysztof Dominiczak and Wojciech Radulski
Alstom Power LTD., ul. Stoczniowa 2, 82-300 Elblag, Poland.

E-mail: krzysztof.dominiczak@power.alstom.com, wojciech.radulski@power.alstom.com

R.Szczepanik
Air Force Institute of Technology, Warsaw, Poland.

Abstract. Considered here are Nonlinear Auto-Regressive neural networks with eXogenous inputs (NARX) as a mathematical model of a steam turbine rotor for controlling steam turbine stress on-line. In order to obtain neural networks that locate critical stress and temperature points in the steam turbine during transient states, an FE rotor model was built. This model was used to train the neural networks on the basis of steam turbine transient operating data. The training included nonlinearity related to steam turbine expansion, heat exchange and rotor material properties during transients. Simultaneous neural networks are algorithms which can be implemented on PLC controllers. This allows for the application neural networks to control steam turbine stress in industrial power plants.

1. Introduction

Thermal stress control of steam turbines has been presented by Busse [1], Dawson [2], Pahl et al. [3]. All of these stress control systems used thermo-physical start-up probes, which were physical models of the steam turbine rotor. The thermo-physical start-up probe measures two temperatures: that of the steam turbine rotor and the mean rotor temperature. More accurate stress control can be achieved by using mathematical models of steam turbine components. Lausterer [4], Lausterer et al. [5], Ehrsam [6] have presented systems in which the mean temperatures of particular turbine components were calculated using a mathematical model and a start-up probe. Sindelar at al. [7] describe a system which assesses stress in a critical turbine component by using only standard power plant measurements. Rusin at al. [8] have presented a steam turbine stress control system based on Duhamel's integral.

This paper for the first time presents a steam turbine stress control system based on a neural network used for the hot start-up and shut down. While neural networks have already been used in power system diagnostics, e.g. by Gluch et al. [9] [10] for diagnostics of power object geometry deterioration, never before have they been used for temperature and stress modeling in on-line steam turbine control systems. The neural networks presented in this paper are able

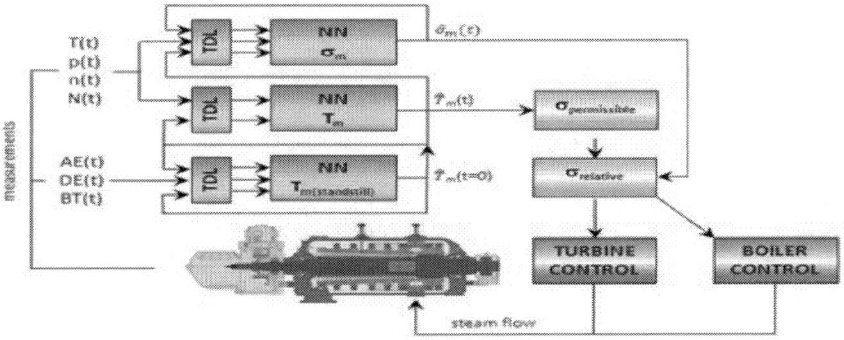

Figure 1. NARX neural network based steam turbine thermal stress control.

to simulate on-line the temperature and stress of a critical turbine component based only on standard power plant measurements, such as speed, power, steam temperature and pressure in front of turbine control valves.

2. Rotor strength modelling with neural network

NARX neural network stress control was implemented on the HP rotor of a 18K390 steam turbine.

2.1. Identification of critical points in a rotor

NARX neural network thermal stress control does not focus on the entire HP rotor, but only on its critical thermal stress points. In order to identify these points, a rotor lifetime assessment (LTA) was performed. This showed that there is only one critical point in the region of the first HP rotor groove.

2.2. Rotor stress control concept using NARX neural

Fig. 1 shows a diagram of the NARX neural network thermal stress control system. Different sets of measurements in the turbine are used in the control system for each period of the steam turbine lifetime.

In our research, seven turbine load cycle regimes were used. In each of these the steam temperature, steam pressure, turbine speed and turbine load were measured. These physical parameters are sufficient to describe rotor boundary conditions during turbine load cycles. Therefore these measurements were chosen as exogenous inputs to neural networks to identify critical point temperature and stress.The neural network steam turbine control system shows actual rotor temperature and stress in the critical point. The control system consists of three NARX neural networks (Fig. 1). The first of these assess critical point temperature during turbine standstill on the basis of the rotor's axial expansion (absolute expansion AE, differential expansion DE, axial bearing float BT). This temperature is used as an initial temperature for the second NARX neural network, which assesses critical point temperature on the basis of turbine speed (n), turbine load (N), steam temperature (T) and pressure (p) before the turbine control valve during turbine start-up. Assuming that rotor is stress free at the beginning of turbine startup, the data concerning critical point temperature, turbine speed, turbine load, steam temperature and pressure before the turbine control valve allows the third NARX network to assess critical point stress.

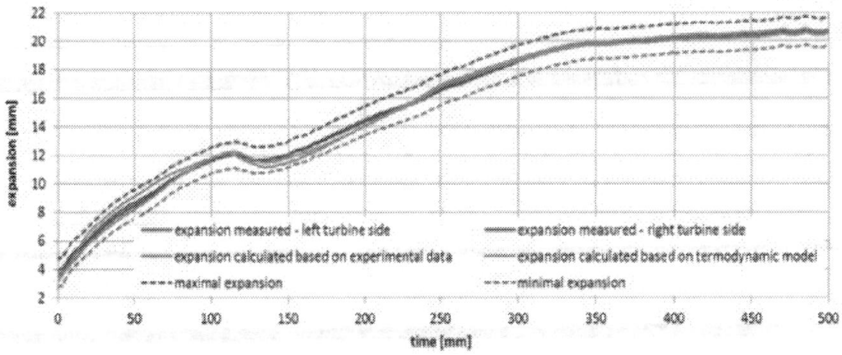

Figure 2. HP rotor axial expansion during cold startup: calculations and measurements.

3. FEM based neural network training

Neural network training is a process which allows to select network weights and biases in order to reflect known output data based on corresponding inputs. Here the neural network training was performed using the Levenberg-Marquardt algorithm, which was possible because of the series-parallel architecture. However, for the NARX neural network the most important factor is the training data.

In order to build a training data set which covers all possible variations of every inputs, 168 FE calculations for transient turbine rotor states were performed. The variation range of each neural network training data set was established on the basis of turbine operational history.

The only experimental possibility of verify the assumed heat exchange and FE models was through expansion of the rotor. Fig. 2 compares the FE analysis with measurements of axial expansion during a cold start up of the turbine. The dotted lines show the expansion, taking into account measurement errors. The results obtained from the numerical calculations and experiments are very close.

4. Neural network testing using experimental data

The HP rotor neural network control system was verified using experimental data. This included each turbine start up category (cold, warm I, warm II, hot), one sliding pressure shut down and one load rejection together with turbine reloading. As a measure of neural network performance, the root mean squared error (RMSE) was used. The maximal RMSE for neural network determined temperature was 5.4 °C and occurred during warm start up I. The other temperatures were: 5.39 °C for load rejection together with turbine reloading, 4.8 °C for cold start up, 4.2 °C for hot start, 4.1 °C for sliding pressure shut down and 2.1 °C for warm start up II.

The maximal RMSE for neural network determined stress was 18.8MPa and occurred during cold start up. The other stress values were: 10,2 MPa, 11,3 MPa, 13,2 MPa, 8MPa and 6.4 MPa. Fig. 3 presents the neural network determined temperature (Tm) and stress (Sig), including relatively errors, in the HP rotor critical location during hot start up. Fig. 3e and 3f show good FE-NET correlation for hot start up. Higher error values were observed in NET determined stress (Fig. 3f). This was because while temperature was determined only on the basis of exogenous inputs and feedback values, stress was not only based on exogenous inputs and feedback values but also on temperatures determined by the other network. Therefore errors in the temperature network affected the stress network (Fig. 3g). Fig. 4 presents neural network determining temperature (Tm) and stress (Sig), with errors, in HP rotor critical location for shut down. Fig. 4e, and 4f show that neural network results for shut down were close to finite element analysis. The relative error for temperature (Fig. 4g) is very small and for stress it is

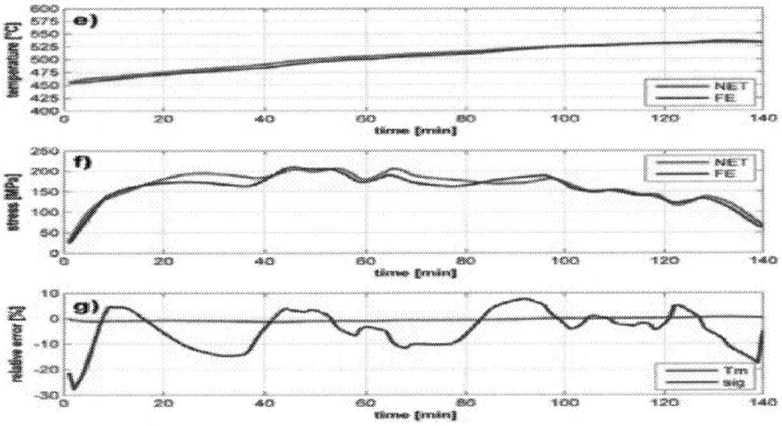

Figure 3. Result of neural network tests for hot start-up.

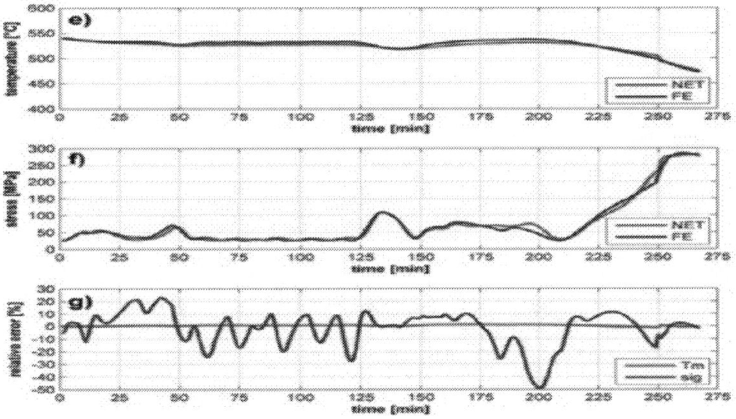

Figure 4. Result of neural network tests for shut down.

no larger than 10% and no lower than -30%.

5. Conclusion

A HP steam turbine stress control system based on a NARX neural network has been presented. The HP rotor neural network control system was verified using experimental data. Neural networks are able to simulate on-line, the temperature and stress of a critical turbine component based only on standard power plant measurements, such as speed, power, steam temperature and pressure in front of turbine control valves. This is a very promising system for controlling various transient thermal stresses of steam turbine and gas turbine rotors. This system can also be applied to control stress in many other power plant components, e.g. boilers, inner and outer casings etc., where it is difficult or impossible to make direct stress measurements on-line. Another advantage of this system are its low hardware requirements, which allow it to be implemented in existing controllers. Temperature and stress analyses show that obtained results from neural networks are very similar to results obtained using the finite element method.The results presented here are for only one critical point (the first groove of an HP rotor), but, after due training, it could be applied in many other points.

Journal of Physics: Conference Series **662** (2015) 012021 doi:10.1088/1742-6596/662/1/012021

References

[1] Busse and L Anfahrsonde 1973 *BBC-Studie* HTGD51147
[2] Dawson 1989 Monitoring and control of thermal stress and component life expenditure in steam turbine England Rugby
[3] Pahl G, Reitze W and Salm 1964 Ueberwachungseinrichtung fuer zulaessige temperaturaenderungen bei dampfturbinen ,*BBC-Nachrichten* **46** 139-147
[4] Lausterer G.K 1997 On-line thermal stress monitoring using mathematical models *Control engineering practice* **5 1** 85-90
[5] Lausterer G.K., Franke J and Eitelberg 1980 Mathematical modeling of a steam generator. Digital computer applications to proccess control *Proc. of 6th IFAC/IFIP International Connference* Duesseldorf
[6] Ehrsam 2011 Steam turbines for solar thermal applications *Proc of ASME Turbo Expo* Vancouver
[7] Sindelar R and Toewe W 2005 TENSOMAX - A retrofit thermal stress monitoring system for steam turbine VGB Power Tech **1** 60-62
[8] Rusin A, Banaszkiewicz M, Lipka M, Lukowicz H and Radulski W 2005 Continuous control and optimization of thermal stresses in the process of turbine start-up *Thermal Stresses Symposium* Vienna University of Technology Vienna
[9] Gluch J and Krzyzanowski J 2004 Application of an for diagnostics of the geometry deterioration of the power system apparatuses *Proc. of ASME Turbo Expo* GT2004-54193 Vienna Austria
[10] Gluch J and Krzyzanowski J 2006 New attempt for diagnostics of the geometry deterioration of the power system based on thermal measurement *Proc. of ASME Turbo Expo 2006* GT2006-90263 Barcelona Spain

Buckling analysis of moderately thick rectangular plates using coupled displacement field method

K.Meera Saheb

Department of Mechanical Engineering, University College of Engineering, JNTUK, Kakinada, Andhra Pradesh, India.

E-mail: meera.aec@gmail.com

K.Aruna

Department of Mechanical Engineering, Ideal Institute of Technology, Kakinada, Andhra Pradesh, India.

E-mail: aruna-kunda23@yahoo.com

Abstract. A simple and efficient coupled displacement field method is developed to study the buckling load parameters of the moderately thick rectangular plates. This method has been successfully applied to study the same for the Timoshenko beams. A single term trigonometric admissible displacement field is assumed for one of the variables, say, the total rotations (in both X, Y directions). Making use of the coupling equations, the spatial variation for the remaining lateral displacement field is derived in terms of the total rotations. The coupled displacement field method makes the energy formulation to contains half the number of unknown independent coefficients, in the case of a rectangular plate, contrary to the conventional Rayleigh-Ritz method. The expressions for the non-dimensional buckling load parameters of the moderately thick rectangular plates with all the edges simply supported are derived. The numerical values of these parameters obtained using the coupled displacement field method match very well with open literature demonstrating the effectiveness of the coupled displacement field method.

1. Introduction

Knowledge of buckling load parameters of moderately thick plates is a necessary design input that has to be considered in the initial design phase. The energy methods provide a convenient means for computing the buckling load parameters and the solutions obtained using this approach are upper bounds and the accuracy of the solution depends on the admissible functions chosen for the lateral displacement and total rotations. The widely used energy method is the classical Rayleigh - Ritz (RR) method where in the displacement field of a structural member is generally approximated by simple trigonometric or algebraic admissible functions. Single or multi-term functions are chosen to satisfy the geometric boundary conditions involved in the problem.

The concept of the coupled displacement field(CDF) which was successfully used in the finite element (FE) structural members analysis [1-3]. A continuum analogue of the FE analysis with the CDF method is not received much attention except in the formulation of Zhou[4] where the two fields (displacement and rotation) are coupled through an equation which is dependent on the applied load. This approach different from the FE method based on the CDF.

A general CDF method is presented here, where the coupling equation is independent of the applied load, to study the buckling and free vibration behavior of uniform Timoshenko beams. The effectiveness of the proposed CDF method is demonstrated successfully by comparing the buckling load and the frequency parameters with those obtained from the RR method [5,6] for the short columns and beams. In the present study, the authors made an attempt to show the applicability of the CDF method to study the buckling of uniform moderately thick rectangular plates.

The proposed CDF method, if generalized to a n term admissible functions, will have n unknown coefficients because of the coupling equation used, where as, the RR method contains 2n unknown coefficients for the square/rectangular plate [12]. On the other hand, if an accurate single term admissible function is used in the CDF method, then a one unknown coefficient problem has to be solved, where as two unknown coefficients are associated in the RR method. Thus the proposed CDF method significantly simplifies the formulation of the buckling problem of moderately thick rectangular plates. In this paper, the coupling equations used are taken from [7]. In general the CDF method reduces the complexity of the problem by a factor two compared to the RR method, as mentioned earlier. The practical utility of the CDF method is demonstrated by solving the buckling problem of the isotropic and uniform moderately thick rectangular plates for all edges simply supported boundary conditions, by using the CDF method. The solution procedure and the numerical results obtained speak for themselves about the simplicity of the CDF method applied to the buckling problem of the moderately thick rectangular plates compared to the RR method. The first order shear deformation theory is briefly given in the following section.

2. First order shear deformation theory of plates

The simplest shear deformation plate theory is the first order shear deformation plate theory (FSDT), also referred to as the Mindlin plate theory [8] where the displacements u,v and w are given by

$$u(x, y, z) = z\theta_x(x, y), \tag{1}$$

$$v(x, y, z) = z\theta_y(x, y), \tag{2}$$

$$w(x, y, z) = w(x, y), \tag{3}$$

where u and v are the inplane displacements in x and y directions, w is the transverse displacement along z direction, θ_x and θ_y denote rotations about the y and x axes respectively. In the FSDT, the shear correction factors are introduced to correct the discrepancy between the actual transverse shear stress distribution and those computed using the kinematic relations of the FSDT. The shear correction factor k depend not only on the geometric parameters, but also on the loading and boundary conditions of the plate. However, a value of k=5/6, the widely used value of the shear correction factor is used in the present study also.

$$U = \frac{D}{2}\int_0^a\int_0^b\left\{\left(\frac{\partial\theta_x}{\partial x}\right)^2 + \left(\frac{\partial\theta_y}{\partial y}\right)^2 + 2\nu\left(\frac{\partial\theta_x}{\partial x}\right)\left(\frac{\partial\theta_y}{\partial y}\right) + 2(1-\nu)\left(\frac{\partial\theta_x}{\partial y}\right)\left(\frac{\partial\theta_y}{\partial x}\right)\right\}dxdy + \frac{kGh}{2}\int_0^a\int_0^b\left\{\left(\frac{\mathrm{d}w}{\mathrm{d}x}+\theta_x\right)^2 + \left(\frac{\mathrm{d}w}{\mathrm{d}y}+\theta_y\right)^2\right\}dxdy, \tag{4}$$

$$W = \frac{N_x}{2}\int_0^a\int_0^b\left(\frac{\partial w}{\partial x}\right)^2 dxdy + \frac{N_y}{2}\int_0^a\int_0^b\left(\frac{\partial w}{\partial y}\right)^2 dxdy, \tag{5}$$

where h is the thickness of the plate, W is the work done because of the biaxial compressive load, N_x is the compressive load per unit length acting on the two edges perpendicular to x axis and N_y is the compressive load per unit length acting on the two edges perpendicular to y axis. Suitable admissible functions satisfying mainly the kinematic boundary conditions (some times the admissible functions may satisfy some or all of the natural boundary conditions and do not violate the variational principles) are assumed for w and θ. In the present study for the sake of simplicity and clarity and easy understanding of the CDF method a single term admissible functions for w and θ are chosen here and is shown that the single term trigonometric admissible functions for the two boundary conditions of the plates considered, given later, are found to be accurate enough for all the practical engineering purposes.

The detailed procedure for CDF method are discussed in this section for the evaluating the buckling load (biaxial and uniaxial compressive load per unit length) of a moderately thick uniform rectangular plate with all edges simply supported for which the exact buckling mode shape for the transverse displacement w is well known(sine waves in the x and y directions).

3. Coupled displacement field (CDF) method

In this method an admissible functions for θ_x and θ_y which satisfies all the geometric boundary conditions on the plate domain are assumed and the field for lateral displacement w is evaluated using the coupling equation, the derivation of which is briefly given below

The expression for the strain energy U is already given in Eq.(4) and the work done by the externally applied lateral loads W_b used in the static analysis is given as

$$W_b = \int_0^a \int_0^b p w dx dy, \tag{6}$$

where p is the lateral load distribution per unit area acting on the plate. Taking the variation of the total potential energy as

$$\delta(U - W_b) = 0, \tag{7}$$

The following static equilibrium equations independent of the externally applied load term [8] are obtained. Note that two coupling equations are obtained for the rectangular plate.

$$\frac{\mathrm{d}w}{\mathrm{d}x} = -\theta_x + \frac{h^2}{3.5}\left[\frac{\partial^2\theta_x}{\partial x^2} + \upsilon\frac{\partial^2\theta_y}{\partial y\partial x}\right] + \frac{h^2}{10}\left[\frac{\partial^2\theta_x}{\partial y^2} + \frac{\partial^2\theta_y}{\partial y\partial x}\right], \tag{8}$$

$$\frac{\mathrm{d}w}{\mathrm{d}y} = -\theta_y + \frac{h^2}{3.5}\left[\frac{\partial^2\theta_y}{\partial y^2} + \upsilon\frac{\partial^2\theta_x}{\partial y\partial x}\right] + \frac{h^2}{10}\left[\frac{\partial^2 y}{\partial x^2} + \frac{\partial^2\theta_x}{\partial y\partial x}\right]. \tag{9}$$

Eqs.(8) and (9) are called as the coupling equations in the CDF method and is used to couple the total rotations θ_x and θ_y and the transverse displacement w, so that the two independent unknown coefficients problem in the RR method reduces to a single unknown coefficient problem in the CDF method. The effectiveness of the CDF method is brought out in the following section.

Though admissible functions θ_x and θ_y can be written in a series form, here a single term admissible functions for θ_x and θ_y is chosen again with the same intention of simplicity and better understanding of the method as

$$\theta_x = \alpha f_1(x, y), \tag{10}$$

$$\theta_y = \alpha f_2(x, y), \tag{11}$$

where α is the undetermined coefficient and $f_i(x,y)$, $i = 1,2$ is the single term admissible function. Note that the functions for θ_x and θ_y are the same as, the rectangular plate is considered in the present study. Substituting the admissible functions for θ_x and θ_y as given in Eq.(10) and (11) in Eq.(8) and (9), the coupled displacement field for the lateral displacement w, after integration is obtained, as

$$w = \alpha f_3(x,y). \tag{12}$$

Note that because of the use of the coupling equation, the transverse displacement distribution w also contains the same undetermined coefficient α as existing in the θ distribution. The lowest buckling load parameter is obtained from the following equation.

$$\frac{d(U-W)}{d\alpha} = 0. \tag{13}$$

In the CDF method the admissible functions for θ_x and θ_y are assumed in the functional form, noting the similarity between $\frac{dw}{dx}$, θ_x and $\frac{dw}{dy}$, θ_y, as

$$\theta_x = \alpha\frac{m\pi}{a}cos\frac{m\pi x}{a}sin\frac{n\pi y}{b}, \tag{14}$$

$$\theta_y = \alpha\frac{n\pi}{b}sin\frac{m\pi x}{a}cos\frac{n\pi y}{b}. \tag{15}$$

Substituting Eqs.(14) and (15) in the coupling Eqs.(8) and (9) and after simplification the slopes are obtained as

$$\frac{dw}{dx} = -\alpha\frac{m\pi}{a}cos\frac{m\pi x}{a}sin\frac{n\pi y}{b}\left\{1+2.8198m^2\left(\frac{h}{a}\right)^2+n^2\left(\frac{h}{b}\right)^2(1.9739+2.8198\upsilon)\right\}, \tag{16}$$

$$\frac{dw}{dy} = -\alpha\frac{n\pi}{b}sin\frac{m\pi x}{a}cos\frac{n\pi y}{b}\left\{1+2.8198m^2\left(\frac{h}{b}\right)^2n^2+m^2\left(\frac{h}{a}\right)^2(1.9739+2.8198\upsilon)\right\}. \tag{17}$$

After integration the lateral displacement field for w, as

$$w = -\alpha sin\frac{m\pi x}{a}sin\frac{n\pi y}{b}\left\{1+2.8198m^2\left(\frac{h}{a}\right)^2+n^2\left(\frac{h}{b}\right)^2(1.9739+2.8198\upsilon)\right\}. \tag{18}$$

Substituting Eqs. (14), (15) and (18) in Eqs. (4) and (5) we get the expression for U and W, as

$$U = \frac{D}{2}\frac{ab}{4}\pi^4\alpha^2\left[\left(\frac{m}{a}\right)^2+\left(\frac{n}{b}\right)^2\right]^2 + \frac{D}{2}\frac{ab}{4}\pi^4\alpha^2h^2 0.35462\left\{7.9512\left[\left(\frac{m}{a}\right)^6+\left(\frac{n}{b}\right)^6\right]+\left[\left(\frac{m}{a}\right)^4\left(\frac{n}{b}\right)^2+\left(\frac{m}{a}\right)^2\left(\frac{n}{b}\right)^4\right](23.8542)\right\}. \tag{19}$$

and

$$W = \frac{N_x}{2}\alpha^2\frac{ab}{4}\pi^2\left(\left(\frac{m}{a}\right)^2\left[1+2.8198m^2\left(\frac{h}{a}\right)^2+n^2\left(\frac{h}{b}\right)^2(2.81984)\right]\right)^2 + \frac{N_x}{2}\alpha^2\frac{ab}{4}\pi^2\left(\left(\frac{n}{b}\right)^2\left[1+2.8198n^2\left(\frac{h}{b}\right)^2+m^2\left(\frac{h}{a}\right)^2(2.81984)\right]\right)^2, \tag{20}$$

where $N_x = N_y$, since the plate is under bi-axial compression load, by minimizing the total potential energy as $(d(U - W)/d\alpha) = 0$, the buckling load parameter is obtained, from one equation only as

$$\lambda = \frac{N_x b^2}{\pi^2 D} = \left\{ \frac{\left[\left(\frac{m}{a}\right)^2 + \left(\frac{n}{a}\right)^2\right]^2 + (1-\nu)\left\{4.0280\left[\left(\frac{h}{a}\right)^2 \frac{m^6}{a^4} + \left(\frac{h}{b}\right)^2 \frac{n^6}{b^4}\right] + \left(\frac{h}{a}\right)^2 \left[\frac{m^4 n^2}{a^2 b^2} + \frac{m^2 n^4}{b^4}\right](12.0844)\right\}}{\left(\frac{m}{ab}\right)^2 \left\{1 + 2.8198 m^2 \left(\frac{h}{a}\right)^2 + n^2 \left(\frac{h}{b}\right)^2 (2.81984)\right\}^2} \right\}. \quad (21)$$

4. Numerical results and discussion

To show the simplicity, effectiveness and ease of the proposed CDF method, buckling problem of the moderately thick uniform rectangular plates with all edges simply supported (Figure.1 and Figure.2) boundary conditions, for different aspect ratios and different loading conditions are given. Single term trigonometric functions are considered to study the influence of the thickness ratio h/b on the buckling load parameters. Table 1 shows values of non dimensional buckling loads for moderately thick square plates under uniaxial and bi axial compressive loads for thin plate. In the same table results obtained by the present method is compared with [7] and the match is very good. Table 2, Table 3 shows values of non dimensional buckling loads for moderately thick rectangular plates under uniaxial compressive loads for different aspect ratios. Table 2 also gives buckling load values for higher modes and for fundamental mode for various plate thickness ratios. Table 2 also shows results obtained by the present method is compared with [7] and the results are matching very closely. Table 4 shows values of non dimensional buckling loads for moderately thick rectangular plates under bi axial compressive loads for different aspect ratios. It is in general observed that non dimensional buckling loads are decreasing with increase of plate thickness ratio.

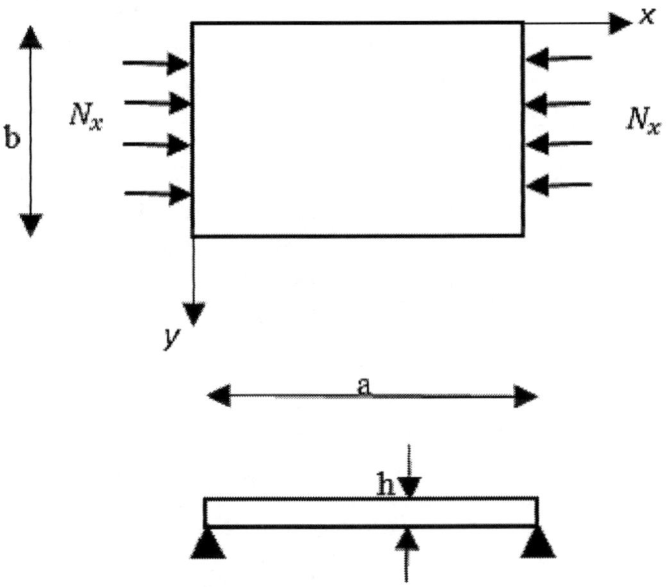

Figure 1. S-S-S-S rectangular plate under uniaxial compression

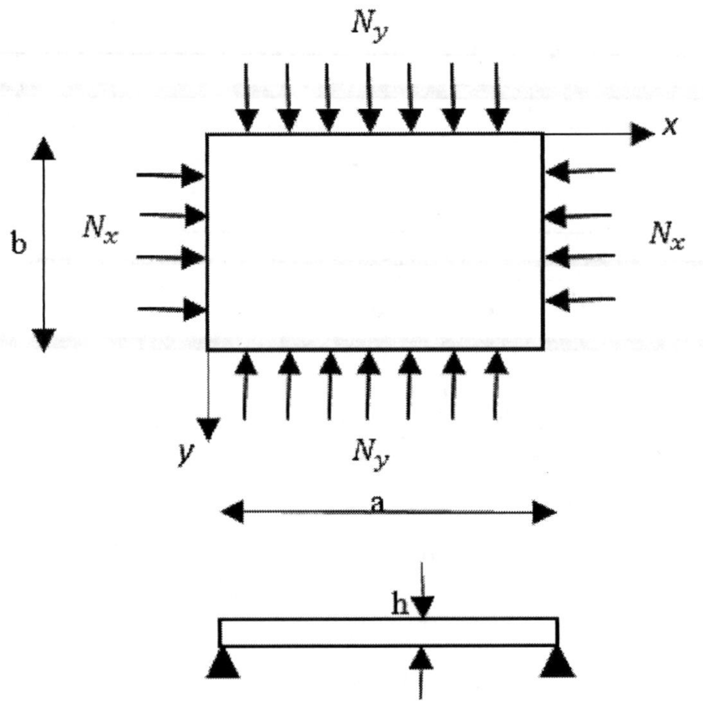

Figure 2. S-S-S-S rectangular plate under biaxial compression

Table 1. Values of non dimensional buckling load parameter λ with shear (h/b) for all edges simply supported moderately thick square plate

γ	a/b	h/b	Present method	Ref[7]
0	1	0	4	4
		0.01	3.9977	3.998
		0.05	3.9443	3.948
		0.1	3.7864	3.8
1	1	0	2	2
		0.01	1.9988	1.9
		0.05	1.972	1.974
		0.1	1.8936	1.999

When γ =0 plate is under uniaxial compression, γ =1 plate is under biaxial compression.

Table 2. Values of non dimensional buckling load parameter λ with shear (h/b) for all edges simply supported moderately thick rectangular plate under uniaxial compression

a/b	h/b	Present method	Ref[7]
0.5	0	6.25	6.25
	0.1	5.4776	5.523
	0.05	6.0369	6.051
	0.01	6.2413	6.25
1.5	0	4.3400 (2,1)	4.34
	0.1 (2,1)	4.0824	4.045
	0.05 (2,1)	4.2734	4.262
	0.01 (2,1)	4.3375	4.337
2	0	4.0000 (2,1)	4
2.5	0	4.1344 (3,1)	4.134
3	0	4.0000 (3,1)	4
	0.1 (3,1)	3.786	3.8
	0.05 (3,1)	3.9446	3.948
	0.01 (3,1)	3.9977	3.998

Numbers in the bracket(m, n) refers at which critical buckling load occurred; (m, n)=(1,1) for all other cases.

Table 3. Values of non dimensional buckling load parameter λ with shear (h/b) for all edges simply supported moderately thick rectangular plate under uniaxial compression

a/b	h/b	Present method	Ref[10]		Ref[11]
			FSDT	HSDT	
0.2	0.5	1.3988	1.3988	1.6851	1.6851
	0.2	6.8755	6.8753	7.0529	7.0529
	0.1	15.6017	15.601	15.658	15.658
	0.05	22.8524	22.851	22.859	22.859
	0.02	26.2698	26.269	26.27	26.27
	0.01	26.8435	26.843	26.843	26.84
0.4	0.5	1.3761	1.3761	1.4455	1.4455
	0.2	4.6264	4.6264	4.6466	4.6466
	0.1	6.9826	6.9824	6.9853	6.9853
	0.05	8.0011	8.001	8.0012	8.0012
	0.02	8.3417	8.3417	8.3417	8.3417
	0.01	8.3934	8.3928	8.3928	8.3928
1	0.5	1.6598	1.6597	1.6759	1.6759
	0.2	3.2633	3.2636	3.2653	3.2653
	0.1	3.7864	3.7864	3.7865	3.7965
	0.05	3.9443	3.9443	3.9443	3.9443
	0.02	3.9909	3.9909	3.9909	3.9909
	0.01	3.9977	3.9977	3.9977	3.9977

Table 4. Values of non dimensional buckling load parameter λ with shear (h/b) for all edges simply supported moderately thick rectangular plate under biaxial compression

a/b	h/b	Present method	Ref[7]
0.5	0.1	4.3821	4.418
	0.05	4.8301	4.841
	0.01	4.993	4.993
	0	5	5
1.5	0.1	1.3878	1.391
	0.05	1.4301	1.431
	0.01	1.4437	1.444
	0	1.4444	1.444
3	0.1	1.0772	1.079
	0.05	1.1025	1.103
	0.01	1.1111	1.111
	0	1.1111	1.111

5. Conclusions

In this paper, a simple and effective CDF method compared with other researchers is proposed to study the buckling of moderately thick rectangular plates. This method because of the use of the static equilibrium equations to couple the total rotations and the transverse displacement, becomes an n undetermined coefficient system when compared to the classical RR method[12] where $2n$ undetermined coefficients are present, thus reducing the computational effort significantly. As a special case, if n=1, which means when a single term admissible function is used, the CDF method contains only one undetermined coefficient which simplifies the solution procedure as has been demonstrated. Numerical results, in terms of the buckling load parameters, for several thickness ratios are presented in this paper for the all edges simply supported boundary condition of the moderately thick rectangular plates show the simplicity of the CDF method when compared to the classical RR method. The results obtained by the other researchers are also compared with the present method wherever possible and are matching very closely.

Acknowledgments
The authors are highly thankful to authorities of University College of Engineering(A), JNTUK-Kakinada for extending the necessary support for publishing the paper.

References

[1] Singh G and Rao G V 2000 Shear flexible finite elements: retrospect and prospects *Journal of the Institute of Engineers (India)* **81** 12-19
[2] Raveendranath P, Sing G and Pradhan B 1999 A two-noded locking-free shear flexible curved beam element *International Journal for Numerical Methods in Engineering* **44** 265-280
[3] Raveendranath P, Sing G and Rao G V 2001 A three-noded locking-free shear flexible curved beam element based on coupled displacement field interpolation *International Journal for Numerical Methods in Engineering* **51** 85-101
[4] Zhou D 2001 Free vibration of multi-span Timoshenko beams using static Timoshenko beam functions *Journal of Sound and Vibration* **241** 725-734
[5] Meera Saheb K, Rao G V, and Rangajanardhana G 2007 Coupled displacement filed formulation for the buckling analysis of shear flexible columns *AIAA J* **33** 413-418
[6] Meera Saheb K, Rao G V and Rangajanardhana G 1998 Free vibration analysis Timoshenko beams using coupled displacement field method *Journal of Structural Engineering* **34** 233-236

[7] Wang C M, Wang C Y and Reddy J N 2005 *Exact solutions for buckling of structural Members* (CRC Press)
[8] Mindlin R D 1951 Influence of rotatory inertia and shear on flexural motions of isotropic, elastic plates *Trans. ASME Journal of Applied Mechanics* **18** 31-38
[9] Shen Hui-shen 1990 Buckling and postbuckling of moderately thick plates *Applied Mathematics and Mechanics* **11** 367-376
[10] Reddy J N and Phan N D 1985 Stability and vibration of isotropic, orthotropic and laminated plates according to higher order shear deformation theory *Journal of Sound Vibration* **98(2)** 157-170
[11] Senthilnathan,Lim N R S P, Lee K H and Chow S T 1987 Buckling of shear deformable plates *AIAA J* **25** 1268-1271
[12] Meerasaheb K, Rao G V and Rangajanardhana G 2009 Buckling and free vibration of moderately thick square plates using coupled displacement field method *Journal of Structural Engineering, India (SERC)* **36** 155-159

Mathematical modeling and multi-criteria optimization of rotary electrical discharge machining process

U Shrinivas Balraj

Department of Mechanical Engineering,
Kakatiya Institute of Technology and Science, Warangal, India.

E-mail: u.shrinivas@gmail.com

Abstract. In this paper, mathematical modeling of three performance characteristics namely material removal rate, surface roughness and electrode wear rate in rotary electrical discharge machining RENE80 nickel super alloy is done using regression approach. The parameters considered are peak current, pulse on time, pulse off time and electrode rotational speed. The regression approach is very much effective in mathematical modeling when the performance characteristic is influenced by many variables. The modeling of these characteristics is helpful in predicting the performance under a given set of combination of input process parameters. The adequacy of developed models is tested by correlation coefficient and Analysis of Variance. It is observed that the developed models are adequate in establishing the relationship between input parameters and performance characteristics. Further, multi-criteria optimization of process parameter levels is carried using grey based Taguchi method. The experiments are planned based on Taguchi's L9 orthogonal array. The proposed method employs single grey relational grade as a performance index to obtain optimum levels of parameters. It is found that peak current and electrode rotational speed are influential on these characteristics.Confirmation experiments are conducted to validate optimal parameters and it reveals the improvements in material removal rate, surface roughness and electrode wear rate as 13.84%, 12.91% and 19.42% respectively.

1. Introduction

RENE80 nickel super alloy is extensively used in aerospace industry due to its high hardness, high strength and creep properties. This super alloy is difficult to machine due to high hardness, low thermal conductivity and high affinity to react with the tool materials at high temperature generated during machining [1]. Hence it requires a non-traditional machining such as electrical discharge machining (EDM). But it has very low metal removal rate (MRR) [2]. The solution to this problem is providing rotation to the electrode. It utilizes the rotary electrode that enhances the flushing of debris formed during machining due to forced circulation of dielectric in the machining zone. This results in better stability of the process. [3-6]. Consequently, rotary EDM (REDM) is adopted for machining RENE80 nickel super alloy in the present study.

Researchers in the past have focused their work on mathematical modeling and performance of EDM/REDM. Pradhan et al. [7] developed mathematical models relating to various significant machining parameters and different machining criteria such as MRR, TWR and overcut during micro-drilling of titanium super alloy by EDM. The models were tested at 95%, 90% and 75% confidence levels and found to be satisfactory. Mohan et al. [3] analyzed the effect of various

REDM parameters on responses like MRR, tool wear rate and SR while machining Al-SiC metal matrix composites. The authors reported that increasing the speed of the rotation of the electrode had positive effect on MRR, tool wear rate and SR than stationary electrode. Aliakbari and Baseri [4] performed REDM of mould steel and found that depending on the geometry of the electrode, the rotational speed of the electrode had different effects on MRR, SR and EWR. Wang et al [5] optimized the blind-hole drilling of $Al_2O_3/6061Al$ composite on REDM using Taguchi methodology. Lin et al [6] pointed out that electrical parameters have significant effect on all considered responses, whereas speed of electrode rotation had significant effect only on MRR. Kao et al. [8] dealt with multi-objective optimization of parameters for machining titanium alloys using GRA and the results showed considerable improvements in MRR, SR and EWR.

It is observed from the literature that limited work has been reported on mathematical modeling and multi-criteria optimization of REDM for RENE80 nickel super alloy. Hence this paper presents the development of mathematical modeling of MRR, SR and using regression approach and multi-criteria optimization of parameters using GRA based Taguchi method.

2. Methodology

2.1. Taguchi method

It is very popular and effective to deal with responses influenced by many parameters and is a systematic approach to determine the optimal process parameters. It mainly focuses on minimization of variation of the response of interest [9]. Further, it reduces the number of experiments drastically and saves time and cost of experimentation. In this method, signal to noise (S/N) ratio is used to measure the deviation of the response from the mean value. There are two types of S/N ratios i.e. Lower-the-better and Higher-the-better types given by Eqns.(1) and (2) respectively,

$$\eta = -10\log_{10}\left[\frac{1}{n}\sum_{i=1}^{n}\frac{1}{y_i^2}\right], \tag{1}$$

$$\eta = -10\log_{10}\left[\frac{1}{n}\sum_{i=1}^{n}y_i^2\right], \tag{2}$$

where η denotes S/N ratio of experimental values, y_i represents experimental value of the ith experiment an n is the total number of experiments.

2.2. Regression approach

Regression analysis is widely used in manufacturing processes as a statistical tool for investigation of relationships between input parameters and output responses. Usually, the researchers are interested to find the causal effect of one input variable on output response for example, the effect of process parameter on machining performance. It is also important to evaluate the 'statistical significance' of the developed mathematical model i.e. the degree of confidence that the true existing relationship among variables is close to the estimated mathematical model. The regression model commonly used is given by,

$$Y = f(A, B, C \quad and \quad D). \tag{3}$$

here Y denotes the response variable like MRR, SR, TWR etc. f is the response function and A, B, C and D are process parameters. In the present study, A, B, C and D are peak current (I), pulse on time (T_{on}), pulse off time (T_{off}) and rotational speed of electrode (S). The general non-linear quadratic model consisting of only linear and quadratic effects is given by the following equation [Montgomery, 2001],

$$Y = \beta_0 + \beta_1 A + \beta_2 B + \beta_3 C + \beta_4 D + \beta_5 A^2 + \beta_6 B^2 + \beta_7 C^2 + \beta_8 D^2 + \epsilon \tag{4}$$

where $\beta_0, \beta_1 ... \beta_8$ are regression coefficients of process parameters and ϵ is the experimental error. β_0 is the free term of the regression model.$\beta_1, \beta_2, \beta_3$ and β_4 are linear coefficients and $\beta_5, \beta_6, \beta_7$ and β_8 are quadratic coefficients. The regression models and regression coefficients are established based on the experimental values of various performance characteristics.

2.3. Grey relational analysis

The grey system theory was proposed by Deng [10] and has evolved gradually to solve certain problems with complicated interrelationships among multiple responses of variety of machining processes. In this, linear normalization of data is usually required in the range 0-1. The optimum parameter levels for complicated multiple responses are converted into optimization of single GRG i.e. this grade is taken as the performance index for multi-criteria optimization [15]. For Larger-the-better type, normalization of the response i is given by Eq.(5),

$$Y_{ij}^* = \left[\frac{y_{ij} - min_j(y_{ij})}{max_j(y_{ij}) - min_j(y_{ij})}\right], \tag{5}$$

where, $Y_{ij}{}^*$ is the normalized value in the jth experiment for $j = 1, 2, ..., q$
For Smaller-the-better type, normalization of the response i is given by Eq.(6),

$$X_{ij}^* = \left[\frac{max_j x_{ij} - x_{ij})}{max_j(x_{ij}) - min_j(x_{ij})}\right]. \tag{6}$$

Later, grey relational coefficient ξ_{ij} is found out by using Eq. (7)

$$\xi_{ij}^* = \left[\frac{min_i min_j(\Delta_{ij}) + \zeta max_i max_j(\Delta_{ij})}{\Delta_{ij} + \zeta max_i max_j(\Delta_{ij})}\right], \tag{7}$$

where, ζ is distinguishing coefficient and lie in the range 0 to 1. Generally, it is taken as 0.5[9].
$\Delta_{ij} = |X_{ij}^* - R|$ and $R = MaxX_{ij}^*$
Now, find the GRG-γ_j for each jth experiment using Eq.(8),

$$\gamma_j = \frac{1}{m}\left(\sum_{i=1}^{m}\sum_{j=1}^{q} w_i \xi_{ij}\right), \tag{8}$$

where, w_i is the weight for i^{th} response and m is the number of responses, $0 < \gamma_j < 1$ and $\sum_{i=1}^{m} w_i = 1$.

3. Experimental details

Pure electrolytic copper of diameter 14.3 mm is used as the electrode. The study material is RENE80 nickel super alloy with hardness 45HRC and density $8.2g/cm^3$ Its chemical composition (wt%) is: Al 5-6; Cr 9.5-12; Ti 2.5-3.2; C 0.13-0.2; Mo 3.5-4.8; W 4.5-5.5; Co 4-4.5; B 0.02 max; Se 0.015 max; Si<0.4; Mn < 0.4; Fe 0.5 max; Ni balance. The rotary setup is fabricated to impart the rotary motion to the electrode and fastened to the ram of die-sink EDM machine (Make -Askar Microns, Model V3525). The work piece dimensions are $70 \times 35 \times 4mm^3$.The work piece and electrode are connected to the positive and negative polarities of power supply respectively. The machining time of each experiment is set for three minutes. The experiments

Table 1. L9 array and experimental results of MRR, SR and EWR.

	Level of parameters				Experimental results		
S.No	I (Amps)	T_{on} (μs)	T_{off} (μs)	S (rpm)	MRR (mg/min)	SR (um)	EWR (mg/min)
1	6	10	10	100	65.035	2.953	12.266
2	6	20	20	300	120.718	3.166	9.833
3	6	50	50	500	176.812	2.903	5.533
4	15	10	20	500	480.495	3.366	60.900
5	15	20	50	100	433.104	3.700	54.050
6	15	50	10	300	467.426	3.833	46.780
7	24	10	50	300	604.060	4.533	73.670
8	24	20	10	500	668.507	4.433	87.866
9	24	50	20	100	618.815	5.633	74.433

are repeated three times at constant gap voltage of 30V for each parametric combination to reduce experimental error and average values of responses are taken for analysis. MRR and EWR are measured by using Eq.9,

$$MRR/EWR = \frac{(W_1 - W_2)}{T}, \tag{9}$$

where W_1 = weight of the work piece/electrode before machining (mg), W_2 = weight of the work piece /electrode after machining (mg), T = machining time (minutes). The weights are measured by digital balance of accuracy of 1mg. SR is measured using surface roughness tester.

The majority of published literature indicates that parameters like peak current, pulse on time and pulse off time are dominant. Also, the comparative study between standard EDM and rotational EDM showed that rotational EDM is superior to standard EDM in machining performance. Hence rotational speed of electrode is taken as the fourth parameter.The selection of an orthogonal array depends on the number of degrees of freedom (DOF). The DOF are equal to number of parameter levels −1. A particular array is selected if the DOF are equal to or less than the DOF of that array. In this paper, four parameters with three levels are selected. Also the pilot experimentation indicates that the interaction among these parameters is negligible. Thus, total DOF are eight. Hence an orthogonal array L9 is selected as this array can has maximum DOF equal to eight. The values of the parameters are taken based on the experimental region and the machine specifications.

4. Results and discussion

The selected L9 orthogonal array based on the Taguchi method and the average experimental values of three trials of MRR, SR and EWR are given in Table 1.

4.1. Mathematical models

Mathematical models are developed for MRR, SR and EWR based on experimental results using statistical software MINITAB15 and are given in Table 2. The validation of developed models is done based on the value of correlation coefficient (R^2 value). It can be seen that R^2 values of these models are more than 95% and hence they are adequate. Another way of checking the adequacy of the model is to carry out ANOVA of these models. The summary of ANOVA results of all models including linear and quadratic terms are shown in Table 2. According to ANOVA, if the calculated F-ratio value is greater than the tabulated F-ratio value at considered

Table 2. Developed mathematical models and their ANOVA.

S.No	Response	Mathematical model	R^2	$R^2_{adj.}$	F-value: Mathematical model	F-value: Linear terms	F-value: Quadratic terms
1	MRR	- 291 + 59.7 A + 3.90 B+ 0.0282 D + - 1.05 A*A - 0.0493 B*B +0.000243 D*D	99.9	99.8	2179.9***	6327.23***	237.07***
2	SR	2.92 - 0.0093 A + 0.0172 B - 0.00113 D + 0.00375 A*A -0.000075 B*B - 0.000000 D*D	96.6	86.40	9.83*	28.56**	0.88
3	EWR	- 4.68 + 1.48 A + 0.763 B- 0.0744D + + 0.0541 A*A - 0.0199 B*B + 0.000169 D*D	99.7	98.7	104.43***	302.5***	10.86*

*** significant at 99% confidence level, **significant at 95% confidence level, *significant at 90% confidence level. $F_{0.01,6,2} = 99.3$, $F_{0.05,6,2} = 19.3$, $F_{0.10,6,2} = 9.33$, $F_{0.05,3,2} = 19.2$, $F_{0.10,3,2} = 9.16$, $F_{0.01,3,2} = 99.3$

Table 3. Normalized values, Grey relational coefficients and GRG of MRR, SR and EWR.

	Normalized values			Grey relational coefficient			
S.No	*MRR*	*SR*	*EWR*	*MRR*	*SR*	*EWR*	*GRG*
1	0.000	0.982	0.918	0.333	0.965	0.859	0.719
2	0.092	0.904	0.948	0.355	0.838	0.905	0.700
3	0.185	1.000	1.000	0.380	1.000	1.000	0.793
4	0.688	0.830	0.328	0.616	0.747	0.426	0.596
5	0.610	0.708	0.411	0.562	0.631	0.459	0.551
6	0.667	0.659	0.499	0.600	0.595	0.500	0.565
7	0.893	0.403	0.172	0.824	0.456	0.377	0.552
8	1.000	0.440	0.000	1.000	0.472	0.333	0.602
9	0.918	0.000	0.163	0.859	0.333	0.374	0.522

confidence level, then the model is considered as statistically significant [7]. The mathematical models for MRR and EWR are strongly significant at 99% confidence level and that of SR is significant at 90% confidence level.

4.2. Multi-criteria optimization of process parameter levels using GRA

The normalized values of responses are measured by using Eqs. 5 and 6 and are shown in Table 3 along with the ideal sequence of value 1. In this study, all the responses are equally weighted i.e.0.33 as all three responses are considered with equal importance. Based on Eqns.7 and 8, GRG for each experiment is calculated. The results of grey relational coefficients, GRGs

Table 4. ANOVA of GRGs.

Parameter	Degrees of freedom	Sum of squares	Variance	F-ratio	Contribution(%)
I	2	0.0598	0.0299	$498.7^{a,b}$	86.73
$T_{on}^{\sharp}$	2	0.00012	0.00006		-0.17
T_{off}	2	0.0012	0.0006	10.01^{b}	1.75
S	2	0.00783	0.0039	$65.28^{a,b}$	11.35
$Error^{\sharp}$	2	0.00012	0.00006		
Total	8	0.0690			100

a significant at 95% confidence level, $F_{0.05,2,2} = 19$,
b significant at 90% confidence level, $F_{0.10,2,2} = 9$,
pooled parameter

Table 5. Results of confirmation experiments.

Initial data	Optimal process parameters			
Response	$I_1T_{on2}T_{off2}S_2$	Prediction $I_1T_{on1}T_{off3}S_3$	Experiment $I_1T_{on1}T_{off3}S_3$	% improvement
MRR	120.718	-	137.433	13.84%
SR	3.166	-	2.580	18.50%
EWR	9.833	-	7.923	19.42%
GRG	0.7097	0.807		13.71%
	Overall average $GRG = 0.6222$			

and their ranks are given in Table 3. The results show that experiment number 3 has the highest GRG. Hence, it can be expected that levels of each process parameter are superior to attain a better multiple responses. ANOVA results of GRGs are shown in Table 4. It depicts that the peak current and electrode rotational speed have percentage contributions as 86.73% and 11.65% respectively. The response graph of the GRGs is shown in Figure 1. From this graph, optimum combinational levels of parameters are identified as $I_1T_{on1}T_{off3}S_3$.Further the confirmation experiments are carried out at optimum parameter levels $I_1T_{on1}T_{off3}S_3$. The results are compared with one of the experiments in orthogonal array $I_1T_{on2}T_{off2}S_2$ [8]. The additive model is used to evaluate the predicted GRG. The corresponding results are given in Table 5. It can be observed that the GRG for multiple responses improved from 0.7097 to 0.807 i.e. 13.71%. MRR increased from 120.718 mg/min to 137.433 mg/min, SR decreased from 3.166μm to 2.580μm and EWR decreased from 9.833 mg/min to 7.923 mg/min. The percentage improvements in MRR, SR and EWR are 13.84%,12.91% and 19.42% respectively.

5. Conclusion

Mathematical models for MRR, SR and EWR are developed using regression approach and the validity of these models is checked by R2 value and ANOVA. Models are found statistically significant and adequate. Multi-criteria optimization of parameter levels is explored using GRA. It is observed that peak current and speed of electrode rotation are significant for multi-criteria optimization of MRR, SR and EWR simultaneously. The optimal levels of parameters are peak current-6A, pulse on time-10μs, pulse off time-50μs and speed of electrode rotation-500rpm. There is an improvement in GRG from 0.7097 to 0.807 i.e.13.71%. MRR increased from 120.718 mg/min to 137.433mg/min. SR decreased from 3.166μm to 2.580μm. The increase in MRR

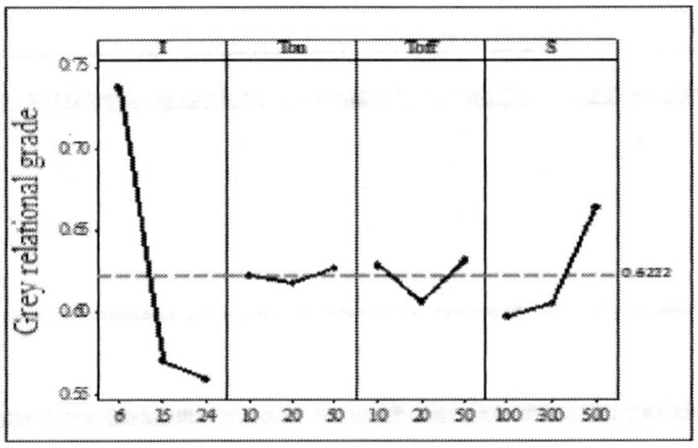

Figure 1. Response graph of GRG.

accompanied by the decrease in SR is attributed to the rotational speed.At higher speed, the debris is completely washed away from the machining zone due to centrifugal action of dielectric fluid. EWR decreased from 9.833 mg/min to 7.923 mg/min. This is due to the fact that at higher speed, the dielectric fluid is forced into machining zone which results in uniform distribution of the heat and cooling of the electrode. Hence EWR is decreased at optimal levels. The percentage improvements in MRR, SR and EWR are 13.84%, 12.91% and 19.42% respectively. This work is useful to the EDM industries dealing with nickel super alloys and provides the guidance to predict the performance and select optimal parameters for overall performance improvement.

References

[1] Choudhary I A and El-Baradie M A 1998 *Journal of Material Processing Technology* B**77** 278-284
[2] Ho K H and Newman S T 2003*International Journal of Machine Tools and Manufacture* B**43** 1287-1300
[3] Mohan B et.al 2002 *Journal of Material Processing Technology* B**124** 297-304
[4] Aliakbari E and Baseri H *International Journal of Advanced Manufacturing Technology* B**10.1007/s00170-011-3862-9**
[5] Wang C C and Yan B H 2002 *Journal of Material Processing Technology* B**102** 90-102
[6] Lin Y C et.al *International Journal of Advanced Manufacturing Technology* B**10.1007/s00170-011-3745-0**
[7] Pradhan B B et.al 2009 *International Journal of Advanced Manufacturing Technology* B**41** 1094-1106
[8] Kao Y C et.al 2010 *International Journal of Advanced Manufacturing Technology* B**47** 395-402
[9] Phadke M S 1989 *Prentice Hall New Jersey*
[10] Deng J 1989 *Journal Grey System* B**1(1)** 1-24

Kinematic analysis of a spatial mechanism for estimating shaking effects

P S S Murthy[1], A Satyadevi[2], A Gopala Krishna[3] and K Eswaraiah[4]

Department of Mechanical Engineering,
[1]KITS,Warangal, [2]GITAM,Hyderabad, [3]JNTUK, Kakinada,[4]KITS,Warangal, India.

E-mail: [1]pmurthy@ymail.com, [2]a.satyadevi@gmail.com

Abstract. Spatial mechanisms are the most general category of kinematic devices. They offer the greatest capability to accomplish any desired kinematic task. A mechanism exerts forces and moments on its supporting frame, which result in vibration. Besides of its effect on efficiency, reducing vibration has become inevitable in the current industrial environment where stern standards on noise and vibration prevail. Balancing of shaking forces and shaking moments in mechanisms is important in order to improve their dynamic performance and fatigue life by reducing vibration, noise and wear. The analysis and synthesis of spatial mechanisms which involves extensive vector mathematics and linear algebra is to be simplified to be taught to engineers in undergraduate education. In the present paper the kinematic analysis of a spatial four-link RSCR mechanism is done to get the velocities and accelerations of its various links which is necessary for the estimation of inertia forces in a mechanism.

1. Introduction

In a world where the standards of living increase every day, today's engineers face an incredible challenge to use technology to answer the growing needs of the population and therefore keep searching for better performing machinery, for higher efficiency. Building high-performance machinery requires limiting machine vibration and the associated dynamic problems, such as fatigue, noise and wear. Besides its effect on efficiency, reducing vibration has become inevitable in the current industrial environment where stern standards on noise and vibration prevail. A possible side effect of machine vibrations is for instance the transmission of significant forces to the machine floor, resulting in disturbing effects in the building in which the machine is placed as well as its surroundings. With this in mind current research focuses on reducing the vibrations caused by the mechanisms in high-speed machinery. A high speed mechanism exerts fluctuating forces and moments on its supporting frame. These forces and moments, of which the resultants are denoted as the shaking force and the shaking moment, induce vibration, on the frame[1,2]. One strategy to suppress these vibrations is to add counterweights to the mechanism moving links. Allowing more counterweight mass, for instance, may give better balancing results, but increases cost or compromises the start procedure of the mechanism and its flexible behavior. Also the space for counterweights is not always available and creating such space can increase the cost. Balancing of shaking forces and shaking moments in spatial mechanisms is important in order to improve their dynamic performances and fatigue life by reducing vibration, noise and wear[3,5]. Spatial mechanisms have become the subject of extensive research interest recently. Robotic devices are becoming increasingly important in manufacturing, materials manipulation in hazardous environment, and in numerous fields of application. Also it is feasible to incorporate complex kinematic requirements in the rational design of such mechanisms. However, prior to wide adaption of spatial mechanisms in high-speed and precision-operating mechanical systems, we must have better knowledge of their dynamic characteristics[6,7]. In the present paper, the kinematic analysis of a spatial four-link RSCR mechanism having two revolute joints, one spherical joint along with one cylindrical joint is done to get the velocities and accelerations of its various links.

2. Spatial mechanisms

Well defined analysis and synthesis techniques of planar mechanisms are readily available and are taught to many mechanical engineers. Additionally, there is a variety of synthesis software available for planar mechanism design. On the other hand, spatial mechanism synthesis and analysis is typically not taught to engineers in undergraduate education. Techniques for analysis and synthesis of spatial mechanisms usually involve extensive vector mathematics and linear algebra techniques. Even for the experienced mechanical designer, it is labor intensive and difficult to design a spatial mechanism. The visualization of these devices can even be difficult. An unbalanced linkage running at high speed transmits shaking forces and shaking moments to its foundation (frame). The shaking force is the resultant inertia force exerted on the frame and is equal to the vector sum of the inertia forces associated with the moving links of the mechanism. The shaking moment about an axis in the frame is the vector sum of the inertia torques and the moments of the inertia forces about the axis. These forces and moments cause vibrations, fluctuations in the input torque and stresses, and therefore impose limitations on the performance of high-speed machinery. There has been a need to develop the optimum balancing of general three-dimensional mechanism. The majority of mechanisms synthesized and found in application are planar devices. These mechanisms have motion such that all elements move in one plane or in parallel planes. While planar mechanisms have been broadly applied, they lack the ability to perform many general motion-control tasks. In fact, planar mechanisms, as well as spherical mechanisms, form a subset of spatial mechanisms. Spatial mechanisms are the most general category of kinematic devices. They offer the greatest capability to accomplish any desired kinematic task. Spatial mechanisms can be composed of any number of links and can include joints with any combination of rotational and translational freedom. It should be noted that robotic manipulators provide controlled actuation of all the joints in the system, which results in a large number of degrees of freedom. This flexibility may be not necessary for the given task, in which case the manipulator could be replaced with a single degree-of-freedom spatial mechanism. If spatial mechanisms could be synthesized quickly and easily, there could be many suitable applications. Some examples include the aerospace industry, exercise equipment, and the rehabilitation medical field. For example, spatial mechanisms are especially useful in satellite design and deployment. Basically, spatial mechanisms could be applied to any task that requires general spatial motion.

3. Kinematic analysis of a spatial four-link RSCR mechanism

The industry is giving top priority to keep up with the current stern norms on noise and vibrations and to limit the workers' exposure to 'physical agents like vibrations and noise. If a choice between two 'equivalent' machines is possible, the less noisy/vibrating machine will be the natural preference. Less vibrating machines will contribute to consolidating and extending market. The present kinematic analysis is useful in estimating the inertia forces acting on the frame which in turn produce vibrations. The method of constant-distance equation for the kinematic analysis of linkages discussed by Rao [2] is based on fact that a constant-distance equation is formulated, wherever the distance between two pair-centers of rigid link remains

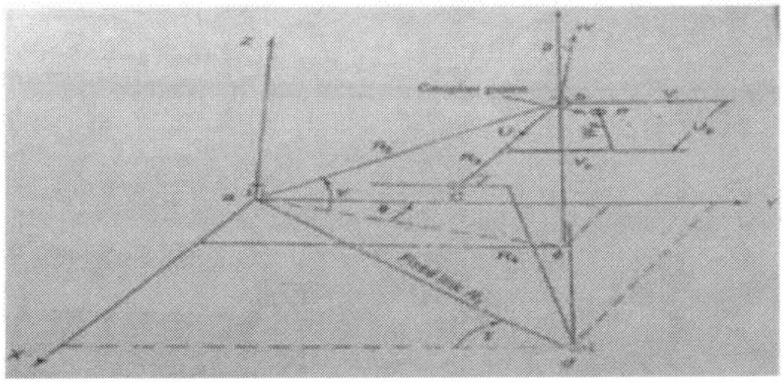

Figure 1. A spatial four bar RSCR linkage with coupler point P.

constant throughout its motions. This results in a much simpler kinematic analysis of the linkages. To illustrate the procedure and the feasibility of the method, the case of a single degree of freedom spatial four-link RSCR mechanism [2] with coupler points as shown in Figure 1 is considered.R_1= 0.148m, R_2= 0.105m, R_3= 0.100m, R_4= 0.105m, ψ= 50 degrees, γ= 35 degrees,U_p= 0.5, V_p= 1, W_p= 1, X_c= 1, Z_c= 1.The set of equations thus form a displacement mathematical model of this mechanism and describes uniquely the linkage configuration. By differentiating the set of equations with respect to time, a complete solution of linkage-kinematics is obtained. In an N-bar spatial linkage ,the pairs of the linkage, which can be of any type, are designed by a, b, i, j, k, m, n. Considering the constant distances between the pair-centers, a set of constant-distance equations can be written in the form

$$Fn(X_a, Y_b, Z_c, X_j, Y_j, Z_j....X_n, Y_n, Z_n) = 0, \tag{1}$$

where $n = 1, 2, ...g$ and g is the number of constant distance equations. Using specified geometrical constraints and known quantities (eg., as demonstrated below by equations 7 and 8 for the case of RSCR Mechanism) equation 1 can be reduced to a set of non-linear, second order algebraic equations with g unknowns in the form

$$[X_1, Y_2, Z_3, Y_4....Z_g]. \tag{2}$$

Solving above two equations and expanding by Taylor's method, we have

$$Fn(X_1, Y_2, Z_3, Z_g) + \delta X_1[\frac{\partial Fn}{\partial X_1}] + \delta Y_2[\frac{\partial Fn}{\partial Y_2}] + \delta Z_g[\frac{\partial Fn}{\partial Z}] = 0. \tag{3}$$

Where δX_1 , δY_2. δZ_g are the correction terms equation 3 is valid only for small increments of the input angle θ of the linkage. In a spatial four-bar RSCR linkage with coupler point P, the driver R_2 rotates about the Z axis. θ is the input angle and ϕ is the output angle.R_3 is the coupler link with the coupler point P, and b is the origin of the moving co-ordinate system, UVW. R_4 is the output link, rotates about the Y- axis. γ and ψ are the constant angles determining the orientation of R_1 and R_2.The following set of constant-distance equation can be written

$$\begin{aligned}
[X_c - X_b]^2 + [Y_c - Y_b]^2 + [Z_c - Z_b]^2 &= [R_3]^2, \\
[X_d - X_c]^2 + [Z_c]^2 &= [R_4]^2, \\
[X_b, Y_b, Z_b] &= [(R_2 cos(\psi) sin(\theta)), (R_2 cos(\theta)), (R_2 sin(\psi))], \\
[X_c, Y_c, Z_c] &= [(X_d + R_4 sin(\phi)), (Y_b), (R_4 cos(\phi))], \\
[X_d, Y_d, Z_d] &= [(R_2 sin(\gamma)), (R_1 cos(\gamma)), (0)].
\end{aligned} \tag{4}$$

By Rearranging

$$F_1(X,Z) = [X_c - R_2cos(\psi)sin(\theta)]^2 + [Z_c - R_2sin(\psi)]^2 - [R_3]^2 = 0, \quad (5)$$

$$F_2(X,Z) = [X_c - R_1sin(\gamma)]^2 + [Z_c]^2 - [R_4]^2 = 0. \quad (6)$$

The output angular displacement θ is computed from

$$\phi = \arctan[X_c - R_1sin(\gamma)]/[Z_c]. \quad (7)$$

The coordinates of the coupler point P are given by the transformation matrix λ for this case

$$[\lambda] = [(\lambda)_{xu}(0)\lambda_{xw};(0)(1)(0);(\lambda)_{zu}(0)(\lambda)_{zw}], \quad (8)$$

$$(\lambda)_{xu} = (\lambda)_{zw} = [X_c - R_2cos(\psi)sin(\theta)]/R_3, (\lambda)_{xw} = -(\lambda)_{zu} = [R_2sin(\psi) - Z_c]/[R_3],$$
$$[X_p;Y_p;Z_p] - [X_b;Y_b;Z_b] = [\lambda][(U_p);(V_p);(W_p)], \quad (9)$$

we have

$$X_p = (\lambda)_{xu})(U_p) + (\lambda)_{xw})(W_p) + X_b,$$
$$Y_p = V_p + Y_b,$$
$$Z_p = (\lambda)_{zu})(U_p) + ((\lambda)_{zw})(W_p) + Z_b. \quad (10)$$

Applying the constant distance equation 3 for the RSCR mechanism then we get the

$$F_1 + \delta(X_c)[\partial F_1/\partial X_c] + \delta Z_c[\partial F_1/\partial Z_c] = 0,$$
$$F_2 + \delta(X_c)[\partial F_2/\partial X_c] + \delta Z_c[\partial F_2/\partial Z_c] = 0. \quad (11)$$

Differentiating equations 4 and 5 with respect to the X_c,and Z_c

$$[\partial F_1/\partial X_c] = 2[X_c - R_2cos(\psi)sin(\theta)],$$
$$[\partial F_1/\partial Z_c] = 2[Z_c - (R_2sin(\psi))],$$
$$[\partial F_2/\partial X_c] = 2[X_c - R_1sin(\gamma)],$$
$$[\partial F_2/\partial Z_c] = 2[Z_c]. \quad (12)$$

Differentiating Equations (6) and (9) and rearranging, we obtain the set of linear equations for velocity and acceleration.

4. Results and discussions

The X, Y and Z coordinates of the linear displacement, linear velocity and linear acceleration of coupler point P of RSCR mechanism for different values of input angle θ from 0 to 360 is calculated by writing a program in MATLAB and variation is shown in Figures 2 to 4. The angular displacement, angular velocity and angular acceleration of output link R_4 of Spatial RSCR mechanism for different values of input angle θ from 0 to 360 is calculated by writing a program in MATLAB and variation is shown in Figure 5.

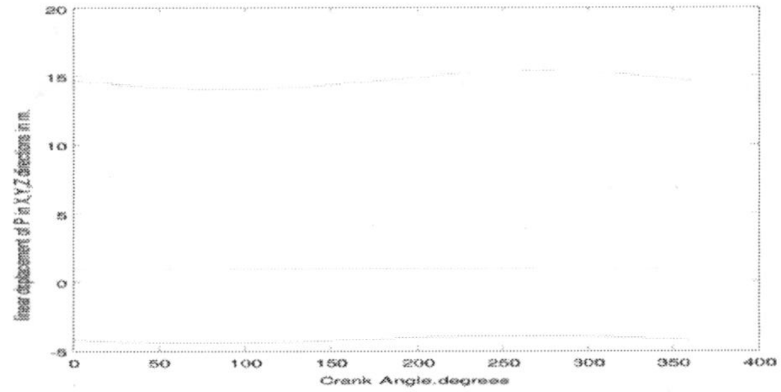

Figure 2. Variation of linear displacement of P with crank angle.

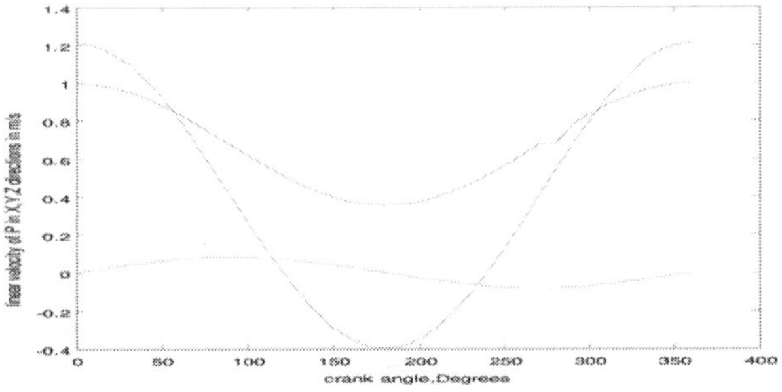

Figure 3. Variation of linear velocity of P with crank angle.

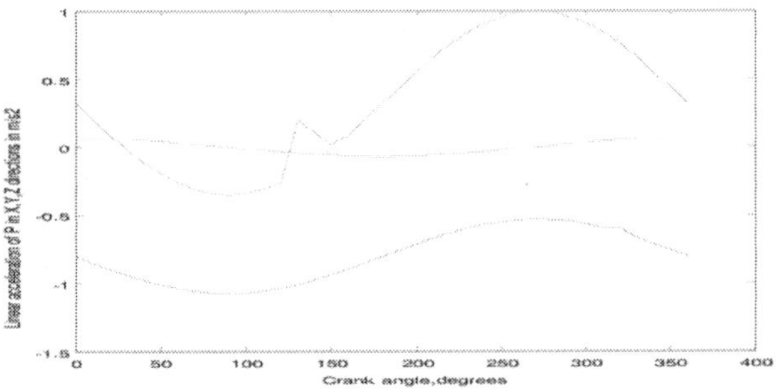

Figure 4. Variation of linear acceleration of P with crank angle.

5. Conclusion

Because of their complexity, it is generally not practical to perform an analysis of spatial linkages by hand computation or by graphical methods. Spatial linkages have therefore attracted much research interest, following the advent of the high speed digital computers. Due to the inability of practicing engineers to design spatial mechanisms, many spatial kinematic tasks are either done by a robotic manipulator, a machine with multiple planar actuators, or are simply ignored for automation and are performed by human operators. The analysis and synthesis of spatial mechanisms which involves extensive vector mathematics and linear algebra is to be simplified to be taught to engineers. The present kinematic analysis is useful for estimating the complete shaking force and shaking moment acting on a spatial mechanism.

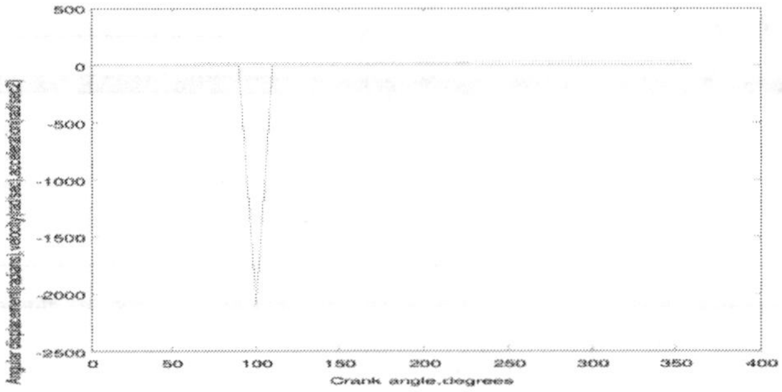

Figure 5. Variation of angular displacement, velocity and acceleration of output link with crank angle.

References

[1] Chaudhary H and Saha S K 2009 Dynamics and Balancing Of Multibody Systems, Berlin, Springer
[2] Rao V Dukkipati 2010 Analysis and Synthesis of spatial mechanism, New Delhi M/s narosa plublising house
[3] Himanshu Chaudhary and Saha S K 2008 *Mechanism and Machine Theory* **43** 506-522
[4] Dresig H and Nguyen phongDien 2010 Germany, TechnischeMechanik publisher
[5] Vigen Arakelian and Sebastien Briot 2010 *Mechanics Research Communications* **37** 265-269
[6] Sebastien Briot Vigen Arakelian and Jean-Paul Le Baron 2012 *Mechanism and Machine Theory* **57** 1-12.
[7] Vinay Gupta Himanshu Chaudhary and Subir K Saha 2013 Proceedings of the *1st International and 16th National Conference on Machines and Mechanisms (iNaCoMM2013)* IIT Roorkee India

Mathematical modeling and analysis of EDM process parameters based on Taguchi design of experiments

J Laxman[1] and K Guru Raj[2]

Department of Mechanical Engineering, Kakatiya Institute of Technology and Science, Warangal, India.

E-mail: [1]Laxmankits9@gmail.com, [2]yadav_56@yahoo.com

Abstract. Electro Discharge Machining is a process used for machining very hard metals, deep and complex shapes by metal erosion in all types of electro conductive materials. The metal is removed through the action of an electric discharge of short duration and high current density between the tool and the work piece. The eroded metal on the surface of both work piece and the tool is flushed away by the dielectric fluid. The objective of this work is to develop a mathematical model for an Electro Discharge Machining process which provides the necessary equations to predict the metal removal rate, electrode wear rate and surface roughness. Regression analysis is used to investigate the relationship between various process parameters. The input parameters are taken as peak current, pulse on time, pulse off time, tool lift time. and the Metal removal rate, electrode wear rate and surface roughness are as responses. Experiments are conducted on Titanium super alloy based on the Taguchi design of experiments i.e. L27 orthogonal experiments.

1. Introduction

Electrical discharge machining is considered as one of the main non-conventional machining processes used for manufacturing geometrically complex or hard material parts that are extremely difficult to machine by conventional machining processes. New developments in the field of material science have lead to new engineering metallic materials, composite materials, and ceramics, having good mechanical properties and thermal characteristics as well as sufficient electrical conductivity so that they can readily be machined by spark erosion. Electrical discharge machine (EDM) technology is increasingly being used in tool, die and mould making industries, for machining of heat treated tool steels and advanced materials (super alloys, ceramics, and metal matrix composites) requiring high precision, complex shapes and high surface finish. Technologically advanced industries such as aeronautics, nuclear reactors and automobiles have been demanding high strength temperature resistant (HSTR) materials having high strength to weight ratio. Today, these alloys have a wide variety of applications in modern aerospace, marine, automobile sector, atomic power plant reactors, and medical implants owing to their high strength to weight ratios, high strength and toughness at elevated temperatures, excellent corrosion resistance, fracture resistance and low modulus of elasticity, good corrosion resistance in most environments. Titanium and Titanium super alloy has been developed in order to satisfy the needs for a class of strong and light weight materials for aircraft engines and airframe manufacturing. Titanium and its alloys are difficult to machine materials due to several inherent properties of the material. In spite of its more advantages and increased utility of titanium alloys, the capability to produce parts with high productivity and good quality becomes challenging. Owing to their poor machinability, it is very difficult to machine titanium and titanium super alloys economically with traditional mechanical techniques. The main difficulties to machine titanium and titanium super alloys are high cutting temperatures and rapid tool wear.

Electrical discharge machining (EDM) is a relatively modern machining process having distinct advantages over other machining processes and can machine Ti alloys effectively. Researchers made attempts to model EDM process to study improvements in performance measures like material removal rate, surface roughness and tool wear rate. Yusuf Keskin et al. investigated the effects of machining parameters on the surface roughness valley machined by EDM on steel using copper tool electrode. The data obtained for response measures had been analyzed using the Design of experiments method multiple regression method. It was observed that the surface roughness had increasing trend with an increase in the discharge duration due to the more discharge energy released during this time [1,2]. Chen and Madhavian [3] suggested a theoretical model to predict the material removal rate and surface roughness. Later compared the theoretical results with the experimental results and found to be in close agreement. Wang and Tsai [4] conducted a study on modeling of surface roughness in electro discharge machining using both response surface methodology and artificial neural networks. Lee and Li [5] studied the influence of operating parameters on tungsten carbide and its machining characteristics. Wang [6] used a hybrid artificial neural network and genetic algorithm methodology to model and optimize EDM process. Mahapatra and Patnaik [7] developed a relationship between control factors and material removal rate and surface roughness using regression analysis and finally genetic algorithm was used to optimize the process as a multi objective function. Kuang-Yuan Kung et al [8] studied the material removal rate and electrode wear ratio on the powder mixed dielectric fluid for machining of cobalt-bonded tungsten carbide on electrical discharge machine. They employed response surface methodology to plan and analyze the experiments. Prasad and Gopalakrishna [9] developed a mathematical model for material removal rate and surface roughness using central composite rotatable design with five factors and finally optimized them as multiobjective optimization using evolutionary algorithms. Chiang [10] proposed mathematical models using response surface methodology for modeling and analysis of the effects of machining parameters on the performance characteristics of the electro discharge machining process of Al2O3+TiC mixed ceramic to study the influence of four machining parameters the discharge current, pulse on time, duty factor and open discharge voltage on material removal rate, electrode wear ratio and surface roughness. Shabgard and Shotorbani [11] established the relation between input parameters the peak current, pulse-on time and voltage and the process outputs using Design of Experiments (DOE), multi linear regression techniques and response surface methodology. In this present paper the work is reported on titanium alloy taking the input parameters, peak current, pulse on time, pulse off time and tool lift time and developed the mathematical model based on regression method.

2. Methodology

2.1. Regression analysis

Regression analysis includes any techniques for modeling and analyzing several variables, when the focus is on the relationship between a dependent variable and one or more independent variables. More specifically, regression analysis helps us to understand how the typical value of the dependent variable changes when any one of the independent variables is varied, while the other independent variables are held fixed. Most commonly, regression analysis estimates the conditional expectation of the dependent variable given the independent variables. In all cases, the estimation target is a function of the independent variables called the regression function.

The Regression analysis is an empirical modeling approach for determining the relationship between various process parameters and responses with the various searches and desired criteria for the significance of these process parameters in the coupled responses. It is a sequential experimentation strategy for optimizing and building the empirical model. The objective of the regression analysis is to develop the mathematical relation between the responses and machining parameters. In many experimental situations, it is possible to represent independent factors in quantitative form. Then these factors can be thought of as having a functional relationship

$$Y_u = f(X_1, X_2, X_3, .X_k) \pm e_r, \tag{1}$$

between the response Y_u and X_1,X_2,X_3,.X_k of k independent quantitative factors as shown in equation 1. The function is called response surface or response function. The experimental error is measured by the residual error e_r. A characteristic surface responds for a given set of independent variables. When the mathematical form is not known, it can be approximated satisfactorily within the experimental region by a polynomial. The higher the degree of the polynomial the better is the correlation.The methodology may be applied for developing the mathematical models in the form of multiple regression equations correlating the dependent parameters such as metal removal rate, Tool wear rate, Surface roughness, etc. with more than two independent parameters, in electric discharge machining process the independent parameters are peak current, pulse on time, pulse off time and tool lift time,. In applying the regression analysis, the dependent parameter is viewed as fitted mathematical model. For the development of regression equations related to various quality characteristics of machined components, the second-order response surface may be assumed as,

$$Y_u = \beta_0 + \sum_{i=1}^{k} \beta_i X_{iu} + \sum_{i=1}^{k}\sum_{j=1}^{k} \beta_{ij} X_{iu} X_{ju} + \sum_{i=1}^{k} \beta_{ii} X_{iu}^2, \tag{2}$$

where Y_u represents the corresponding response, contains linear, squared and cross-product terms of X_{iu}'s variables. In the present paper, the code values of $i^t h$ machining parameters for u $^t h$ response are represented byX_{iu}. The values of k indicate the number of machining parameters. The terms β_i, β_{ii} and β_{ij} are the second order regression co-efficient. The second term under the summation sign of this polynomial equation attributes to linear effects, whereas the third term of the above equation corresponds to the higher order effects and lastly the fourth term of the equation includes the interactive effects of the parameters [13].

2.2. Design of experiments

Experimental design methods are too complex and are not easy to use. When the number of process parameters increases a large number of experiments have to be carried out. To solve this problem, the Taguchi method uses a special design of orthogonal arrays to study the entire parameter space with only a small number of experiments. According to the Taguchi method, a robust design and an orthogonal array are employed for the experimentation. Four machining parameters are considered as controlling factors namely, peak current, pulse on time, pulse off time and tool lift time and each parameter has three levels. Table.1 shows the parameters and their levels considered for the experimentation. The experimental design considered for the investigation based on the L27 orthogonal array. Based on this, 27 experiments are conducted; each having a different combination of levels of factors as shown in Table 3 was carried out. Orthogonal arrays are special standard experimental design that requires only a small number of experimental trials to find the main factor effects on output.The standard Orthogonal arrays available are L4, L8, L12, L16, L18, L9, L27, L32 etc. In this work L27 array is selected based on main factors, interaction effects between variables, the variables are assigned at columns,

Figure 1. Machined workpiece and tool.

Table 1. Selected process parameters in EDM process.

Parameter	Symbol	units	Level 1	Level 2	Level 3
Peak current	Ip	Amp	9	12	15
Pulse on time	Ton	μs	10	20	50
Pulse off time	Toff	μs	20	50	100
Tool lift time	Tlift	μs	5	10	20

Table 2. Chemical composition of work material.

Titanium	Al	C	Mo	Zr	Si	Fe	P	V
Base	5.5-6.8%	0.13%	0.5-2%	1.5-2.5%	0.15%	0.3%	0.8-2.5%	0.3%

as stipulated by orthogonal array. Some columns can be kept dummy, but no row should be left out. Once the orthogonal array is selected, the experiments are selected as per the level combinations; all the experiments must be conducted. The interaction effect columns can be kept dummy while conducting experiments, but to be considered for analysis.

3. Experimental work

3.1. Experimental setup

We constructed an appropriate mathematical model to predict the value of MRR, EWR and SR. Experiments were carried out on die sinking EDM of type Askar microns, model V3525 with servo head constant gap voltage, positive polarity. Commercial 30 Grade EDM Oil was used as Dielectric fluid. The investigation was done on Titanium super alloys with Brinell hardness number 35 HB. The chemical composition of work material is given in Table.1.The size of work pieces is 50mmX30mmX6mm and the diameter copper electrode is 12mm have been taken for the study. Preliminary experiments were conducted by varying the process input parameters for finding range of parameters. L27 orthogonal array was used to design the number of experiments. The results obtained were analyzed and the models were produced using MINITAB Software. Tool material is of copper because of easy available in market and lower in cost.

3.2. Selection of input parameters

The input parameters are selected based on the preliminary experiments and the operational limitations of the EDM machine. The selected input parameters (peak current, pulse on time and pulse off time and tool lift time) are shown in Table. 1, and the output parameters are

Metal removal rate (MRR), Tool wear rate(TWR) and Surface roughness(SR) have been taken for the study.

3.3. Calculation of MRR and TWR

To calculate the value of MRR and TWR, the work piece and tool are to be weighted, for this purpose we used SHIMADZU PHILLIPPINES, Digital sensitive weighting balance which is of 1mg accuracy. The work piece sample is weighted before machining and after machining and the MRR is obtained using the following equation.

$$MRR = \frac{wp_1 - wp_2}{T}, \tag{3}$$

where wp_1 and wp_2 are the weights of the workpiece before and after machining in mg/min respectively, T is the machining time in minutes. Similarly TWR is calculated taking the weights of the tool before machining and after machining and is obtained by using following equation

$$TWR = \frac{wt_1 - wt_2}{T}, \tag{4}$$

where wt_1 and wt_2 are the weights of the tool electrode before and after machining in mg/min respectively. The SR was measured on HANDYSURF. Three measurements of SR had been taken at different locations and the average values are taken for the analysis. The experiments are conducted according to L27 orthogonal array and the results are shown in Table 3.

4. Mathematical models

Regression analysis combines mathematical and statistical techniques for empirical model building. A model of the response to certain independent input variables can be obtained by conducting experiments and applying regression analysis,. The mathematical models commonly used are represented by,

$$Y = \Phi(I_p, T_{on}, T_{off}, T_{lift}) \pm e_r, \tag{5}$$

where Y is the machining response, Φ is the response function and I_p,T_{on},T_{off} andT_{lift} are machining independent variables and e_r is the error. The general second-order polynomial response equation is mostly used. It also confirms that this model provides an excellent explanation of the relationship between the Independent factors and the responses. The second order response surface representing the MRR, TWR and SR can be expressed as a function of machining parameters I_p,T_{on},T_{off} andT_{lift}.The relationship between the response and machining parameters has been expressed based on equation 2 as follows.

$$\begin{aligned} Y_u = \beta_0 + \beta_1 I_p + \beta_2 T_{on} + \beta_3 T_{off} + \beta_4 T_{lift} + \beta_5 I_p T_{on}, \\ +\beta_6 I_p T_{off} + \beta_7 I_p T_{lift} + \beta_8 T_{on} T_{off} + \beta_9 T_{on} T_{lift}, \\ +\beta_{10} T_{off} T_{lift} + \beta_{11} I_p^2 + \beta_{12} T_{on}^2 + \beta_{13} T_{off}^2 + \beta_{14} T_{lift}^2. \end{aligned} \tag{6}$$

To obtain predictive quantitative relationships, it is necessary to model the machining responses and the process variables. In the present work, the mathematical models were developed on the basis of EDM experimental results using MINITAB-17. From the observed data for MRR, TWR and SR, the response function has been determined using polynomial second order equations, as given below.

$$\begin{aligned} MRR = 16.96 - 3.25 I_p + 0.945 T_{on} - 0.0822 T_{off} + 0.089 T_{lift} - 0.01335 I_p T_{on} \\ +0.00045 I_p T_{off} - 0.0576 I_p T_{lift} + 0.001547 T_{on} T_{off} - 0.00356 T_{on} T_{lift} \\ +0.00270 T_{off} T_{lift} + 0.2039 I_p^2 - 0.01213 T_{on}^2 - 0.000445 T_{off}^2 + 0.0101 T_{lift}^2, \end{aligned} \tag{7}$$

Table 3. Experimental results.

	Level of parameters				Experimental results		
S.No	Ip (Level)	Ton (Level)	Toff (Level)	Tlift (Level)	MRR (mg/min)	TWR (mg/min)	SR (μm)
1	1	1	1	1	8.233	1.247	5.830
2	1	1	2	2	4.100	0.607	4.432
3	1	1	3	3	1.666	0.577	3.651
4	1	2	1	2	11.570	0.453	5.642
5	1	2	2	3	8.520	0.280	6.100
6	1	2	3	1	5.333	0.517	6.763
7	1	3	1	3	8.661	0.156	7.641
8	1	3	2	1	12.100	0.416	8.745
9	1	3	3	2	7.933	0.223	5.910
10	2	1	1	2	8.331	0.760	7.010
11	2	1	2	3	4.000	0.320	6.352
12	2	1	3	1	3.166	0.830	8.485
13	2	2	1	3	7.367	0.743	7.219
14	2	2	2	1	12.133	0.710	8.531
15	2	2	3	2	5.833	0.303	7.118
16	2	3	1	1	12.333	0.396	8.607
17	2	3	2	2	10.667	0.326	7.514
18	2	3	3	3	5.200	0.210	7.479
19	3	1	1	3	8.367	1.303	8.454
20	3	1	2	1	10.967	1.720	9.923
21	3	1	3	2	4.833	0.913	7.031
22	3	2	1	1	20.667	2.583	10.178
23	3	2	2	2	16.667	1.893	9.264
24	3	2	3	3	7.600	0.837	6.777
25	3	3	1	2	11.443	0.970	10.869
26	3	3	2	3	8.330	1.063	8.580
27	3	3	3	1	14.103	1.300	11.010

$$\begin{aligned} TWR = 6.12 - 0.900I_p - 0.0117T_{on} - 0.0001T_{off} - 0.10441T_{lift} + 0.00029I_pT_{on} \\ -0.0011I_pT_{off} - 0.0511I_pT_{lift} + 0.000183T_{on}T_{off} + 0.000665T_{on}T_{lift} \\ +0.000234T_{off}T_{lift} + 0.0485I_p^2 - 0.000313T_{on}^2 - 0.000014T_{off}^2 + 0.00414T_{lift}^2, \end{aligned} \tag{8}$$

$$\begin{aligned} SR = -0.64 + 0.785I_p + 0.1175T_{on} + 0.0302T_{off} - 0.311T_{lift} - 0.00220I_pT_{on} \\ -0.00130I_pT_{off} - 0.01460I_pT_{lift} + 0.000034T_{on}T_{off} + 0.00065T_{0n}T_{lift} \\ -0.0016521T_{off}T_{lift} + 0.0010I_p^2 - 0.000973T_{on}^2 - 0.000052T_{off}^2 + 0.01780T_{lift}^2, \end{aligned} \tag{9}$$

R^2- values of models for MRR, TWR and SR are 95.36%, 90.55% and 93.43% respectively. This shows the second order regression model is accurate. The R^2 is defined as the ratio of variability explained by the model to the total variability in the actual data and is used as a measure of the goodness of fit [12].

5. Results and discussion

As shown in the main effects plot for MRR shown in figure 2, the MRR increases with increase in peak current because of the depth of craters more due to the higher intensity of the sparks,

Table 4. Response table for means of MRR.

Level	Ip	Ton	Toff	Tlift
1	7.568	5.963	10.775	11.004
2	7.670	10.632	9.720	9.042
3	11.442	10.086	6.185	6.635
Delta	3.873	4.670	4.589	4.369
Rank	4	1	2	3

Table 5. Response table for means of TWR.

Level	Ip	Ton	Toff	Tlift
1	0.4973	0.9197	0.9568	1.0799
2	0.5109	0.9243	0.8150	0.7164
3	1.3980	0.5622	0.6344	0.6099
Delta	0.9007	0.3621	0.3223	0.4700
Rank	1	3	4	2

Table 6. Response table for means of SR.

Level	Ip	Ton	Toff	Tlift
1	6.079	6.796	7.939	8.675
2	7.591	7.510	7.716	7.199
3	9.121	8.484	7.136	6.917
Delta	3.041	1.687	0.803	1.758
Rank	1	3	4	2

similar trend was observed with pulse on time up to 20sec then it causes to decrease because of the working current increases due to the increase in pulse on time which causes to form a carbon layer on tool surface at higher working currents. Working current is calculated based on equation 10.

$$IW = (IP) * D, \tag{10}$$

$$D = \frac{T_{on}}{T_{on} + T_{off}}. \tag{11}$$

where IW= Working current in Amperes; IP = Peak current in Amperes; D= Duty ratio, T_{on} =Pulse on time, T_{off} =Pulse off time. MRR decreases with increase in pulse off time and tool lift time because of the idle time increases. As shown in the main effects plot for TWR shown in figure 3. TWR increases with increase in current because of the depth of craters increases due to the higher intensity of the sparks which causes higher temperatures, however, from the response tables for means of the MRR, TWR and SR, It is found that the percentage of variation of MRR by varying the input factors , peak current, pulse on time, pulse off time and and tool lift time are 8.33% , 2.20% , 1.25% and 6.66% respectively. The percentage of variation of TWR by varying the input factors , peak current, pulse on time, pulse off time and tool lift time are 8.33% , 3.33% , 1.26% and 6.67% respectively. The percentage of variation of SR by varying the input factors , peak current, pulse on time, pulse off time and and tool lift time are 8.33% , 2.50% , 1.25% and 6.66% respectively.The variation of the responses MRR and SR are higher with respect to the variation of the current, and least variation of the responses is observed with the variation of the pulse off time. The main effects plots are generated based on the response

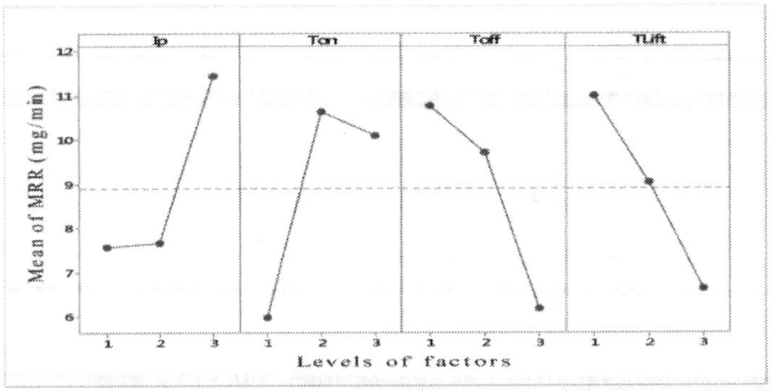

Figure 2. Effects of process parameters on Metal Removal Rate.

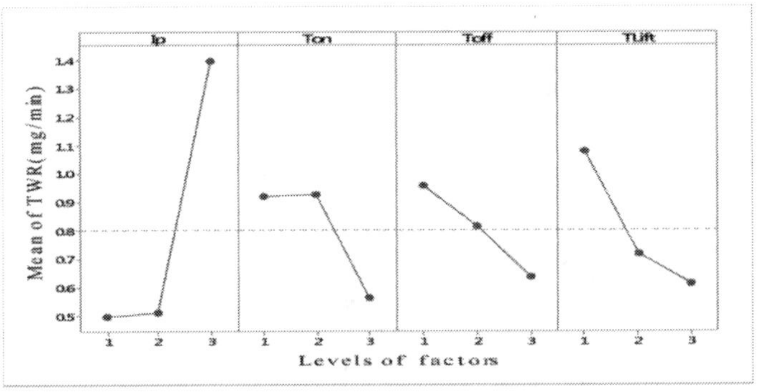

Figure 3. Effects of process parameters on Tool Wear Rate.

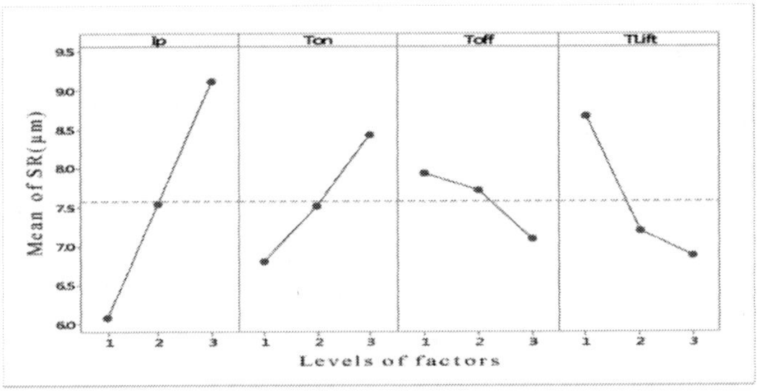

Figure 4. Effects of process parameters on Surface Roughness.

tables 4 , 5 and 6 for the means of responses MRR, TWR and SR and are shown in figures 2, 3 and 4 respectively. TWR increases with increase in pulse on time up to 20sec then it causes to decrease because of the working current increases due to the increase in pulse on time which causes to form a carbon layer on tool surface at higher working currents. TWR decreases with increase of the other parameters . It is observed from the main effects plot for SR shown in figure 4, SR increases with increase in peak current and pulse on time because of higher depths of craters formed on the machined surface where as increase in pulse off time and tool lift time causes SR to decrease the surface finish of the machined surface is more with increase in pulse off

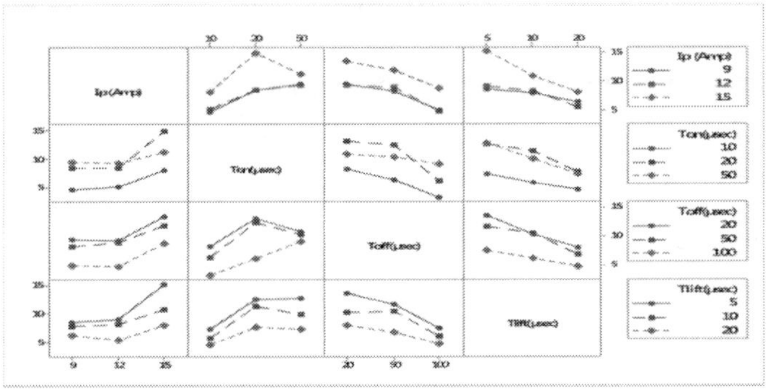

Figure 5. Interaction effects of process parameters on metal removal rate.

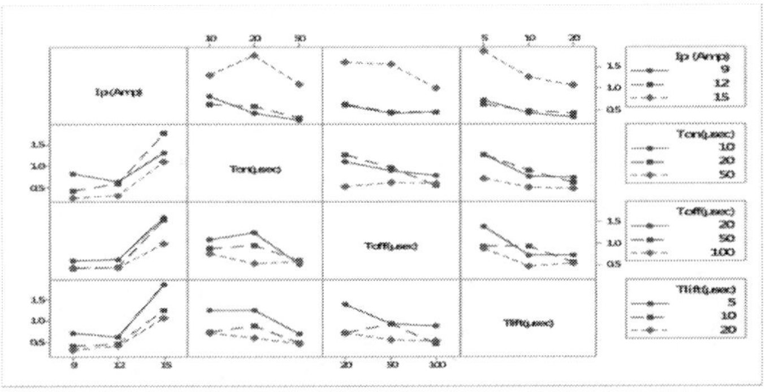

Figure 6. Interaction effects of process parameters on Tool wear rate.

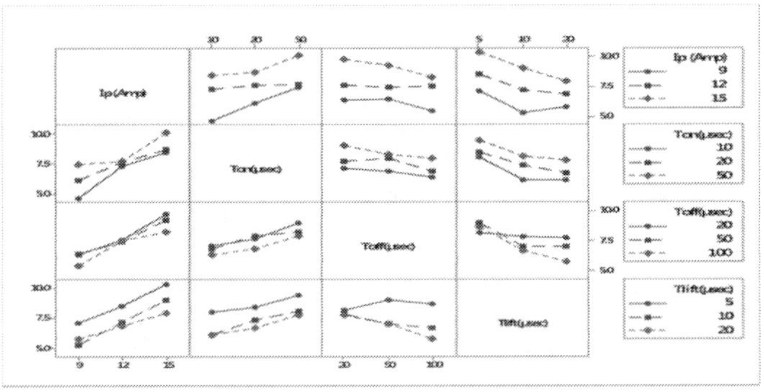

Figure 7. Interaction effects of process parameters on surface roughness.

time and tool lift time but the MRR is less and peak current is the found to the most important factor effecting the responses.

It is found from the figure 6 that there is an interaction between the process parameters in affecting the tool wear rate since the responses at different levels of process parameters for a given level of parameter value are not parallel.It is seen from the figures 5 and 7 that there is weak interaction between the process parameters in affecting the metal removal rate and surface roughness since the responses at different levels of process parameters for a given level of parameter value are almost parallel.

6. Conclusion

In this study, the MRR, TWR and SR in EDM process of Titanium alloy using copper electrode were modeled and analyzed based on the Taguchi design of experiments and regression analysis. The input parameters peak current, pulse on time, pulse off time and tool lift time is employed to carry out the experimental study. The following conclusions were drawn based on the study.

- Peak current was found to be the most important factor effecting the MRR, TWR and SR.
- Increase in peak current causes MRR, TWR and SR to increases continuously.
- Increase in pulse on time causes MRR and TWR to increase up to second level then started to decrease but increase in pulse on time causes SR to increase.
- Increase in tool lift time causes MRR, TWR and SR to decrease continuously.
- Increase in pulse off time causes MRR, TWR to decrease, however, SR to decrease up to level 2, followed to increase till level 3.
- It is found that the percentage of variation of responses (MRR, TWR and SR) by varying the input factors , peak current, pulse on time, pulse off time and tool lift time are 8.33
- It is found that the percentage of variation of responses MRR, TWR and SR by varying the input factor pulse on time are 2.201 % , and 3.33% and 2.50% respectively.
- There is weak interaction between the process parameters affecting the metal removal rate and surface roughness as compared to affecting the Tool wear rate.
- R^2 - values of models for MRR, TWR and SR are 95.36%, 90.55% and 93.43% respectively.

The research findings along with various mathematical models will provide effective guide line to result parameter settings for achieving required MRR, TWR and SR during the machining of titanium super alloy on EDM.

References

[1] Yusuf Keskin, H Seicuk Halkaci and Meluut Kzil 2006 *I. J. of Manuf. Technol.* **28** 1118-21
[2] R K Garg , K K Singh, Anish Sachdeva, Vishal S. Sharma Kul deep Objha and Sharanjit Singh 2010 *I. J. of adv. Manuf. Technol.* **50** 611-624
[3] Chen Y, Madhavian S M 2000*Journal of Material Processing Technology* 150-157.
[4] Wang P J, Tsai K M 2001 *Journal of Material Processing Technology* **102** 1455-57
[5] Lee S H and Li X P 2001 *Journal of Material Processing Technology* **102** 344-353
[6] Wang K 2003 *International Journal of Machine Tools and Manufacture* 995-999
[7] Yusuf Keskin and Selcuk H 2006 *International Journal of Advanced Manufacturing Technology* **41** 1118-21
[8] Swarup Mahapatra and Amar Patnaik 2006 *J. of Braz. Society of Mech. Sci. Eng.* **47** 4 423
[9] Kuang-Yuan Kung, Jenn-Tsong Horng and Ko-Ta Chiang 2009*I. J. of Adv. Manuf. Technol.* **43** 95-104
[10] Prasad D V S S S V and Gopala Krishna A 2008 *I. J. of Adv. Manuf. Technol.* **43** 914-925
[11] Ko-Ta Chiang 2008 *I. J. of Adv. Manuf. Technol.* B**37** 523-533
[12] Shabgard M R and Shotorbani R M 2009 *World Academy of Sci. Engg. and Technol.* **3** 04-29
[13] Montgomery D C 2001 *Wiley New York*
[14] Ross P J 2005 *Tata McGraw Hill New Delhi*

Experimental studies on semi active suspension systems with various dampers

B Anil Kumar, Ch Bharath Raj and K Rajendar

Department of Mechanical Engineering, Kakatiya University College of Engineering, Warangal - 506004, India.

E-mail: anil.kucet@gmail.com

Abstract. Some results of an experimental probe on semi Active suspension systems with various Dampers have been studied in the present paper. Dampers with different configurations were studied by using Magneto Rheological gel Damper. A comparison have been done for different configurations of Dampers viz.MR gel Damper with Magnetic field, MR gel Damper without magnetic field, Damper with conventional oil, Natural vibration without damper.

1. Introduction

The experimental setup was fabricated for studying the performance of MR gel under various circumstances. The experimental results on the gel suggest that a significant increase in storage modulus is possible under the influence of magnetic field. This is an important observation in the context of application of the magnetic gel to damp the vibrations produce during ride, as this indicates that it is possible to tune the magnetic composite gels by changing magnetic fields. Also, the significant deformations of the gels and the magnetic shape retaining ability noticed during the conduction of the experiments is a very important aspect in understanding the property of the magneto rheological gel properly. The improvements made in the existing paper i.e., use of variable stiffness springs in suspension system also helps in regulating ride comfort and stability of the vehicle. Both the systems combined together will produce greater ease, flexibility of control and performance. Such, control system is close to active suspension systems and provide large domain to control the ride stability and comfort without compromising according the road conditions. The result obtained from the study proves that, using MR dampers and variable stiffness springs will provide a smoother ride then conventional suspension system.

2. Literature review

Literature provides vibration isolation systems have been studied broadly and in great depth. The vibration control systems can be categorized as passive, active and semi-active systems. Semi-active control systems fill the gap between passive and active control system and they represent a compromise between performance improvement and simplicity of implementation. They only expend a small amount of energy to change system parameters, such as damping and stiffness. The basic idea of variable damping systems has been proposed by many researchers to provide effective vibration control [1-4]. However, there is room for further improvement because variable spring stiffness systems have not been thoroughly investigated in terms of their practical implementation, despite the fact that vibration systems with variable stiffness control were proposed by a few researchers [5,6]. Kooris proposed a variable stiffness system to suppress buildings response to earthquakes [6]. The aim of Kooris work was to achieve a non-stationary and non-resonant state during earthquakes. Youn and Hack used an air spring in a suspension system to vary the stiffness among three discrete values [5]. In the proposed vehicle suspension system, A Voigt element and a spring in series are used to control the system stiffness. The Voigt element is comprised of a controllable damper and a constant spring. The equivalent stiffness of the whole system is changed by controlling the damper in the Voigt element and the second damper which is parallel with the other elements provides variable damping for the system, a new variable stiffness and variable damping system in which the stiffness and damping can be independently and easily controlled.

3. Experimental set-up

In order to facilitate vibrations with MR gel and to accommodate the Damper, apparatus has been designed and fabricated. This Equipment consists of a Base built up with Mild steel Angles of dimensions 3x1.5 feet. A vertical frame of 2feet height is mounted at one of its ends. A cantilever beam of length 3.6 feet is hinged at a height of 1.5 feet from the base to facilitate the vibrations to occur only in vertical plane. This beam is attached to compression spring, which is fixed to the base. The gap between the spring and beam is compensated by GI rod which is welded to the spring and the beam. The Damper is located adjacent to the spring to show the variations amplitude & damping, a frame is welded to the initial setup. Two metal rods were attached above the frame at a distance from center, perpendicular to the base. PVC pipes with elastic bands wound around them are inserted onto the rods. So that, it supports the drawing sheet & facilitates rolling without slipping. The marker is placed on the bar and pointed towards the sheet to trace the amplitude of vibrations produced. DC motor with a pulley is fixed in a wooden block and it is attached to one of the PVC pipes with connecting belt. This pipes acts as Rollers, such that rotation of one pipe makes the other to rotate and drags the sheet against the marker. When this bar is made to oscillate for a given load, the marker traces the amplitude of vibration & damping effect produced. DC motor is operated by a DC adapter if 12volts and wattage of 12.To produce Magnetic field from AC supply, a step-down transformer with 36 wattage capacity is coupled with AC to DC converter so as to produce a maximum voltage of 12 volts. The output from the converter is connected to MR gel damper. At first the beam, mass and spring assembly is tested to find its natural frequency and

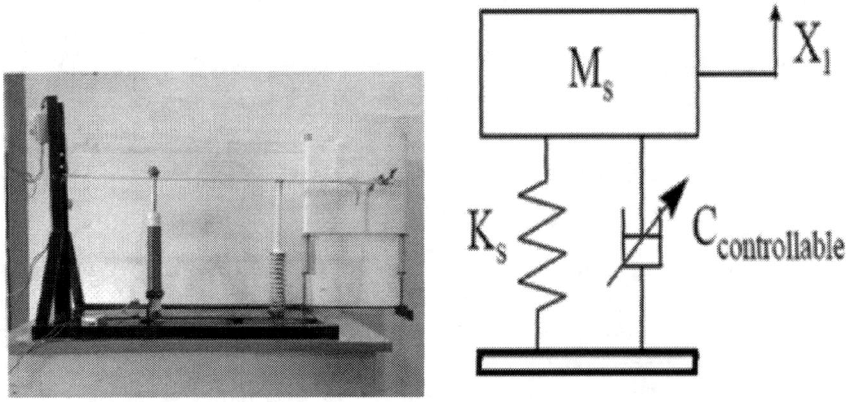

Figure 1. The experimental setup and Schematic diagram.

damping ratio by exciting the end of the beam under different loads and then leaving it for free vibration. The total mass of the beam and GI rod is 300gms. A known mass is now attached at the end of cantilever beam (without any damper) and excited to produce vibrations which are analyzed to find the natural frequency of vibration. Preliminarily the initial height of the beam is measured using a vernier height gauge placed on flat plate. A weight of 700gms added to the bar, and then the resultant height is measured to determine the deflection produced. Similar process is repeated by varying the loads, incremented one by one and deflection is noted down for loads 1400gram, 2100gram, and 2800gram. Values are noted and average stiffness of the spring is determined. The damper is placed in C-clamps which are welded at the bottom side of the frame. The clamps are holding the damper with the help of nut and bolt system. The piston rod is passed through the holes which are drilled to the beam and nut is used to hold the piston firmly with the beam. The damper and spring are parallel to each other this acts as a shock absorber system. Now the damper is placed in respective position and time period of 5 oscillations is noted down with 2.8 kg mass added to the beam. Beam is excited and the amplitude and decay of vibrations is traced by the system.

3.1. MR Gel preparation procedure

Selecting the appropriate proportions of the gel matrix is the first task in the sample preparation. The plain non-magnetic gel matrix is prepared using the copolymer, White lithium grease like Lubricant, a non-aerosol with mass fractions of the mineral oil as plasticizer. Upon particular proportions of linking agent and plasticizer mass fraction for the gel matrix we get the required softness for the matrix. The MR gel is prepared by interspersing carbonyl iron particles in this matrix. Carbonyl iron powder is used in this study show irregular shape of the particles with sizes ranging from 2 to 10μm. Isotropic magnetic composite gels are prepared with a proportion of 50% by weight or carbonyl iron particles mixed with a constant ratio 1:12 of copolymer to mineral oil mass fraction. The current scope of preparation is limited to isotropic magnetic gel. The magnetic composite gel is prepared with proportions as follows:

(i) Mineral Oil 60g / 65.5ml (density 918 kg/m^3)
(ii) Grease 13g / 10 ml (density 1300 kg/m^3)
(iii) carbonyl iron particles 92g /25 ml (density 3670 Kg/m^3)

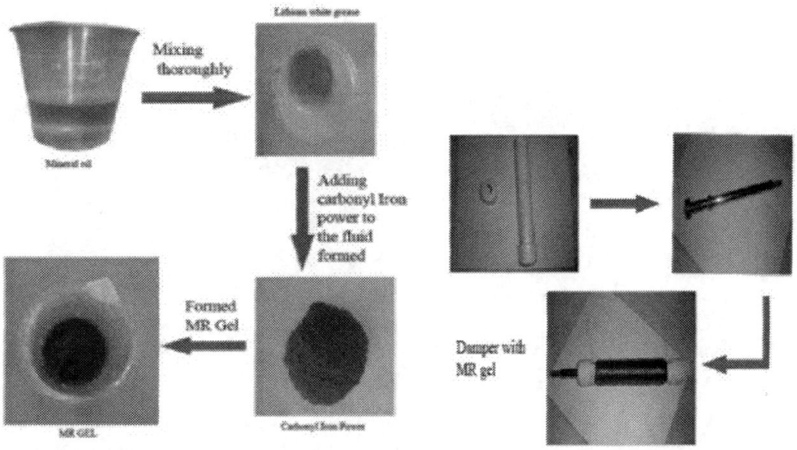

Figure 2. Preparation of MR gel.

Initially 60 ml of mineral oil is taken into a Glass beaker and a proper proportion of lithium (white grease) is mixed thoroughly until entire grease and oil form a uniform suspension. Then carbonyl iron particles are introduced into the mixture and stirred so as to form required MR Gel as shown in figure below.

3.2. Damper with the mineral oil

In this system the damper is filled with mineral oil. When weights are added to the beam, the spring starts oscillating with its natural frequency which is damped by the damper. The damping takes place in 5 seconds. With the help of the graph tracing system the damping graph is traced. The amplitude of vibration decreases compared to the graph without damper.

3.3. Damper with MR-GEL and without magnetic field

In system the oil is replaced with the MR GEL. This gel behaves like Newtonian fluid or normal oil when no field is applied. Weights are added to the beam and excited, graph is traced by system.

3.4. Damper with MR-GEL and with magnetic field

Now a Magnetic field is applied to the gel by applying direct current to copper wires which are wounded around the damper. The Field is parallel to the direction of applied load and magnetic particles align along the same direction increasing the stiffness. When bar is made to oscillate by giving external force, as stiffness of MR gel is greater than mineral oil. It damp outs vibrations very quickly within few seconds (2.2 sec). Graph is traced with the help of graph tracing system. The dynamic response of the system for the given initial excitation depends primarily on the

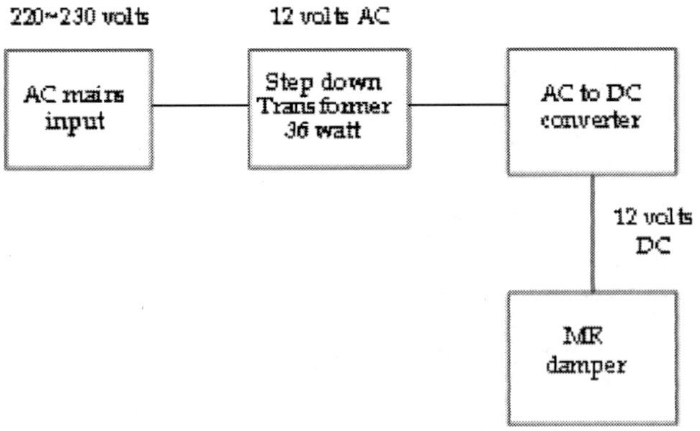

Figure 3. Circuit diagram representing generation of magnetic field.

three dynamic attributes of the system the stiffness of spring elements, the dissipation due to hysteresis damping (internal damping) of the system and the inertia due to mass. The testing system is assumed to be linear. Tests on the system were conducted to validate this assumption. With the mass of the system known, a simple straightforward procedure is adopted to estimate the other two attributes of the system, namely, the storage part, i.e. stiffness of the spring and the dissipation part, i.e. damping coefficient. Using a free decay test, i.e. by introducing a small initial disturbance excitation to the system, these attributes are then calculated from the measured response data obtained from experimentation. The procedure and the principle used to calculate the above mentioned system attributes are discussed below.

4. Mathematical formulation and solution methodology

Free undamped vibration occurs when a mechanical system is set off with an initial input and then allowed to vibrate freely. The mechanical system will then vibrate at one or more of its "natural frequency" and damp down to zero due to hysteresis damping. The fundamentals of vibration analysis can be understood by studying the simple mass–spring–damper model. The force applied to the mass by the spring is proportional to the amount the spring is stretched x (we will assume the spring is already compressed due to the weight of the mass). The proportionality constant, k, is the stiffness of the spring and has units of force/distance (N/m). The negative sign indicates that the force is always opposing the motion of the mass attached to it:

$$F_s = kx. \tag{1}$$

The force generated by the mass is proportional to the acceleration of the mass as given by Newtons second law of motion:

$$\sum F = ma = m\ddot{x} = m\frac{d^2x}{dt^2}. \tag{2}$$

The sum of the forces on the mass then generates this ordinary differential equation:

$$m\ddot{x} + kx = 0. \tag{3}$$

The natural frequency of vibration is given by

Table 1. The experimental data.

S.No	Load mass	Initial height h1cm	Final height h2cm	Deflection H=h1-h2 cm	Stiffness (k)N/m= F/H	Average stiffness of the spring N/m	Natural frequency (f) Hz
1	No load (0.3Kg)	0	0	0	-	2850	15.512
2	0.7	18.62	18.37	0.246	2790	2850	8.496
3	1.4	18.62	18.15	0.469	2920	2850	6.516
4	2.1	18.62	17.91	0.709	2905	2850	5.485
5	2.8	18.62	17.635	0.985	2785	2850	4.825

$$f_n = \frac{1}{2\pi}\sqrt{\frac{k}{m}}. \tag{4}$$

Natural frequency of load of mass 2.8 kg (f_n) = 4.825 Hz. Angular velocity (ω_n)= 2π f_n= 30.31 rad/s. Therefore the stiffness of the spring is 2.88 kN/m Amplitude 5.5 cm, decay period 32 sec The amplitude of vibration traced by the system is shown figure 4.

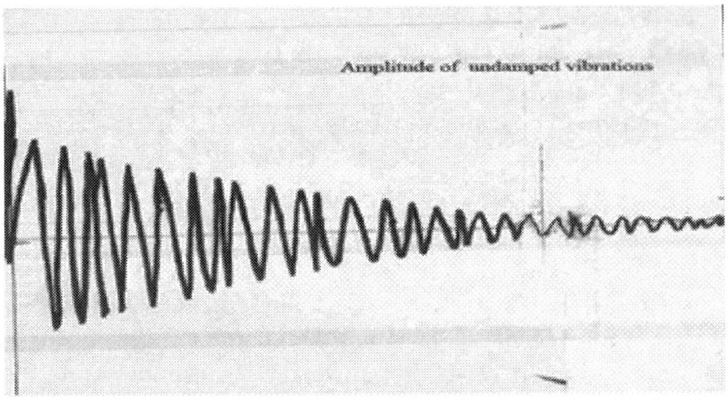

Figure 4. Pattern of vibration occurred and traced for natural vibration of the beam.

4.1. Free vibration with damping

We now add a "viscous" damper to the model that outputs a force that is proportional to the velocity of the mass. The damping is called viscous because it models the effects of an object within a fluid. The proportionality constant C is called the damping coefficient and has units of Force over velocity (N s/m).

$$F_d = -cv = -c\dot{x} = -c\frac{dx}{dt}. \tag{5}$$

By summing the forces on the mass we get the following ordinary differential equation:

$$m\ddot{x} + c\dot{x} + kx = 0. \tag{6}$$

The solution to this equation depends on the amount of damping. If the damping is small enough the system will still vibrate, but eventually, over time, will stop vibrating. This case is called under damping this case is of most interest in vibration analysis. If we increase the damping just to the point where the system no longer oscillates it reaches the point of critical damping (if the damping is increased past critical damping the system is called over damped). The value that the damping coefficient needs to reach for critical damping in the mass spring damper model is:

$$C_c = 2\sqrt{km}. \tag{7}$$

To characterize the amount of damping in a system a ratio called the damping ratio (also known as damping factor and critical damping) is used. This damping ratio is just a ratio of the actual damping over the amount of damping required to reach critical damping. The formula for the damping ratio ζ the mass spring damper model is:

$$\zeta = \frac{C}{2\sqrt{km}}. \tag{8}$$

Metal structures (e.g. airplane fuselage, engine crankshaft) will have damping factors less than 0.05 while automotive suspensions in the range of 0.2–0.3. The solution to the under damped system for the mass spring damper model is the following:

$$x(t) = Xe^{-\zeta\omega_n t}cos(\sqrt{1-\zeta^2\omega_n t - \phi}), \tag{9}$$

$$\omega_n = 2\pi f_n. \tag{10}$$

The value of X, the initial magnitude, and ϕ the phase shifts, are determined by the amount the spring is stretched. The formulas for these values can be found in the references.

4.2. Damped vibration with MR gel without Magnetic field.

From the table the damped frequency of spring mass and MR damper without Magnetic Field is given by Then $\zeta = 0.90$, Damping coefficient = 169.19 (N. sec/m) ,Amplitudes 5.5cm , decay 5 sec The amplitude of vibration traced by the system is given by figure

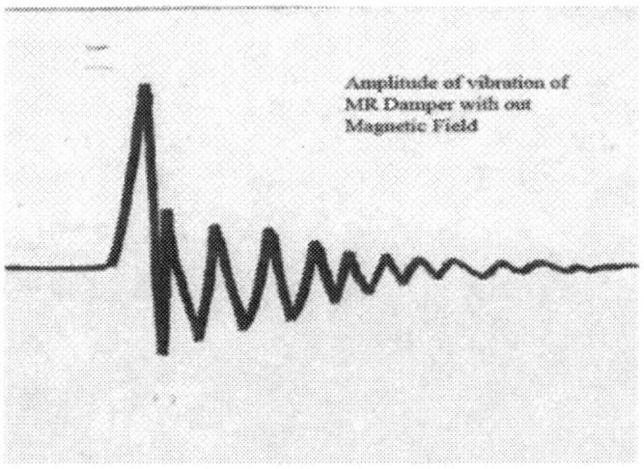

Figure 5. Pattern of vibration occurred and traced far damped vibrations with damper of MR Gel without Magnetic Field.

4.3. Damped vibration with MR Gel Magnetic field.

The magnetic fields generated by currents and calculated from Ampere's Law or the Biot-Savarts Law are characterized by the magnetic field B measured in Tesla. Using Ampere's law, one can determine the magnetic field associated with a given current or current associated with a given magnetic field, providing there is no time changing electric field present. Ampere's law is a mathematical statement of the relationship between currents and the magnetic fields they generate. Field at Centers of Current Loop, The form of the Magnetic field from a current element in the loop is given by

$$dB = \frac{\mu_0 Id\vec{L} \times \vec{r}}{4\pi R^2} = \frac{\mu_0 IdLsin\theta}{4\pi R^2}. \tag{11}$$

Which in this case simplifies greatly because the angle 90 degrees far all points along the path and the distance to the field point is constant. The integral becomes

$$B = \mu_0 \frac{I}{4\pi R^2} \oint dL = \frac{\mu_0 I}{4\pi R^2} 2\pi R = \frac{\mu_0 I}{2R}. \tag{12}$$

$$\mu_0 = 4\pi \times 10^{-7}\ T.m/A.$$

Now the current in the copper coil is determined by determining the resistance as voltage is constant. The resistance of a wire is given by

$$R = \rho \frac{L}{A}. \tag{13}$$

Resistance = $resistivity \times \frac{length}{area}$. For a wire of length **L** =2π R_p x no of turns. R_p=outer radius of CPVC pipe=.0225m. R_p=2π x 0.0225 x 1450 =205 meters.

$$Area(A) = 3.142 \times 10^{-6} m^2,$$

resistance R= 4.38 Ω. Current I= voltage/ resistance = 12/4.38 = 2.74 Amperes. Therefore the magnetic field produced is given by B =0.11 Tesla. The amplitude of vibration traced by the system is given by figure. From table 5 the damped frequency of spring made and MR damper is given by ζ =0.92. Damping coefficient = 172.96 (N. sec/m). Amplitude 5.5cm, decay period 3sec.

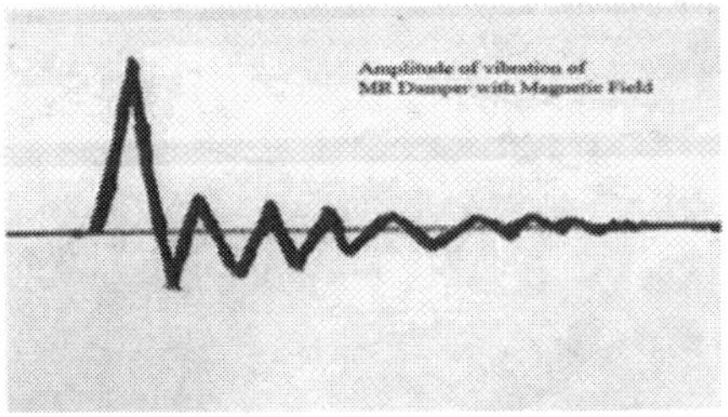

Figure 6. Pattern of vibration occurred and traced for damped vibrations with damper of MR Gel with applied Magnetic Field.

Table 2. The experimental data of various configurations with dampers.

S.No	Load (W)	Time for 5 oscillations (sec)	Average time sec	Time for 1 oscilla-tion(t) sec	Damped frequency(f) Hz F_d	Damped Angular velocity (rad/s) ω_d
1. Damper With Oil	2.8 Or 27.45 N	4.42 4.62 4.23 4.33	4.4	.44	2.27	14.26
2. Damper with MR Gel & No Magnetic Field	2.8 Or 27.45 N	2.39 2.57 2.36 2.42	2.44	0.487	2.05	12.88
3. Damper with MR Gel	2.8 Or 27.45 N	2.55 2.69 2.61 2.58	2.61	.522	1.91	12

5. Results and discusion

Comparing various systems with respect to amplitude of vibration and decay period is shown in below figure This clearly shows the effect of damping produced ba MR gel damper with

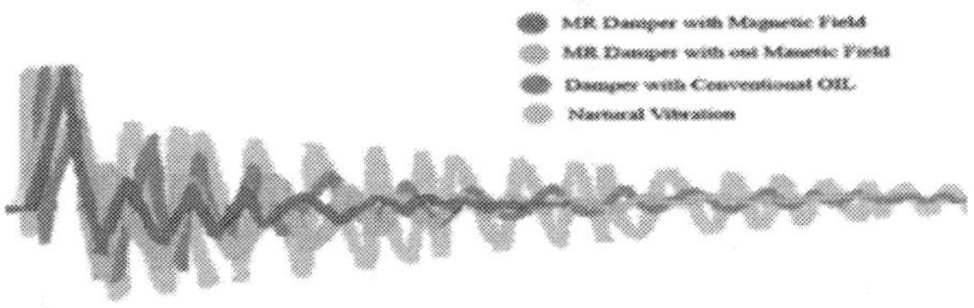

Figure 7. Pattern of vibrations

Magnetic Field is more effective than the other systems. So this provides added flexibility to control damping according to the road conditions and The effect of variation of current on the magnetic field generation is linear. The observations made during the experimentation of various systems and comparing of various systems with respect to amplitude of vibration, by un damped system are as follows, the damped system with mineral oil is 47% (decrease of 53 %), the damped system with MR gel and without magnetic field is 42% (decrease of 58%) and the damped system with MR gel and with magnetic field is 39% (decrease of 61%). The improvements made in the existing work i.e., use of variable stiffness springs in suspension system also helps in regulating ride comfort and stability of the vehicle. Both the systems combined together will produce greater ease, flexibility of control and performance. Such, control system is close to active suspension systems and provide large domain to control the ride stability and comfort without compromising according the road conditions. The results obtained from the study prove that, using MR dampers will provide a smoother ride then conventional suspension system.

References

[1] Venkateswara Rao P and Maniprakash S 2010 Functional behaviour of isotropic magnetorheological gels Smart Mater *Struct* **19** 085019

[2] Gheorghe Ghita and Marius Giuclea 2008 Semi-active Suspension Control

[3] Daniel Fischer and Rolf Isermann 2003 Mechatronic semi-active and active vehicle suspensions (0967-0661)

[4] Bing Wei and Xinglong Gong 2010 Study on the Properties of Magnetorheological Gel Based on Polyurethane.**118** 2765-71

[5] Cherry alex .(2011). Determination of model parameters of MR gel subjected to magneto mechanical loading.

[6] Yanqing Liu Hiroshi Matsuhisa 2008 Semi-active vibration isolation system with variable stiffness and damping control (0022-460X)

[7] Ly H V, Reitich F, Jolly M R, Banks H T and Ito K 1999 Simulations of Particle Dynamics in Magnetorheological Fluids *Journal of Computational Physics* **155** 160-177

[8] Chen L, Gong X L , Jiang W Q, Yao J J, Deng H X and Li W H 2007 Investigation on magneto rheological elastomers based on natural rubber *J. mater.sci.* **42** 5483-5489

[9] Fuchs A, Hu B, Gordaninejad F, Evrensel C 2005 Synthesis and Characterization of Magnetorheological Polyimide Gels, *Journal of Applied Polymer Science* **98** 2402-13

[10] Shokrollahi H and Janghorban K 2007 Different annealing treatments for improvement of magnetic and electrical properties of soft magnetic composites *Journal of Magnetism and Magnetic Materials* **317** 61-67

[11] Melek Yalcintas and Heming Dai 2004 Vibration suppression capabilities of magneto rheological materials based adaptive structures *Smart Mater.Struct.***243** 21-23

[12] Xiaojie Wang, Faramarz Gordaninejad, Mert Calgar, Yanming Liu, Joko Sutrisno, Alan Fuchs 2009 Sensing Behavior of Magnetorheological Elastomers *Journal of Mechanical Design* **131** 910-41

Initial value method for general singular perturbation problems

Loka Pavani

Assistant Professor, Department of Mathematics, OUCW, Koti, O.U., T.S., India

E-mail: lokapavani@hotmail.com

Abstract. In general, the numerical solution of a boundary value problem will be more difficult than the numerical solution of the corresponding initial value problem. Hence, we prefer to convert the given second order problem into a first order problem. In this paper we present an initial value method. It is distinguished by the following fact: the original second order problem is replaced by an asymptotically equivalent first order problem and then solved as an initial value problem. Classical RungeKutta method is used to solve the first order equation. The method is first described for solving problems with left end boundary layer. This is extended for solving singular perturbation problems with right end, internal and terminal layers. We solve one problem to demonstrate the applicability of the method. The numerical results are compared with the exact solution.

1. Introduction

Singular perturbation problems containing a small positive parameter ϵ have appeared in many fields such as fluid mechanics, chemical kinetics, elasticity, aerodynamics, plasma dynamics and magneto hydrodynamics. A few notable examples are boundary layer problems, WKB problems. For small values of ϵ, it is well known that standard numerical methods for solving such problems are unstable and fail to give accurate results. Therefore, it is important to develop suitable numerical methods for these problems, whose accuracy does not depend on the parameter value, i.e. methods that are uniformly convergent. A wide variety of papers and books have been published in the recent years, describing various methods for solving singular perturbation problems, among these we mention C.M Bender, and S.A.Orszag [1], E.Issacson, H.B.Keller [2], P.W.Hemker and J.J.H Miller [3], J.Kevorkian and Cole [4], Nayfeh [5], O'Malley [6], Y.N.Reddy [7], M.K.Kadalbajoo and Y.N.Reddy[8] ,R.Vulanovic [9], F.Mazzia and D.Trigiante[10] Van Dyke [11], J.J.H.Miller [12], [13]. Usually, the numerical solution of a boundary value problem will be more difficult than that of initial value problem. Hence, we prefer to convert the given second order problem into a first order problem.In this paper we present an initial value method. It is distinguished by the following fact: the original second order problem is replaced by an asymptotically equivalent first order problem and then solved as an initial value problem. Classical RungeKutta method is used to solve the first order equation. The method is first described for solving problems with left end boundary layer. This is extended for solving singular perturbation problems with right end, internal and terminal layers. We solve one problem to demonstrate the applicability of the method. The numerical results are compared with the exact solution.

2. Left end boundary layer problems

Let us consider a class of linear singularly perturbed two point boundary value problems of the form

$$\epsilon y''(x) + [a(x)y(x)]' + b(x)y(x) = h(x) \quad x \in [0,1], \tag{1}$$

$$\text{with } y(0) = \alpha, \tag{2.a}$$

$$\text{and } y(1) = \beta. \tag{2.b}$$

where ϵ is a small positive parameter ($0 < \epsilon \leq 1$), α, β are given constants, $a(x)$, $b(x)$, and $h(x)$ are assumed to be sufficiently continuously differentiable functions in $[0,1]$. Furthermore, we assume that $a(x) \geq M > 0$ throughout the interval $[0,1]$ where M is some positive constant. This assumption merely implies that the boundary layer will be in the neighborhood of $x = 0$. The initial value method consists of the following steps.

Step 1: Preliminary step we obtain the reduced problem of (1) by setting $\epsilon = 0$ in equation (1) and solve it for the solution with appropriate boundary condition. Let $y_0(x)$ be the solution of the reduced problem of $(1-2)(i.e)$

$$[a(x)y_0(x)]' + b(x)y_0(x) = h(x), \tag{3}$$

$$\text{and } y_0(1) = \beta. \tag{3.a}$$

Step 2: Set up the approximate equation to equation (1) as follows

$$\epsilon y''(x) + [a(x)y(x)]' + b(x)y_0(x) = h(x), \tag{4}$$

where we have simply replaced the $y(x)$ term by $y_0(x)$, the solution of the reduced problem $(3-3a)$.

Step 3: Replace the approximated second order problem $(4-2)$ by an asymptotically equivalent first order problem as follows:

Integrate equation (4) to get

$$\epsilon y'(x) + a(x)y(x) = f(x) + k, \tag{5}$$

where $f(x) = \int (h(x) - b(x)y_0(x))dx$ and k is a constant to be determined.

Step4: To determine the constant k, we introduce the condition that the reduced equation of (5) should satisfy the boundary condition $y(1) = \beta$ $(i.e)$ $y(1) = \frac{(f(1)+k)}{(a(1))} = \beta$,

$$k = a(1)\beta - f(1). \tag{6}$$

Remark 1: This choice of k ensures that the solution of the reduced problem of $(1-2)$ satisfies the reduced equation of (5).

Step 5: We now adjoin the condition (which we drop, whenever we formulate the reduced problem of the equation $(1-2)$, $y(0) = \alpha$ to the equation (5) to obtain first order problem as follows

$$\epsilon y'(x) + a(x)y(x) = f(x) + k \quad x \in [0,1], \tag{7}$$

$$\text{with } y(0) = \alpha, \tag{7.a}$$

k is given by (6). Thus in a manner of speaking we have replaced the original second order problem $(1-2)$ with the asymptotically equivalent first order problem $(7-7a)$. We solve this

initial value problem to obtain the solutions over the interval $0 \leq x \leq 1$. There now exist a number of efficient methods for the solution of initial value problems. In order to solve the initial value problems in our numerical experimentation we make use of the classical fourth order Runge-Kutta method. In fact any standard analytical or numerical methods can be used.
Remark 2: For the case $b(x) = 0$ we do not require step (1)&(2) because we can integrate directly the given equation. Detailed discussion and numerical examples can be found in Kadalbajoo and Reddy [8].

3. Right end boundary layer problems

We now describe this initial value method for the singularly perturbed two point boundary value problems with the right end boundary layer of the underlying interval. To be specific we consider a class of linear singularly perturbed two point boundary value problems of the form

$$\epsilon y''(x) + [a(x)y(x)]' + b(x)y(x) = h(x) \quad x \in [0,1], \tag{8}$$

$$\text{with } \; y(0) = \alpha, \tag{9.a}$$

$$\text{and } \; y(1) = \beta, \tag{9.b}$$

where ϵ is a small positive parameter $(0 < \epsilon \leq 1)$, α, β are given constants, $a(x)$, $b(x)$, and $h(x)$ are assumed to be sufficiently continuously differentiable functions in $[0,1]$. Furthermore, we assume that $a(x) \geq M > 0$ throughout the interval $[0,1]$ where M is some positive constant. This assumption merely implies that the boundary layer will be in the neighborhood of $x = 1$. The initial value method consists of the following steps.
Step 1: Preliminary step:
We obtain the reduced problem by setting $\epsilon = 0$ in equation (8) and solve it for the solution with appropriate boundary condition. Let $y_0(x)$ be the solution of the reduced problem of $(8-9)(i.e)$

$$[a(x)y_0(x)]' + b(x)y_0(x) = h(x), \tag{10}$$

$$\text{and } \; y_0(1) = \alpha. \tag{10.a}$$

Step 2: Set up the approximate equation to equation (8) as follows

$$\epsilon y''(x) + [a(x)y(x)]' + b(x)y_0(x) = h(x), \tag{11}$$

where we have simply replaced the $y(x)$ term by $y_0(x)$, the solution of the reduced problem $(10-10a)$.
Step 3: Replace the approximated second order problem $(11-9)$ by an asymptotically equivalent first order problem as follows:
Integrate equation (11) to get

$$\epsilon y'(x) + a(x)y(x) = f(x) + k, \tag{12}$$

where $f(x) = \int (h(x) - b(x)y_0(x))dx$ and k is a constant to be determined.
Step 4: To determine the constant k, we introduce the condition that the reduced equation of (12) should satisfy the boundary condition $y(0) = \alpha$

$$y(0) = \frac{1}{(a(0))}[f(0) + k] = \alpha,$$

$$k = a(0)\alpha - f(0). \tag{13}$$

Remark 3: This choice of k ensures that the solution of the reduced problem of $(8-9)$ satisfies the reduced equation of (12).
Step 5: We now adjoin the condition (which we drop, whenever we formulate the reduced problem of equation $8-9$, $y(1)=\beta$ to (12) to obtain the first order problem as follows.

$$\epsilon y^{'}(x) + a(x)y(x) = f(x) + k \quad x \in [0,1], \tag{14}$$

$$\text{with} \;\; y(1) = \beta, \tag{14.a}$$

where k is a constant given by (13)

Thus in a manner of speaking we have replaced the original second order problem $(8-9)$ with the asymptotically equivalent first order problem $(14-14a)$. We solve this initial value problem to obtain the solution over the interval $0 \leq x \leq 1$.There now exist a number of efficient methods for the solution of initial value problem. In our numerical experimentation, we make use of classical fourth order Runge - Kutta method. In fact any standard analytical or numerical method can be used.
Remark 4: For the case $b(x) = 0$, we do not require the preliminary step, because we can directly integrate the given equation.

4. Internal layer problems

We will now extend the initial value method to singular perturbation problem with an internal layer of the underlying interval. In this case $a(x)$ changes sign in the domain of interest. Without loss of generality we can take $a(0)=0$ and the interval to be $[-1,1]$. To describe the method, we again consider a class of linear singularly perturbed two point boundary value problems of the form

$$\epsilon y^{''}(x) + [a(x)y(x)]^{'} + b(x)y(x) = h(x) \quad x \in [0,1], \tag{15}$$

$$\text{with} \;\; y(-1) = \alpha, \tag{16.a}$$

$$\text{and} \;\; y(1) = \beta. \tag{16.b}$$

where ϵ is a small positive parameter $(0 < \epsilon << 1)$, α, β are given constants $a(x)$, $b(x)$, $h(x)$ are assumed to be sufficiently continuously differentiable functions in $[-1,1]$. Furthermore we assume that $a(x) \leq M \leq 0$ in $[-1,0]$ where M is some negative constant and $a(x) \geq M > 0$ in $[0,1]$ where M is some positive constant. This assumption implies that the boundary layer is in the nbd of $x=0$. We now proceed as follows.
Step1: We first find the approximate solution at $x=0$. Without loss of generality we can take $a(0)=0$. At $x=0$ equation (15) becomes

$$\epsilon y^{''}(x) + [a(x)y(x)]^{'} + b(x)y(x) = h(x) \quad \text{at} \;\; x = 0. \tag{17}$$

The reduced problem of (17) obtained by putting $\epsilon = 0$ gives us the approximation to $y(0)$.

$$b(0)y(0) = h(0), \tag{18}$$

is the reduced problem. Therefore

$$y(0) = \frac{h(0)}{b(0)} = \gamma. \tag{19}$$

Step 2: We now divide the interval $[-1,1]$ into two subintervals $[-1,0]$ and $[0,1]$ so that equation (15) has a right layer in $[-1,0]$ and a left layer in $[0,1]$.

Step 3: We now use initial value method for right end boundary layer as described in section (3) in the interval $[-1,0]$. We have

$$\epsilon y''(x) + [a(x)y(x)]' + b(x)y(x) = h(x) \quad x \in [-1,0], \tag{20}$$
$$\text{with } \; y(-1) = \alpha, \; \text{ and } \; y(0) = \gamma. \tag{21}$$

The corresponding initial value problem is

$$\epsilon y'(x) + a(x)y(x) = f(x) + k; \quad x \in [-1,0], \tag{22}$$
$$\text{with } \; y(0) = \gamma. \tag{22.a}$$

where k is a constant given by

$$k = a(-1)\alpha - f(-1). \tag{23}$$

Step 4: We use initial value method for left end boundary layer for the other half interval $[0,1]$ as described in section(2). We have

$$\epsilon y''(x) + [a(x)y(x)]' + b(x)y(x) = h(x) \quad x \in [0,1], \tag{24}$$
$$\text{with } \; y(0) = \gamma, \tag{25.a}$$
$$\text{and } \; y(1) = \beta. \tag{25.b}$$

The corresponding initial value problem is

$$\epsilon y'(x) + a(x)y(x) = f(x) + k; \quad x \in [0,1], \tag{26}$$
$$\text{with } \; y(0) = \gamma. \tag{26.a}$$

where k is a constant given by

$$k = a(1)\beta - f(1). \tag{27}$$

Thus, in a manner of speaking we have replaced the original second order problem $(15-16)$ with asymptotically equivalent first order problems $(22-22a)$ and $(26-26a)$. We solve these initial value problems to obtain solution over $[-1,0]$ and $[0,1]$ respectively.

5. Numerical example

With the help of one model example, we shall demonstrate the applicability of initial value method for internal layer problems.

Example 5.1: Consider the following SPP

$$\epsilon y''(x) + xy'(x) - y(x) = 0; \quad x \in [-1,1],$$
$$\text{with } \; y(-1) = 1, \text{and } \; y(1) = 2.$$

For this example we have $a(x) = x$, $b(x) = -1$ and $f(x) = 0$. Further we have an internal layer of width $o(\sqrt{\epsilon})$ at $x = 0$ (for details, see O'Malley $[6, pp168-172, eq8.1case(i)]$ and Kevorkian and Cole $[4, pp41-43$, eqs (2.3.76) and (2.3.77)]
The given equation can be written as $\epsilon y''(x) + (xy)' - 2y(x) = 0$;
Here $a(x) = x$; $b(x) = -2$; $h(x) = 0$
Step1: $\gamma = 0$
Step 2: In the interval $[-1,0]$ we have a right layer. The solution of the reduced problem

$(xy)' - 2y(x) = 0$ with $y(-1) = 1$ is $y_0(x) = -x$, $k = 0$, $f(x) = -x^2$
The initial value problem is

$$\epsilon y'(x) + xy(x) = -x^2; \quad x \in [0,1]$$
$$\text{with } y(0) = 0.$$

Step 3: In the interval $[0,1]$ we have a right layer. The solution of the reduced problem $(xy)' - 2y = 0$ with $y(1) = 2$ is $y_0(x) = 2x$, $k = 0$, $f(x) = 2x^2$. We solve these initial value problems using classical RungeKutta method. The numerical results are presented in **Table 1** for $\epsilon = 10^{-3}$

Table 1. Computational results for Example 5.1 with $\epsilon = 10^{-3}$ and $h = 0.01$.

x	y(x)	Exact Solution
-1.000	1.0000000	1.0000000
-0.500	0.4979919	0.4999964
-0.100	0.0884604	0.1032475
-0.080	0.0642969	0.0866667
-0.060	0.0384793	0.0727863
-0.040	0.0152044	0.0629293
0.000	0.0000000	0.0606750
0.020	0.0049300	0.0785011
0.040	0.0316510	0.1029288
0.060	0.0777414	0.1327857
0.080	0.1287349	0.1666600
0.100	0.1768589	0.2032466
0.500	0.9959838	0.9999939
1.000	2.0000000	2.0000000

6. Two boundary layers problems

The suggestions given for internal layers problems can be extended mutatis mutandis to problems with two boundary layers. Consider the class of linear singularly perturbed problems of the form

$$\epsilon y''(x) + [a(x)y(x)]' + b(x)y(x) = h(x) \quad x \in [-1,1], \tag{28}$$
$$\text{with } y(-1) = \alpha, \tag{29.a}$$
$$\text{and } y(1) = \beta. \tag{29.b}$$

where ϵ is a small positive parameter $(0 < \epsilon \leq 1)$ α, β are given constants, $a(x)$, $b(x)$, $h(x)$ are assumed to be sufficiently continuously differentiable function in$[1,1]$. Furthermore we assume $a(x) \geq M > 0$ in $[-1,0]$ and $a(x) \leq M < 0$ in $[0,1]$. This assumption implies that the boundary layer is at $x = -1$ and $x = 1$. With out loss of generality we take $a(x) = 0$ at $x = 0$ since $a(x)$ changes sign in the domain of interest. We now proceed as follows:
Step1: We first find the approximate solution at $x = 0$. Equation (28) is now

$$\epsilon y''(x) + b(x)y(x) = h(x). \tag{30}$$

The reduced problem of (28) gives us the approximation to $y(0)$

$$\text{where} \qquad y(0) = \frac{h(0)}{b(0)} = \gamma. \tag{31}$$

Step 2: We now divide the interval $[-1, 1]$ into two subintervals $[-1, 0]$ and $[0, 1]$ so that equation (28) has a left layer in $[-1, 0]$ and a right layer in $[0, 1]$.
Step 3: We now use initial value technique for left end boundary layer as described in section (2) in the interval $[-1, 0]$. We have

$$\epsilon y''(x) + [a(x)y(x)]' + b(x)y(x) = h(x) \quad x \in [-1, 0], \tag{32}$$
$$\text{with } \ y(-1) = \alpha, \tag{33.a}$$
$$\text{and } \ y(0) = \gamma. \tag{33.b}$$

The corresponding initial value problem is

$$\epsilon y'(x) + a(x)y(x) = f(x) + k \quad x \in [-1, 0], \tag{34}$$
$$\text{with } \ y(-1) = \alpha, \tag{34.a}$$

where k is constant given by

$$k = a(0)\gamma - f(0). \tag{35}$$

Step 4: We use initial value technique for right end boundary layer as described in section (3) above for the interval $[0, 1]$. We have

$$\epsilon y''(x) + [a(x)y(x)]' + b(x)y(x) = h(x), \tag{36}$$
$$\text{with } \ y(0) = \gamma, \tag{37.a}$$
$$\text{and } \ y(1) = \beta. \tag{33.b}$$

The corresponding initial value problem is

$$\epsilon y'(x) + a(x)y(x) = f(x) + k \quad x \in [0, 1], \tag{38}$$
$$\text{with } \ y(1) = \beta, \tag{38.a}$$

where k is constant given by

$$k = a(0)\gamma - f(0). \tag{39}$$

Thus in a manner of speaking we have replaced the original second order problem $(28 - 29)$ with asymptotically equivalent first order problems (34-34a) and (38-38a). We solve these initial value problems to obtain solutions over $[-1, 0]$ and $[0, 1]$ respectively. There now exist a number of efficient methods for the solutions of initial value problems. We use classical Runge-Kutta method in our numerical experimentation. Any other method can also be used.

7. Numerical example

To demonstrate the applicability of the method we solve one problem.
Example 7.1: Consider the following SPP

$$\epsilon y''(x) - xy'(x) - y(x) = 0 \quad x \in [-1, 1],$$
$$\text{with } \ y(-1) = 1\text{and } \ y(1) = 2.$$

Journal of Physics: Conference Series **662** (2015) 012027 doi:10.1088/1742-6596/662/1/012027

For this example we have $a(x) = -x$, $b(x) = 1$ and $f(x) = 0$. Further we have two boundary layers one at $x = -1$ and one at $x = 1$ (for details, see O'Malley [6,pp168-173,eq 8.1 case (i)] The given equation can be written as

$$\epsilon y^{''}(x) - (xy)^{'}(x) = 0,$$

here $a(x) = -x$, $b(x) = 0$, and $h(x) = 0$.
Step 1: $\gamma = 0$.
Step 2: In the interval $[-1, 0]$ we have a left layer. The solution of the reduced problem $(xy)^{'} = 0$ with $y(0) = 0$ is $y_0(x) = 0$, $k = 0$, $f(x) = 0$.
The initial value problem is

$$\epsilon y^{'}(x) - xy(x) = 0 \quad x \in [-1, 0],$$
$$\text{with } \; y(-1) = 1.$$

Step 3: In the interval $[0, 1]$ we have a right layer. The solution of the reduced problem $(xy)^{'} = 0$ with $y(0) = 0$ is $y_0(x) = 0$, $k = 0$, $f(x) = 0$.
The initial value problem is

$$\epsilon y^{'}(x) - xy(x) = 0 \quad x \in [0, 1],$$
$$\text{with } \; y(1) = 2.$$

We solve these initial value problems using classical Runge-Kutta method. The numerical results are presented in **Table 2** for $\epsilon = 10^{-3}$.

Table 2. Computational results for Example 5.1 with $\epsilon = 10^{-3}$ and $h = 0.01$.

x	y(x)	Exact Solution
-1.000	1.0000000	1.0000000
-0.980	0.0000000	0.6393120
-0.960	0.0000000	0.0042141
-0.940	0.0000000	0.0002865
-0.920	0.0000000	0.0000201
-0.900	0.0000000	0.0000014
-0.700	0.0000000	0.0000000
-0.300	0.0000000	0.0000000
0.300	0.0000000	0.0000000
0.900	0.0000000	0.0000029
0.920	0.0000000	0.0000402
0.940	0.0000000	0.0005730
0.960	0.0000000	0.0084281
0.980	0.0000000	0.1278624
1.000	2.0000000	2.0000000

8. Discussion and conclusion

We have presented and illustrated an initial value method for solving singularly perturbed two point boundary value problem. We have replaced the original second order problem by an equivalent first order problem and then solved it as an initial value problem. We used Runge-Kutta method to solve the first order equations. We extended the method to solve problems with internal and terminal layers. We solved one problem each to demonstrate the applicability of the method. It is observed that the method compares well with the known method.

References

[1] Bender C M and Orszag S A 1978 *Advanced mathematical methods for scientists and engineers* (New York: McGraw-Hill)
[2] Isaacson E and Keller H B 1966 *Analysis of numerical methods* (New York: John Wiley & Sons)
[3] Hemker P W and Miller J J H (Editors) 1979, *Numerical analysis of singular perturbation problems* (New York: Academic Press)
[4] Kevorkian J and Cole J D 1981 *Perturbation methods in applied mathematics,* (New York: Springer-Verlag)
[5] Nayfeh A H 1973 *Perturbation methods* (New York: Wiley)
[6] O'Malley R E 1974 *Introduction to singular perturbations* (New York: Academic Press)
[7] Y N Reddy 1986 *Numerical treatment of singularly perturbed two point boundary value problems* (IIT, Kanpur, India:Ph.D. thesis)
[8] Kadalbajoo M K and Y N Reddy 1989 An initial value technique for a class of non-liear single perturbation problem *Applied Mathematics and computation* **30** 223
[9] Vulanovic R 1991 A second order numerical method for non-linear singular perturbation problems without turning points *Comput.Maths.Math.Phys.* **31** 28
[10] Mazzia F and Trigiante D 1992 Numerical methods for second order singular perturbation problems*Computers. Math.Applic.* **23** 81
[11] Van Dyke M 1964 *Perturbation methods in fluid mechanics* (New York:Academic Press)
[12] Miller J J H (Editor) 1980 *Boundary and interior layers - Computational and Asymptotic Methods Proceedings of BAIL Int. Conference* Dublin: Boole Press
[13] Miller J J H, O'Riordan E, and Shishkin GI 1996*Fitted numerical methods for singular perturbation problems, error estimates in te maximum norm for linear problems in one and two dimensions* (World Scientific Publishing Company Pvt. Ltd)

International Conference on Vibration Problems (ICOVP-2015) IOP Publishing
Journal of Physics: Conference Series **662** (2015) 012028 doi:10.1088/1742-6596/662/1/012028

Numerical solution of fourth order boundary value problem using sixth degree spline functions

P Kalyani[1], A S Madhusudhan Rao[2], P S Rama Chandra Rao[3]

Department of Mathematics, Hawassa University, Hawassa, Ethiopia[1]

Department of Physics, Hawassa University, Hawassa, Ethiopia[2]

Kakatiya Institute of Technology and Science, Warangal-506001, India[3]

E-mail: kk.parcha@yahoo.com, madhuammiraju@yahoo.co.in

Abstract. In this communication, we developed sixth degree spline functions by using Bickley's method for obtaining the numerical solution of linear fourth order differential equations of the form $y^{(4)}(x)+f(x)y(x)=r(x)$ with the given boundary conditions where f(x) and r(x) are given functions. Numerical illustrations are tabulated to demonstrate the practical usefulness of method.

1. Introduction

Spline functions play an important role in finding numerical solution of boundary value problems. Splines are types of curves, originally developed for shipbuilding in the days before computer modeling. Bickley[1] has suggested a different approach of solving linear two-point boundary value problem. To model the solution curve he used cubic spline interpolation and applied the differential equation as well as the boundary conditions for obtaining unknown constants. Khan [2] has considered the applications of cubic spline functions for the solution of two point boundary value problems. Some of the books which discuss splines include Ahlberget et al.[3], deBoor [4], Prenter[5], Schumaker [6], Shikin and Plis [7] and Spath [8]. Chawla and Katti [9] employed finite difference method for a class of singular two point BVPs. A class of BVPs has been solved by Rama Chandra Rao[10] using numerical integration. Caglar et al. [11] have solved third order linear and nonlinear BVPs using fourth degree B-spline. Usmani and Sakai [12] solved the special case of linear third order BVPs using the quartic spline.Usmani [13] considered the problem of bending a rectangular clamped beam resting on an elastic foundation.He derived numerical techniques of order 2, 4 and 6 for the solution of a fourth order linear boundary value problem. Papamichael and Worsey [14] have developed a cubic spline method for the solution of a linear fourth order two point boundary value problem. Siddiqi and Akram [15] described a quintic spline method for the solution of fourth order boundary value problems and derived end conditions for quintic spline interpolation, at equally spaced knots. We used Bickley's method for obtaining the numerical solution of linear fourth order boundary value problems by constructing sixth degree spline function. Numerical illustrations are summarized in tabular form and shown graphically.

2. Cubic spline-Bickley's method

Suppose the interval $[x_0, x_n]$ is divided into n subintervals with knots $x_0, x_1, x_2, ..., x_n$ starting at x_0, the function u(x) in the interval $[x_0, x_1]$ is represented by a cubic spline in the form

$$s(x) = a + b(x - x_0) + c(x - x_0)^2 + d(x - x_0)^3. \tag{1}$$

For the next interval $[x_1, x_2]$ the spline function s(x) is supposed in the form

$$s(x) = s(x) \quad \text{on} \quad [x_0, x_1] + d_1(x - x_1)^3,$$

proceeding in to the next interval $[x_2, x_3]$ we add another term $d_2(x - x_2)^3$ and so until we reach x_n. Thus the function s(x) is represented in the form

$$s(x) = a + b(x - x_0) + c(x - x_0)^2 + \sum_{i=0}^{n-1} d_i(x - x_i)^3, \tag{2}$$

$$s'(x) = b + 2c(x - x_0) + 3\sum_{i=0}^{n-1} d_i(x - x_i)^2, \tag{3}$$

$$s''(x) = 2c + 6\sum_{i=0}^{n-1} d_i(x - x_i). \tag{4}$$

2.1. *The two-point second order boundary value problem*

First, we consider the linear differential equation

$$p(x)u'' + q(x)u' + r(x)u = v(x), \tag{5}$$

with the boundary conditions

$$\alpha_0 u + \beta_0 u' = \gamma_0 \quad \text{at} \quad x = x_0, \alpha_n u + \beta_n u' = \gamma_n \quad \text{at} \quad x = x_n. \tag{6}$$

The number of coefficients in (2) is (n+3), the satisfaction of the differential equation by the spline function at (n+1) nodes gives (n+1) equations in the (n+3) unknowns. Also the boundary conditions (6) give two more equations in the unknowns. Thus we get $(n+3)$ equations in $(n+3)$ unknowns $a, b, c, d_0, d_1, ..., d_{(n-1)}$, after determining these unknowns we substitute them in (2) and thus we get the cubic spline approximation of $u(x)$. Putting $x = x_0, x_1, x_2, ..., x_n$ in the spline function thus determined, we get the solution at the nodes. The system of equations to be satisfied by the constants $a, b, c, d_0, d_1, ..., d_{(n-1)}$ is derived below. Substituting (2), (3), (4) in (5), at $x = x_m$ we get

$$\begin{aligned} &ar_m + b[r_m(x_m - x_0) + q_m] + c[r_m(x_m - x_0)^2 + 2q_m(x_m - x_0) + 2p_m] \\ &+ \sum_{i=0}^{m-1} d_i[r_m(x_m - x_i)^3 + 3q_m(x_m - x_i)^2 + 6p_m(x_m - x_i)] = s_m, m = 0, 1, 2, ..., n \end{aligned} \tag{7}$$

where $p_m = p(x_m), q_m = q(x_m), r_m = r(x_m)$ and $s_m = s(x_m)$. Applying boundary conditions (6), we get

$$\begin{aligned} &\alpha_0 a + \beta_0 b = \gamma_0; \alpha_n a + [\alpha_n(x_n - x_0) - \beta_n]b + [\alpha_n(x_n - x_0)^2 - 2\beta_n(x_n - x_0)]c + \\ &\textstyle\sum_{m=0}^{n-1}[\alpha_n(x_n - x_0)^3 - 3\beta_m(x_n - x_m)^2]d_m = \gamma_n. \end{aligned} \tag{8}$$

If these equations are taken in the order (8),(7) with $m = n, \ n-1, \ ..., \ 0$ the coefficient matrix of unknowns $d_{(n-1)}, d_{(n-2)} d_1, d_0$, c, b, a is of the Hessenberg form, namely an upper triangle with a single lower sub-diagonal. The forward elimination is then simple with only one multiplier at each step and the back substitution is correspondingly easy.

3. Construction of sixth degree spline

Suppose the interval $[x_0, x_n]$ is divided into subintervals with grid points $x_0, x_1, x_2, ..., x_n$. Starting at x_0, the function y(x) in the interval $[x_0, x_1]$ is represented by a sixth degree spline.

$$s(x) = a + b(x - x_0) + c(x - x_0)^2 + d(x - x0)^3 + e(x - x_0)^4 + g(x - x_0)^5 + h_0(x - x_0)^6.$$

Proceeding in to the next interval $[x_1, x_2]$, we add a term $h_1(x - x_1)^6$, proceeding in to the next interval $[x_2, x_3]$ we add another term $h_2(x - x_2)^6$ and so until we reach x_n. Thus the function y(x) is represented in the form

$$s(x) = a + b(x - x_0) + c(x - x_0)^2 + d(x - x_0)^3 + e(x - x_0)^4 + g(x - x_0)^5 + \sum_{i=0}^{n-1} h_i(x - x_i)^6. \tag{9}$$

It can be shown that s(x) and its first five derivatives are continuous across nodes.

3.1. Method of obtaining the solution of fourth order boundary value problem by sixth degree spline

Consider the linear fourth order differential equation

$$y^{(4)}(x) + f(x)y(x) = r(x), \tag{10}$$

with the boundary conditions

$$y(x_0) = \alpha, y(x_n) = \beta, y'(x_0) = \alpha', y'(x_n) = \beta'. \tag{11}$$

From (11, and taking spline approximation in (10) at $x = x_i$ for $i = 0, 1, 2, \ldots, n$, we get $(n+5)$ equations in $(n + 6)$ unknowns $a, b, c, d, e, g, h_0, h_1, h_2, ..., h_{n-1}$. To have the solution for the unknowns, one more equation is required. So we assume that $h_{n-1} = h_{n-2}$. After determining these unknowns we substitute them in (9) and thus we get the sixth degree spline approximation of $y(x)$. Putting $x = x_1, x_2, x_3, ..., x_n$ in the spline function thus determined, we get the solution at the nodes.

4. Numerical illustrations

In this section we consider a linear boundary value problem. The numerical solution and absolute errors are obtained by sixth degree spline function when $h = 0.2$. The approximate solution, exact solutions and absolute errors at the grid points are given in the tabular form. Further the approximate solution and exact solution are shown graphically.

Example 1

Consider the following fourth order linear boundary value problem

$$y^{(4)}(x) + xy(x) = -(8 + 7x + x^3)e^{(x)}, 0 \leq x \leq 1, \tag{12}$$

With the boundary conditions

$$y(0) = y(1) = 0, y'(0) = 1, y'(1) = -e. \tag{13}$$

Solution of BVP (12)-(13) when h = 0.2:

The interval [0,1] is divided in to 5 equal subintervals , we denote the knots by x_0, x_1, x_2, x_3, x_4

and x_5 where $x_0 = 0, x_1 = 0.2, x_2 = 0.4, x_3 = 0.6, x_4 = 0.8, x_5 = 1$ The sixth order spline s(x) which approximate y(x) is given by

$$s(x) = a + b(x - x_0) + c(x - x_0)^2 + d(x - x_0)^3 + e(x - x_0)^4 + g(x - x_0)^5 + \sum_{i=0}^{4} h_i(x - x_i)^6,$$

There are 11 unknowns in s(x) which are to be determined from the following conditions by assuming $h_3 = h_4$

$$\begin{aligned} &s(x_0) = 0, s(x_5) = 0, s'(x_0) = 1, s'(x_5) = -e, \\ &s^{(4)}(x_i) + x_i s(x_i) = -(8 + 7x_i + x_i^3)e^{(x_i)}, \quad \text{for} \quad i = 0, 1, 2, 3, 4, 5. \end{aligned} \tag{14}$$

From the conditions $s(x_0) = 0, s'(x_0) = 1$, we have a = 0, b=1, hence the spline s(x) reduces to the form

$$s(x) = (x - x_0) + c(x - x_0)^2 + d(x - x_0)^3 + e(x - x_0)^4 + g(x - x_0)^5 + \sum_{i=0}^{4} h_i(x - x_i)^6, \tag{15}$$

From (16)

$$s'(x) = 1 + 2c(x - x_0) + 3d(x - x_0)^2 + 4e(x - x_0)^3 + 5g(x - x_0)^4 + 6\sum_{i=0}^{4} h_i(x - x_i)^5, \tag{16}$$

and

$$s^{(4)}(x) = 24e + 120g(x - x_0) + 360\sum_{i=0}^{4} h_i(x - x_i)^2. \tag{17}$$

Substituting (16), (17) in (15) we get a system of equations and solving those equations we get

$$\begin{aligned} &c = -0.000025, \quad d = -0.499905, \quad e = -0.033333, \\ &g = -0.1250311, \quad h_0 = -0.03676, \quad h_1 = -0.016080, \\ &h_2 = -0.013420, \quad h_3 = -0.025155, \quad h_4 = -0.025155. \end{aligned} \tag{18}$$

Substituting these values in (16) we get the spline approximation s(x) of y(x). The values of s(x),y(x) and the corresponding absolute errors at x_1, x_2, x_3, x_4, x_5 have been given in the Table-1 and the comparison has been shown in figure 1.

Table 1. Approximate solution s(x), exact solution y(x) and absolute errors

x	s(x)	y(x)	Absolute error
0.2	0.19542406	0.19542440	3.7715E-07
0.4	0.35803683	0.35803793	1.0954E-06
0.6	0.43730732	0.43730851	1.1955E-06
0.8	0.35608594	0.35608655	6.0648E-07

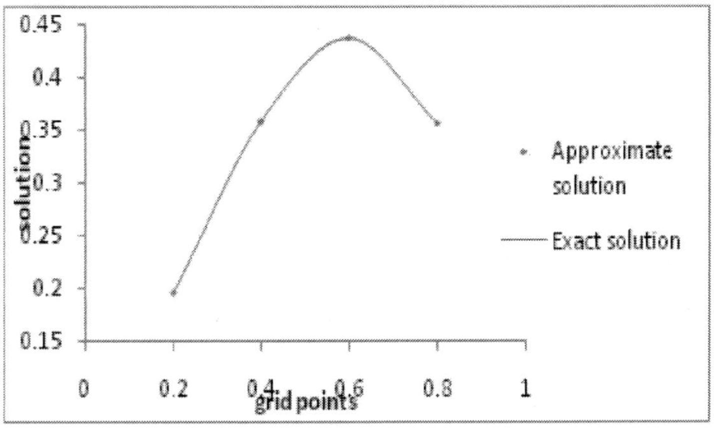

Figure 1. Comparison of approximate solution and exact solution with h=0.2.

5. Conclusion

The numerical method is developed to obtain the solution of fourth order boundary value problems using sixth degree spline. Approximate solution, exact solution and absolute errors with $h = 0.2$ are summarized in the table. The comparison has been shown graphically. It is observed that the approximate solutions are in good agreement with the exact solution. It is also observed that the approximate solution is more close to the exact solution when h is small. Further it is also noted that the numerical solutions obtained are remarkably accurate and have negligible absolute errors even for the values of h as large as 0.2.

References

[1] Bickely W G 1998 *Computer J.* **202** 11
[2] Khan A 2004 *Applied Mathematics and Computation* **175154**
[3] Ahlberg J H et al 1967 *Academic press Inc*
[4] de Boor C 2001 *(New York: Springer-Verlag))*
[5] Prenter P M 1975 *(New York: Wiley)*
[6] Schumaker L L 1981 *(New York: John Wiley and Son)*
[7] Shikin E V and Plis A I 1995 *CRC press*
[8] Spath H 1995 *A.K. Peters*
[9] Chawla M M and Katti P 1984 *IMA .J. Number. Anal* **457**
[10] Rama Chandra Rao P S 2006 *Indian Journal of Mathematics and Mathematical Sciences* **137** 2
[11] Caglar H N et al 1999 *Int. J. Comput. Math* **71373**
[12] Usmani R A and Sakai M 1984 *J. Math. and Phys .Sc.* **365**
[13] Usmani Riaz A 1978 *Mathematics of Computation* **108** 732
[14] Papamichael N and Worsey A J 1981 *J. Comptut. Appl. Math.* **187** 7
[15] Siddiqi S S and Akram G 2008 *International Journal of Numerical analysis and Modeling* **101** 5

Journal of Physics: Conference Series **662** (2015) 012029 doi:10.1088/1742-6596/662/1/012029

Modified generalized Chaplygin gas model in Bianchi type-V space-time geometry with dynamical G and Λ

S Kotambkar[1], G P Singh[2] and R Kelkar[3]

[1]Department of Applied Mathematics, L I T, RTM Nagpur University, Nagpur, India.
[2]Department of Mathematics, V N I T, Nagpur, India.
[3]Department of Applied Mathematics, S B Jain ITMR, Nagpur, India.

E-mail: shubha.kotambkar@rediffmail.com

Abstract. In this paper a new set of exact solutions of Einsteins field equations have been obtained for Bianchi type V space-time geometry. The first part of the paper deals with perfect fluid cosmological models in the presence of variable gravitational and cosmological constants. The second part of the paper is devoted to study the modified generalized Chaplygin gas model. The physical and dynamical behaviors of the models have been discussed with the help of graphical representation.

1. Introduction

It has been well established in the literature [1–5] that the universe is currently accelerating. However, final satisfactory explanation about physical mechanism and driving force of accelerated expansion of the universe is yet to achieve. To understand the accelerating behavior of the universe, role of varying cosmological constant is also considered as very important. The current interest in stems mainly from observations of type Ia high red shift supernovae which indicate that the universe is accelerating fueled perhaps by a small cosmological - term [6]. Cosmological constant problem and consequences on cosmology with a time varying cosmological constant are investigated by some researchers [7–11].

A variation of G has many interesting consequences both in geology and astrophysics. Canuto and Narlikar [12] have shown that the G varying cosmology is consistent with what so ever cosmological observations presently available. Singh and Kotambkar [13, 14] have discussed cosmological models with G and Λ in higher dimensional space-time. Vishwakarma [15] has investigated a Bianchi type I model with variable G and Λ . Singh et. al. [16] have studied a new class of cosmological models with variable G and Λ . Singh and Kale [17] have discussed anisotropic bulk viscous cosmological models with variable G and Λ .

The observed universe is homogeneous and isotropic, hence space- time is usually described by Fridman- Lemaitra-Robertson-Walkar (FLRW) cosmology. But it is widely believed that FLRW model does not give a correct matter description in the early stage of universe. The inconsistencies found in the cosmic microwaves background (CMB) and the large structure observations are stimulations to grow interest in the study of anisotropic cosmological model of the universe. Recent experimental data support the existence of an anisotropic phase, whch turns in to an isotropic one during the evolution of the universe. A number of authors have investigated Bianchi type V cosmological model in general relativity

Journal of Physics: Conference Series **662** (2015) 012029 doi:10.1088/1742-6596/662/1/012029

in different context [18–21]. Recently Yadav et. al. [22], Yadav and Sharma [23] have discussed about transit universe in Bianchi type V space-time with variable G and Λ.

There is mounting evidence from supernovae [24], WMAP [25–29], BAO oscillation data that the universe at present is dominated by a smooth component with negative pressure, the so called dark energy, leading to accelerated expansion. To avoid the problems associated with cosmological models with and quintessence models, recently it has been shown that Chaplygin gas (CG) may be useful for describing dark energy because of its positive energy density and negative pressure. Due to effectiveness of CG in explaining the evolution of the universe, several generalization of Chaplygin gas have been proposed in the literature [30, 32]. The generalized Chaplygin gas (GCG), described by equation of state

$$p = \frac{-B}{\rho^{\alpha}}$$

is considered, where constants B and α satisfy respectively $B > 0$ and $0 < \alpha \leq 1$ [33, 34]. State finder diagnostic for modified Chaplygin gas in Bianchi type V universe has been discussed by Adhav [35]. Number of researchers investigated cosmological models with modified generalized Chaplygin gas [36–39]. Observational constraints on modified Chaplygin gas in Horava-Lifschitz gravity with dark radiation have been studied by Paul et. al. [40].

2. Field equations

The Einstein field equation with time dependent G and Λ may be written as

$$R_i^j - \frac{1}{2}Rg_i^j = -8\pi G T_i^j + \Lambda g_i^j. \tag{1}$$

The energy momentum tensor of cosmic fluid is given by

$$T_i^j = (\rho + p)u_i u^j - p g_i^j, \tag{2}$$

where ρ is the energy density, p represents equilibrium pressure and u^i is the fluid four velocity of the fluid satisfying the condition $u_i u^i = 1$.
The spatially homogeneous and anisotropic Bianchi type V space-time metric is given by

$$ds^2 = dt^2 - R_1^2(t)dx^2 - R_2^2(t)e^{2kx}dy^2 - R_3^2(t)e^{2kx}dz^2. \tag{3}$$

The Einstein field equation (1) and the energy momentum tensor (2) for the space-time metric (3) yield the following equations:

$$\frac{\dot{R}_1}{R_1}\frac{\dot{R}_2}{R_2} + \frac{\dot{R}_2}{R_2}\frac{\dot{R}_3}{R_3} + \frac{\dot{R}_3}{R_3}\frac{\dot{R}_1}{R_1} - \frac{3k^2}{R_1^2} = 8\pi G\rho + \Lambda. \tag{4}$$

$$\frac{\ddot{R}_1}{R_1} + \frac{\ddot{R}_2}{R_2} + \frac{\dot{R}_1}{R_1}\frac{\dot{R}_2}{R_2} - \frac{k^2}{R_1^2} = -8\pi G\rho + \Lambda. \tag{5}$$

$$\frac{\ddot{R}_2}{R_2} + \frac{\ddot{R}_3}{R_3} + \frac{\dot{R}_2}{R_2}\frac{\dot{R}_3}{R_3} - \frac{k^2}{R_1^2} = -8\pi G\rho + \Lambda. \tag{6}$$

$$\frac{\ddot{R}_1}{R_1} + \frac{\ddot{R}_3}{R_3} + \frac{\dot{R}_1}{R_1}\frac{\dot{R}_3}{R_3} - \frac{k^2}{R_1^2} = -8\pi G\rho + \Lambda. \tag{7}$$

$$2\frac{\dot{R}_1}{R_1} - \frac{\dot{R}_2}{R_2} - \frac{\dot{R}_3}{R_3} = 0. \tag{8}$$

By combining equations (4)-(7) one can easily obtain the continuity equation as

$$\dot{\rho}+(\rho+p)\left(\frac{\dot{R}_1}{R_1}+\frac{\dot{R}_2}{R_2}+\frac{\dot{R}_3}{R_3}\right)+\rho\frac{\dot{G}}{G}+\frac{\dot{\Lambda}}{8\pi G}=0. \tag{9}$$

The energy momentum conservation equation ($T^{ij}_{;j}$) suggests

$$\dot{\rho}+3(\rho+p)H=0, \text{ where Hubble parameter } H=\frac{1}{3}\left(\frac{\dot{R}_1}{R_1}+\frac{\dot{R}_2}{R_2}+\frac{\dot{R}_3}{R_3}\right). \tag{10}$$

Equations (9) and (10) implies

$$\rho\frac{\dot{G}}{G}+\frac{\dot{\Lambda}}{8\pi G}=0. \tag{11}$$

The physical quantities of observational interest viz. the expansion scalar Θ, shear scalar σ^2, the relative anisotropy and deceleration parameter q are defined by

$$\Theta=3H=\frac{\dot{V}}{V}, \text{ (Volume } V=R_1R_2R_3), \tag{12}$$

$$\sigma^2=\frac{1}{2}\left[\left(\frac{\dot{R}_1}{R_1}\right)^2+\left(\frac{\dot{R}_2}{R_2}\right)^2+\left(\frac{\dot{R}_3}{R_3}\right)^2\right]-\frac{\Theta^2}{6}, \tag{13}$$

$$\text{Relative anisotropy}=\frac{\sigma^2}{\rho}, \tag{14}$$

$$q=-1-\frac{\dot{H}}{H^2}. \tag{15}$$

3. Solutions of field equations

It can be easily seen that the system have five equations from equation (4)-(8) with seven unknowns R_1, R_2, R_3, ρ, p, G and Λ. Hence in order to obtain complete set of exact solutions, two additional physically plausible relations among these variables are required. Equation (5) and equation (7) suggest a relation between scale factors R_2 and R_3 as

$$\frac{\dot{R}_2}{R_2}-\frac{\dot{R}_3}{R_3}=\frac{k}{(R_2R_3)^{\frac{3}{2}}}. \tag{16}$$

We consider the anasatz

$$R_1=t^{\alpha}e^{t^{\beta}}. \tag{17}$$

Here α and β are constants can be determined on the basis of observational limits on cosmological parameters.

Under this condition equation (8) may be written as

$$\frac{\dot{R}_2}{R_2}-\frac{\dot{R}_3}{R_3}=2\left(\frac{\alpha}{t}+\beta t^{\beta-1}\right). \tag{18}$$

With help of equations (8) and (16) and (17) takes the form

$$\frac{\dot{R}_2}{R_2}-\frac{\dot{R}_3}{R_3}=\frac{k}{(t^{\alpha}e^{t^{\beta}})^3}. \tag{19}$$

From equations (18) and (19), one can easily obtain

$$R_2 = C_1 t^{\alpha} e^{t^{\beta}} exp\left(\int \frac{k}{2(t^{\alpha}e^{t^{\beta}})^3}\right) dt. \tag{20}$$

$$R_3 = C_2 t^{\alpha} e^{t^{\beta}} exp\left(\int \frac{-k}{2(t^{\alpha}e^{t^{\beta}})^3}\right) dt. \tag{21}$$

By use of the aforesaid equations, we have computed expressions for physical parameters viz. spatial volume V, Hubble parameter H, expansion scalar Θ, shear scalar σ^2 and deceleration parameter q

$$V = Ct^{3\alpha}e^{3t^{\beta}},\ H = \frac{\alpha}{t} + \beta t^{\beta-1},\ \Theta = 3\left(\frac{\alpha}{t} + \beta t^{\beta-1}\right),\ \sigma^2 = \frac{k^2}{4(t^{\alpha}e^{t^{\beta}})^6},\ q = -1 - \frac{[-\alpha + \beta(\beta-1)t^{\beta}]}{(\alpha + \beta t^{\beta})^2}, \tag{22}$$

where $C = C_1C_2$. The model has a singularity at the origin of the universe. The expression for expansion scalar Θ clearly shows that for all $\beta < 1$, expansion ceases and shear dies out for when cosmic tine t takes very large value.If $\beta > 1$ the universe expands forever and for all $\alpha > 1, \beta > \frac{1}{2}$ cosmological model shows accelerated expansion of the universe.

3.1. Case I

In this case we consider the equation of state connecting the pressure and the energy density as

$$p = \gamma\rho,\ 0 \leq \gamma \leq 1, \tag{23}$$

$\gamma = 0$, $\gamma = \frac{1}{3}$ and $\gamma = 1$ represents dust model, radiation dominated model and matter dominated model respectively. Using equations (10), (22) and (23), we can obtain the expression for energy density and by use of equations (4)-(5), (17) , (21)-(22) we can obtain expression for gravitational constant as

$$\rho = \frac{D}{(t^{\alpha}e^{t^{\beta}})^{3(1+\gamma)}}, \tag{24}$$

where D is a constant of integration, and

$$G = \frac{-1}{8\pi D(1+\gamma)}\left[\frac{k^2}{2}(t^{\alpha}e^{t^{\beta}})^{3(\gamma-1)} + 2\beta(\beta-1)t^{\beta-2}(t^{\alpha}e^{t^{\beta}})^{3(1+\gamma)} + 2k^2(t^{\alpha}e^{t^{\beta}})^{3\gamma+1} - \frac{2\alpha}{t^2}(t^{\alpha}e^{t^{\beta}})^{3(1+\gamma)}\right]. \tag{25}$$

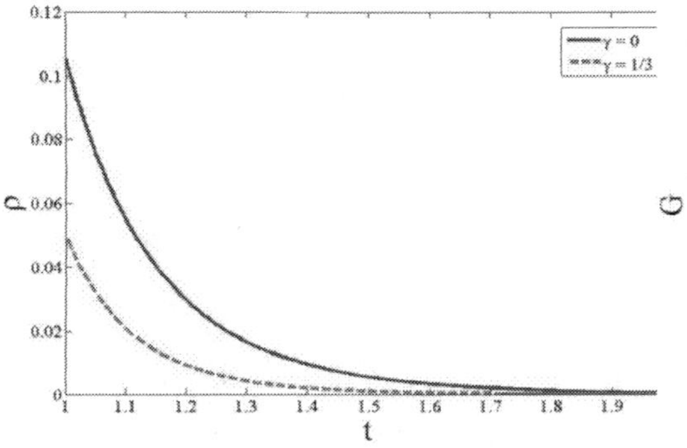

Figure 1. Variation of energy density with cosmic time t.

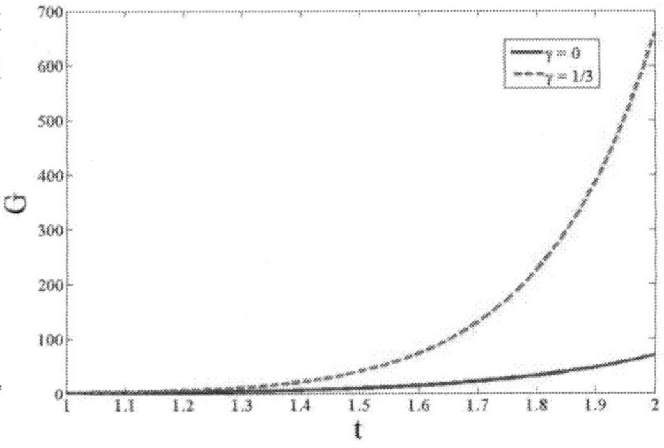

Figure 2. Variation of gravitational constant with cosmic time t.

By use of equations (22) and (24) expression (14) for relative anisotropy takes the form

$$\text{Relative anisotropy} = \frac{k^2}{4D(t^\alpha e^{t^\beta})^{3(1-\gamma)}}. \tag{26}$$

Using equations (5), (23) and (25), we get the expression for cosmological constant as

$$\Lambda = 3\left(\frac{\alpha}{t} + \beta t^{\beta-1}\right)^2 + \frac{1}{1+\gamma}\left[2\beta(\beta-1)t^{\beta-2} - \frac{2\alpha}{t^2} + \frac{k^2}{4}(1-\gamma)(t^\alpha e^{t^\beta})^{-6} - k^2(3\gamma+1)(t^\alpha e^{t^\beta})^{-2}\right]. \tag{27}$$

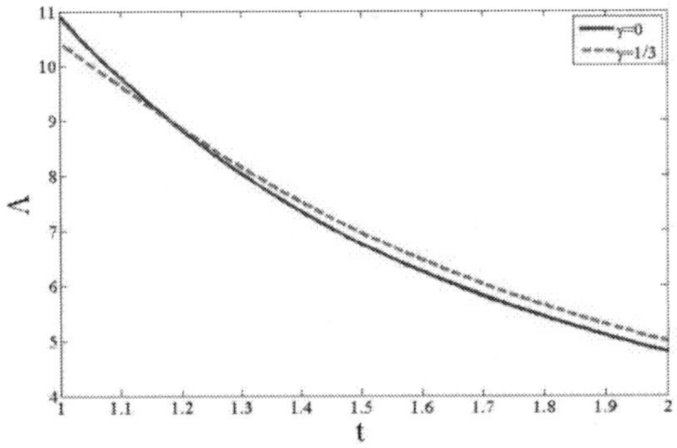

Figure 3. Variation of Cosmological "constant" with cosmic time t.

Anisotropy parameter of expansion is given by

$$\Delta \equiv \frac{1}{3}\sum_{i=1}^{3}\left(\frac{H_i - H}{H}\right)^2 = D_1(t^\alpha e^{t^\beta})^{-6}\left(\frac{\alpha}{t} + \beta t^{\beta-1}\right)^{-2}, \text{ where } D_1 = \frac{k^2}{6}.$$

The state finder parameters for this model takes the form

$$r = \frac{\dddot{R}}{RH^3} = \frac{(\alpha^3 - 3\alpha^3 + 2\alpha) + [3\alpha\beta(\alpha+\beta-2) + \beta(\beta^2 - 3\beta + 2)]t^\beta + 3\beta^2(\alpha+\beta-1)t^{2\beta} + \beta^3 t^{3\beta}}{(\alpha+\beta t^\beta)^3}, \tag{28}$$

$$s = \frac{r-1}{3(q-\frac{1}{2})} = \frac{2\alpha\left[(\alpha^3 - 3\alpha^3 + 2\alpha) + [3\alpha\beta(\alpha+\beta-2) + \beta(\beta^2 - 3\beta + 2)]t^\beta + 3\beta^2(\alpha+\beta-1)t^{2\beta} + \beta^3 t^{3\beta} - (\alpha+\beta t^\beta)^3\right]}{(6-9\alpha)(\alpha+\beta t^\beta)^3}. \tag{29}$$

The critical energy density and the vacuum energy density are respectively given by

$$\rho_c = \frac{3H^2}{8\pi G} = \frac{-3D(1+\gamma)(\alpha+\beta t^\beta)^2}{(t^\alpha e^{t^\beta})^{3(1+\alpha)}\left[-2\alpha + 2k^2t^2(t^\alpha e^{t^\beta})^{-2} + \frac{k^2t^2}{2}(t^\alpha e^{t^\beta})^{-6} + 2\beta(\beta-1)t^\beta\right]}. \tag{30}$$

$$\rho_v = \frac{3(\alpha+\beta t^\beta)^2 + \frac{1}{1+\gamma}\left(-2\alpha + 2\beta(\beta-1)t^\beta + \frac{k^2t^2(1-\gamma)}{4}(t^\alpha e^{t^\beta})^{-6} + k^2(3\gamma+1)t^2(t^\alpha e^{t^\beta})^{-2}\right)}{\frac{-1}{D(\gamma+1)}(t^\alpha e^{t^\beta})^{3(1+\gamma)}\left[-2\alpha + 2k^2t^2(t^\alpha e^{t^\beta})^{-2} + \frac{k^2t^2}{2}(t^\alpha e^{t^\beta})^{-6} + 2\beta(\beta-1)t^\beta\right]}. \tag{31}$$

Mass density parameter and density parameter of the vacuum are given by

$$\Omega_M = \frac{\rho}{\rho_c} = \frac{-1}{3(1+\gamma)(\alpha+\beta t^\beta)^2\left[\frac{k^2t^2}{2}(t^\alpha e^{t^\beta})^{-6} + 2k^2t^2(t^\alpha e^{t^\beta})^{-2} - 2\alpha + 2\beta(\beta-1)t^\beta\right]}. \quad (32)$$

$$\Omega_\Lambda = \frac{\rho_v}{\rho_c} = 1 + \frac{-1}{3(\alpha+\beta t^\beta)^2}\left[\frac{1}{1+\gamma}\left(\frac{k^2(\gamma-1)t^2}{4}(t^\alpha e^{t^\beta})^{-6} - k^2(3\gamma+1)t^2(t^\alpha e^{t^\beta})^{-2} + 2\beta(\beta-1)t^\beta - 2\alpha\right)\right]. \quad (33)$$

From equation (30) to (33) one can see that the critical energy density, critical vacuum energy density, mass density parameter and density parameter of vacuum are decreasing with evolution of the universe.

3.2. Case II

It is known that ordinary matter fields available from standard model of particle physics in general relativity, fails to account the present observation. Therefore modifications of the matter sectors of the Einstein-Hilbert action with exotic matter is considered in the literature, Chaplygin gas is considered to be one of the such candidates for dark energy. The generalized Chaplygin gas equation is given by $p = -\frac{B}{\rho^a}$, $0 < a \leq 1$. As we can see from equation of GCG is corresponding to almost dust at high density which is not agreeing completely with present universe. Therefore in this case we consider modified Chaplygin gas (MCG) with the following equation of state

$$p = \gamma\rho - \frac{B}{\rho^a}, \quad (34)$$

where γ is positive constant. To have constant negative pressure at low energy density and high pressure at high energy density, MCG model is appropriate choice. By use of equations (10), (17) and (34) one can obtain energy density of MCG and by use of equations (4)-(5), (17) , (20)-(21) , (34) one can obtain gravitational constant of MCG as,

$$\rho = \left[\rho_0 + C(t^\alpha e^{t^\beta})^{-3n}\right]^{\frac{1}{a+1}}, \quad (35)$$

where $\rho_0 = \frac{B}{\gamma+1}$, $n = (\gamma+1)(a+1)$ and C is a constant of integration.

$$G = \frac{(t^\alpha e^{t^\beta})^{3n}}{8\pi C(\gamma+1)}\left[\rho_0 + C(t^\alpha e^{t^\beta})^{-3n}\right]^{\frac{a}{a+1}}\left[\frac{2\alpha}{t^2} - 2\beta(\beta-1)t^{\beta-2} - \frac{k^2}{2}(t^\alpha e^{t^\beta})^{-6} - 2k^2(t^\alpha e^{t^\beta})^{-2}\right]. \quad (36)$$

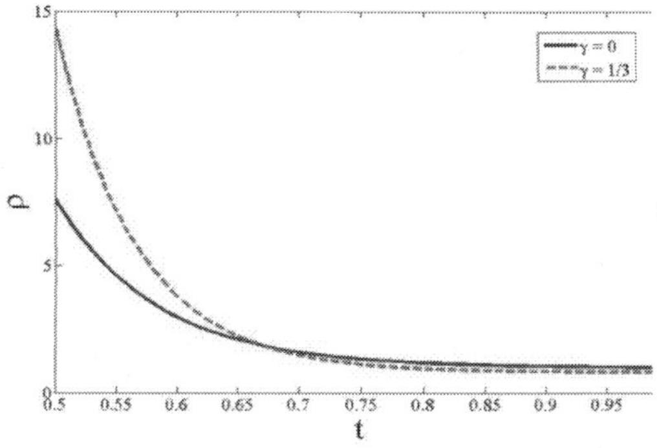

Figure 4. Variation of energy density with cosmic time t.

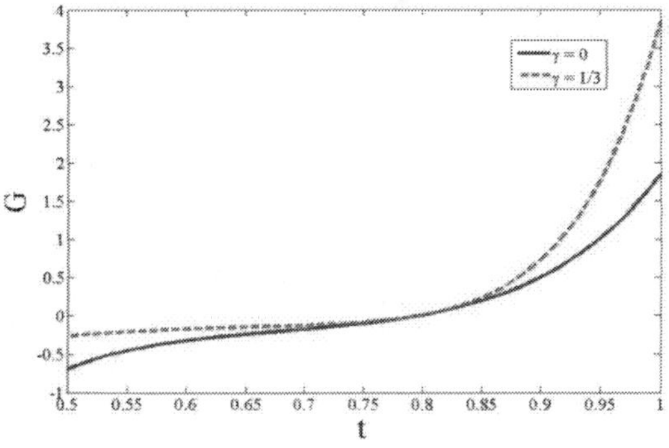

Figure 5. Variation of gravitational constant with cosmic time t.

Using equations (5), (35) and (36) we get the expression for cosmological constant as

$$\Lambda = 3\left(\frac{\alpha}{t}+\beta t^{\beta-1}\right)^2+\frac{2M_1}{t^2}(t^\alpha e^{t^\beta})^{3n}\left\{\alpha-\beta(\beta-1)t^\beta-\frac{k^2t^2}{4}(t^\alpha e^{t^\beta})^{-6}-k^2t^2(t^\alpha e^{t^\beta})^{-2}\right\}$$
$$+\frac{2}{(\gamma+1)t^2}\left\{-\alpha+\beta(\beta-1)t^\beta+\frac{k^2t^2(1-\gamma)}{8}(t^\alpha e^{t^\beta})^{-6}-\frac{k^2t^2(3\gamma+1)}{2}(t^\alpha e^{t^\beta})^{-2}\right\}, \quad (37)$$

where $M_1=\frac{\gamma\rho_0-B}{C(\gamma+1)}$.

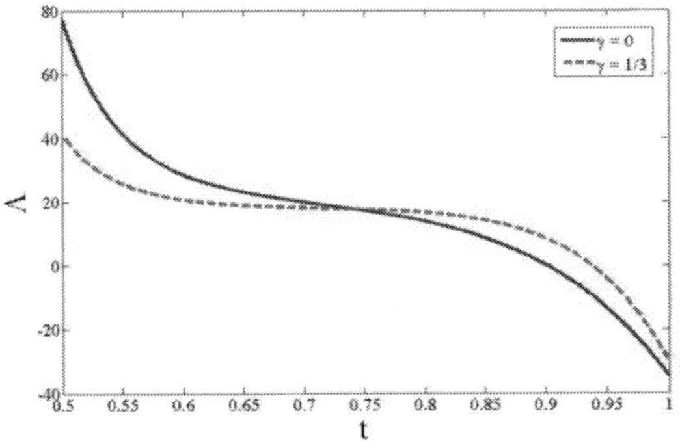

Figure 6. Variation of Cosmological "constant" with cosmic time t.

The critical energy density and the vacuum energy density ρ_c and ρ_v in this case are given by

$$\rho_c=\frac{3C(\gamma+1)(\alpha+\beta t^\beta)^2}{(t^\alpha e^{t^\beta})^{3n}\left[\rho_0+C(t^\alpha e^{t^\beta})^{-3n}\right]^{\frac{a}{a+1}}\left[2\alpha-2\beta(\beta-1)t^\beta-k^2t^2(t^\alpha e^{t^\beta})^{-6}-2k^2t^2(t^\alpha e^{t^\beta})^{-2}\right]}, \quad (38)$$

$$\rho_v=\frac{3(\alpha+\beta t^\beta)^2+2M_1(t^\alpha e^{t^\beta})^{3n}\left\{\alpha-\beta(\beta-1)t^\beta-\frac{k^2t^2}{4}(t^\alpha e^{t^\beta})^{-6}-k^2t^2(t^\alpha e^{t^\beta})^{-2}\right\}+\frac{2}{\gamma+1}\left\{-\alpha+\beta(\beta-1)t^\beta+\frac{k^2t^2(1-\gamma)}{8}(t^\alpha e^{t^\beta})^{-6}-\frac{k^2t^2(3\gamma+1)}{2}(t^\alpha e^{t^\beta})^{-2}\right\}}{\frac{1}{C(\gamma+1)}(t^\alpha e^{t^\beta})^{3n}\left[\rho_0+C(t^\alpha e^{t^\beta})^{-3n}\right]^{\frac{a}{a+1}}\left[2\alpha-2\beta(\beta-1)t^\beta-\frac{k^2t^2}{2}(t^\alpha e^{t^\beta})^{-6}-2k^2t^2(t^\alpha e^{t^\beta})^{-2}\right]}, \quad (39)$$

whereas mass density parameter and the density parameter of the vacuum are given by

$$\Omega_M=\frac{1}{3C(\gamma+1)(\alpha+\beta t^\beta)^2}\left[\rho_0(t^\alpha e^{t^\beta})^{3n}+C\right]\left[2\alpha-2\beta(\beta-1)t^\beta-\frac{k^2t^2}{2}(t^\alpha e^{t^\beta})^{-6}-2k^2t^2(t^\alpha e^{t^\beta})^{-2}\right], \quad (40)$$

$$\Omega_\Lambda = 1+\frac{1}{3(\alpha+\beta t^\beta)^2}\Bigg[2M_1(t^\alpha e^{t^\beta})^{3n}\left\{\alpha-\beta(\beta-1)t^\beta-\frac{k^2t^2}{4}(t^\alpha e^{t^\beta})^{-6}-k^2t^2(t^\alpha e^{t^\beta})^{-2}\right\}$$
$$+\frac{2}{\gamma+1}\left\{-\alpha+\beta(\beta-1)t^\beta+\frac{k^2t^2(1-\gamma)}{8}(t^\alpha e^{t^\beta})^{-6}-\frac{k^2t^2(3\gamma+1)}{2})(t^\alpha e^{t^\beta})^{-2}\right\}\Bigg]. \quad (41)$$

From equation (38) to (41) one can see that the critical energy density, vacuum energy density, mass density parameter and density parameter of vacuum are decreasing with time.

4. Discussion

In this paper we have discussed Bianchi Type V space-time model with equilibrium pressure and modified generalized Chaplygin gas. The energy density ρ and cosmological constant Λ are decreasing with evolution of the universe whereas G becomes infinite as $t \to \infty$ in all the models. If $\beta > 1$ the universe expands forever and for $\frac{1}{2} < \beta < 1$ the universe expansion vanishes when $t \to \infty$. Shear dies out with evolution of the universe for large value of t. For all $\alpha > 1, \beta > \frac{1}{2}$ equation (30) shows that present cosmological model suggests accelerated expansion of the universe. Considering present day limit for deceleration parameter $q = -0.53^{+0.17}_{-0.13}$, equation (30) suggests $0.6097 < \beta < 0.7462$. In order to have clear idea of variation in behavior of cosmological parameters relevant graphs have been plotted. In first case for $\gamma = 0$ energy density is rapidly decreasing than for $\gamma = \frac{1}{3}$ whereas G is increasing slowly for $\gamma = 0$ than for $\gamma = \frac{1}{3}$ which is in fair agreement with observations. In both the cases cosmological constant is decreasing fastly for $\gamma = 0$ than for $\gamma = \frac{1}{3}$ which goes with observations.

Acknowledgments

S. K. and G. P. S. would like to thank U. G. C. New Delhi for providing financial support under the scheme of major research project F. No. 41-765/2012(SR). S. K. and R. K. would like to thank Inter University Centre for Astronomy and Astrophysics for providing facilities.

References

[1] Riess A. G. Et al., 1998 Observational evidence from supernovae fpr an acceletating universe and a cosmological constant *Astron J.* **116** 1009
[2] Perlmutter S. Et al., 1999 Measurements of Omega and Lambda from 42 high red shift supernovae *Astrophys. J.* **517** 565
[3] Percival W J 2001 The 2 dF galaxy redshift survey: The power spectrum and the matter content of the universe *Mon. Not. R. Astron. Soc.* **327** 1297
[4] Jimenez R, Verde L, Tren T and Stern D 2003 Constraints on the equation of state of dark energy and the Hubble constant from stellar ages and the CMB *Astrophys. J.* **593** 622
[5] Stern R, Jimenez R Verde L, Kamion Kowski M and Stanford S A 2010 Cosmic Chronometers: Constraining the equation of state of dark energy I: H(z) measurements *J. Cosmolo. Astropart.Phys.* **1002** 8
[6] Sahni V and Starobinsky A 2000 The case of positive cosmological lambda term *Int. J. Mod. Phys. D* **9** 373
[7] Ratra B and Peeble P J E 1988 Cosmological consequences of a Rolling homogeneous scalar field *Phys. Rev. D* **37** 3406
[8] Dolgov A D 1997 Higher spin fields and the problem of cosmological constant, *Phys. Rev. D* **55** 5881
[9] Sahni V and Strarobinsky A 2000 The case of positive cosmological lambda term, *Int. J. Mod. Phys. D* **9** 373
[10] Padmanabhan T 2003 Cosmological constant: The weight of the vacuum *Phys. Rept.* **380** 235
[11] Peeble P J E 2003 The cosmological constant and dark energy,*Rev. Mod. Phys.***75**599
[12] Canuto V M and Narlikar J N 1980 Cosmological tests of the Hoyley- Narlikar conformal gravity *Astrophys. J.* **6** 236
[13] Singh G P and Kotambkar S 2001 Higher dimensional cosmological models with gravitational and cosmological constant *Gen. Rel. Grav.* **33** 621
[14] Singh G P and Kotambkar S 2003 Higher dimensional dissipative cosmology with varying G and Lambda *Grav. and Cosmol.* **9** 206
[15] Vishwakarma R G 2005 A model of explain varying Λ, G and σ^2 simultaneously *Gen. Rel. Grav.* **37** 1305
[16] Singh G P Kotambkar S and Pradhan A 2007 A new class of higher dimensional cosmological models of universe with variable G and lambda- term *Romanian J. Phys.* **53** 607
[17] Singh G P and Kale A Y 2009 Bulk viscous Bianchi type V cosmological models with variable gravitational and cosmological constant *Int. J. Theor Phys.* **48** 1177
[18] Bali R and Meena B L 2004 Conformally Flat Tilted Bianchi Type-V Cosmological Models in General Relativity *Pramana* **62** 1007
[19] Bali R and Yadav M K 2005 Bianchi Type-IX Viscous Fluid Cosmological Model in General Relativity *Pramana* **64** 187
[20] Bali R and Tinkar S 2008 Bianchi type V bulk viscous barotropic fluid cosmological model with variable G and Lambda, *Chin. Phys. Lett.* **25** 3090
[21] Bali R and Tinker S 2009 Bianchi type III bulk viscous barotropic fluid cosmological model with variable G and Lambda *Chin. Phys. Lett.* **26** 029802
[22] Yadav A K Pradhan A and Singh A K 2012 Bulk viscous LRS Bianchi type I universe with variable G and decaying Lambda *Astophys. Space Sci.* **337** 379
[23] Yadav A K and Sharma A 2013 A transitioning universe with time varying G and decaying Lambda *Res. Astr. Astrophys.* **13** 501

[24] Torny J L et al 2003 Cosmological Results from High-z Supernovae *Astrophys. J.* **594** 1
[25] Bridle S Lahav O Ostriker J P and Steinhardt P J 2003 Precision cosmology? Not just yet *Science* **299** 1532
[26] Bennet C Et al 2003 First year Wilkinson Microwave Anisotropic Probe (WMAP) observations: Preliminary maps and basic results *Astrophys. J. Suppl.* **148** 1
[27] Hinshaw G Et al 2003 First year Wilkinson Microwave Anisotropic Probe (WMAP) observations: The angular power spectrum *Astrophys. J. Suppl.* **148** 135
[28] Spergel D N Et al 2003 First year Wilkinson Microwave Anisotropic Probe (WMAP) observations: Determination of cosmological parameters *Asrtophys. J. Suppl.* **148** 175
[29] Kogut A Et al 2014 Results from the Wilkinson Microwave Anisotropic Probe *PTEP-2014* **06B** 102
[30] Benaoum H B hep-th 0205140 Dev A Alcaniz J S and Jain D 2003 Cosmological consequences of a Chaplygin gas dark energy *Phys. Rev. D* **67** 023515
[31] Sen A A and Scherrer R J 2005 Generalizing the generalized Chaplygin gas *Phys. Rev. D* **72** 063511
[32] Debnath U 2011 Modified Chaplygin gas with variable G and Lambda *Chin. Phys. Lett.* **28** 19801
[33] Kamenshchik A Gorini V Moschella U and Pasquier V 2001 An alternative to quintessence *Phys. Lett. B***511** 265
[34] Bento M C Bertolami O and Sen A A 2003 WMAP constraints on the generalized Chaplygin gas model *Phys. Lett. B* **575** 172
[35] Adhav K S 2011 State finder diagnostic for modified Chaplygin gas in Bianchi type - V universe *Eur.Phys. J. plus* **126** 52
[36] Liu D J and Li X H 2005 CMBR constraint on a modified Chaplygin gas model *Chin. Phys. Lett.* **22** 1600
[37] Debnath U Banerjee A and Chakraborty S 2004 Role of modified Chaplygin gas in accelerated universe *Class. Quant. Grav.* **21** 5609
[38] Thakur P Ghose S and Paul B C 2009 Modified Chaplygin gas and constraints on its B parameter from CDM and UDME cosmological models *Mon. Not. Astron. Soc.* **397** 1935
[39] Paul B C and Thakur P 2013 Observational constraints on modified Chaplygin gas from cosmic growth *JCAP* **052** 1311
[40] Paul B C Thakur P and Verma M K 2013 Observational constraints on modified Chaplygin gas in Horava- Lifshitz gravity with dark radiation *Pramana* **81** 691

A study on photon attenuation coefficients of different wood materials with different densities

B Saritha and A S Nageswara Rao

Department of Physics, Kakatiya University, Warangal 506009, Telangana, India.

E-mail: `saruphysics@gmail.com`

Abstract. A study on the variation of linear attenuation coefficients with the densities of the wood samples is under taken. The soft wood and hard wood samples were collected from the forest area of Pakal in Warangal district. The linear and mass attenuation coefficients are measured using gamma ray spectrometry based on NaI (Tl) scintillation detector with energies of 662 KeV and 59.5 KeV respectively. The mass attenuation coefficient values measured from experiment and are compared with theoretical methods using XCOM program. The plots of density versus linear attenuation coefficient for different wood materials correspond to higher order polynomial are presented. It is observed that variation of linear attenuation coefficient depends on densities of materials. The Chloroxylon swietenia with more density has more linear attenuation coefficient at 59.5 KeV and 662 KeV. The variation in attenuation coefficient attributed to chemical composition of wood used in the experiment.

1. Introduction

Shielding from gamma radiation requires large amounts of density. The radiation is better absorbed by materials with high atomic numbers density, although neither effect is important compared to the total mass/area in the path of the gamma ray. For this reason, a lead shield is 25 percent more efficient to a shield gamma radiation than an equal mass of another shielding material such as aluminum, concrete, or soil, or same quantity of water. Lead's capability of shielding radiation is associated with density. The higher the energy of the gamma rays, the thicker the shielding required. Materials for shielding gamma rays are typically measured by the thickness required to reduce the intensity of the gamma rays by one half (the half value layer or HVL). However, the mass of this much concrete or soil is only $20 - 30$ percent larger than that of this amount of lead. Depleted uranium is used for shielding in portable gamma ray sources, but again the savings in weight over lead is modest, the main effect being reducing the shielding bulk. Now a days wood, as one of the main materials, cheap as compared to lead and other better shields, used in modern building construction, and its method of production in the subject of detailed scrutiny, particularly where the buildings are of critical importance such as a hospital.

Some investigations in the period of 1900 - 2000 is presented by Mortatti and Fillo [1]. They have used a method to study ten different samples using 662 KeV and 59.54 KeV gamma-rays. Barros-Ferraz and Rezenda [2] used the method to study the density variability in two different samples of pine wood using Am^{241} at 59.54 KeV, with a view to understand their efficiency and adaptation. A new method of water content gradation in wood based on low-energy gamma ray attenuation in pine wood samples are developed by Aguiar and Barros-Ferraz [3]. They showed that the method is quick, nondestructive and has reliably good sensitivities for moisture contents varying from 30 to 90 percent. Miranda, Regina and Pascholati [4] studied gamma attenuation in natural wood, dry wood and dry leaves using Am^{241} gamma rays and they found that the variation is small among different species. The challenge of the research activity in wood science and technology must be oriented to the development of non-contact, nondestructive techniques which can be used in situ and in vivo and not induce any structural damage in samples of various sizes ranging from small clear laboratory specimens to trees, structural elements, and industrial-size wood-based panels [5,6]. The gamma-ray attenuation technique is used to determined the wood density and moisture content and is recommended by the investigators [7,8]. The gamma ray source Cs^{137}, the photon peak which has energy of 662 KeV for determination of attenuation coefficients is used in research work [9,11].

Attenuation coefficient is an important parameter for the study of interaction of radiation with matter and indicates the fraction of energy scattered or absorbed [12]. The probability of a photon interaction with a given material, per unit path length, is called the 'Linear attenuation coefficient'. This coefficient is of great importance in radiation shielding materials. Linear attenuation coefficients depend on the shielding material. The density often does not have a unique value but depends on the physical state of the material such as, in the case of wood, the moisture content. To prevent the effects of variations in the density of a material, the linear attenuation coefficient is expressed as mass attenuation coefficient which is the linear attenuation coefficient per unit density of the material [13]. Recently, Saritha and Rao [14-15] have made experimental and theoretical investigations to determine mass attenuation coefficients of complex biological molecules such as wood samples having H, C, N and O elements.The present work is to estimate attenuation of the gamma-ray of different wood samples with different densities. The mixture rule (see eq. (1) below) has been used for measuring mass attenuation coefficients. Gamma ray sources Am^{241} and Cs^{137} with energies of 59.54 KeV and 662 KeV have been used in these studies.

2. Experimental details

Gamma ray spectrometry system with NaI(Tl) detector high efficiency is used with present study. The spectrum is analyzed using the software PHASTE. The detector is shielded using 4 to 5 inches lead on all sides to reduce the back ground level of the system. The spectrum calibration is carried out using gamma ray sources Am^{241} and Cs^{137} at energies 662 KeV and 59.5 KeV for the wood sample. Attenuation values are plotted against the density of wood samples for particular gamma ray narrow beam geometry. In the present investigation, two points are systematically located on either side of the centered of the peak. In this method we can correct the multiple scattering. After noting the counts without sample (I_0), samples of known thickness (I) are placed in between the collimators placed before the detector and the mass attenuation coefficient is calculated using the well known Lambert's law:

$$\frac{I}{I_0} = exp(-\mu t). \qquad (1)$$

The experimental setup as shown in the figure consists of two collimators, one at the front of the detector and another at the front of source that is placed in between the detector and source having 2° of scattered angle. Fig-1 shows the experimental set up. A direct measurement of source is first taken and then each sample in between the collimators is taken. By this procedure, the intensity of gamma ray with out sample I_0 and that of with wood I is determined and the ration I/I_0 is calculated for each sample. The linear and mass attenuation coefficient for different wood materials are measured with narrow beam geometry (mono energetic transmission beam) for energies 59.5KeV and 662KeV. The linear attenuation coefficients with corresponding

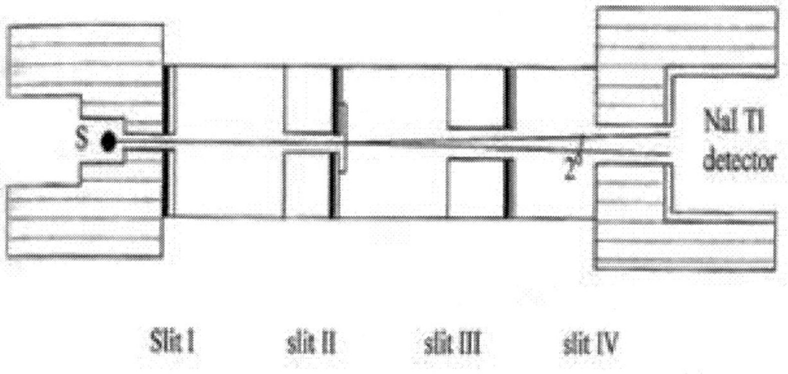

Figure 1. Schematic diagram problem.

Table 1. Density related mass attenuation coefficient of different wood samples for 59.5 KeV.

S.No	Wood sample name	Mass attenuation coefficient	Density(g/cm^3)
1	Sterculia urens	0.209 ± 0.004	0.340
2	Albiziz amara	0.194 ± 0.003	0.721
3	Lagerstroemila parviflora	0.195 ± 0.003	0.975
4	Ziziphus Xylopyras	0.199 ± 0.003	0.695
5	Albiziz adara	0.194 ± 0.003	0.859
6	Pterocarpus marsupium	0.192 ± 0.003	0.747
7	Chloroxylon swietenia	0.193 ± 0.003	1.186
8	Terminalia alata	0.194 ± 0.003	0.929
9	Wrightia tinctoria	0.201 ± 0.004	0.818
10	Cochlospermum religiosus	0.206 ± 0.004	0.306

Table 2. Density related mass attenuation coefficient of different wood samples for 662 KeV.

S.No	Wood sample name	Mass attenuation coefficient	Density(g/cm^3)
1	Sterculia urens	0.080 ± 0.002	0.340
2	Albiziz amara	0.081 ± 0.002	0.721
3	Lagerstroemila parviflora	0.079 ± 0.002	0.975
4	Ziziphus Xylopyras	0.081 ± 0.002	0.695
5	Albiziz adara	0.080 ± 0.002	0.859
6	Pterocarpus marsupium	0.080 ± 0.002	0.747
7	Chloroxylon swietenia	0.081 ± 0.002	1.186
8	Terminalia alata	0.081 ± 0.002	0.929
9	Wrightia tinctoria	0.083 ± 0.002	0.818
10	Cochlospermum religiosus	0.082 ± 0.002	0.306

densities of samples are given in tables 1 and 2. After collection, the samples dried in air up to one year and samples were cut into different dimensions having different densities.

3. Results and discussion

The mass attenuation coefficient (μ/ρ) for different wood samples is measured for the two energies 59.5 KeV and 662 KeV. For each energy, the measurements for all types of samples are carried out in three dimensions. In this investigation, ten wood samples namely Sterculia urens, Albizia amara, Largerstroemia parviflora, Ziziphus xylopyrus, Albizia odaratisma, Pterocarpus marsupium, Chloroxylon swietenia, Terminalia alata, Wrightia tinctoria, and Cochlospermum religious are used with their absolute uncertainties.

For sample 1, the theoretical error for wood samples rangeses from 0.003 % to 0.004 % where as experimental error is 0.004 % and no difference is observed in between com values and experimental values in sample 1. In sample 2, the difference is 0.003.

Linear attenuation coefficients versus different densities of wood samples are plotted and depicted graphically in figure 2. The best line of fit in the form of equation is given below.

$$Y = -0.000x^4 + 0.011x^3 - 0.087x^2 + 0.263x - 0.116,\ R^2 = 0.829,$$
$$Y = -0.000x^4 + 0.005x^3 - 0.0367x^2 + 0.111x - 0.051,\ R^2 = 0.838.$$

The R^2 is the coefficient of determination is a statistical measurement of how will the regression line approximately the real data points. From the numerical results, the value 1.00 of R^2 shows that the regression line perfectly fits the data. From the figure 2, it is clear that the

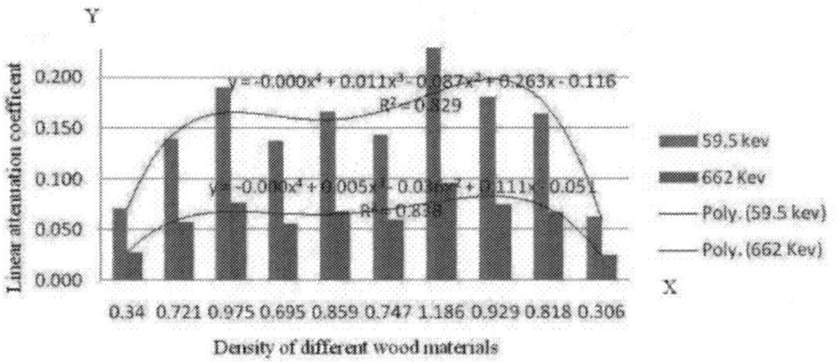

Figure 2. Linear attenuation coefficient versus density of different wood materials.

trend follows higher order polynomial. Here, it is observed that linear attenuation coefficient increases as density increases. Also linear attenuation coefficient decreases as density decreases as shown in figure. The density of Chloroxylon swietenia has more linear attenuation coefficient at 59.5 KeV and 662 KeV. This might be due to chemical composition present in the wood. On examination of the measured photon linear attenuation coefficient for softwood and hard wood materials are across the energies 59.5 KeV and 662 KeV. There are slight differences between the results; there may be due to the composition variation among the different types of soft and hard wood materials. In Chloroxylon swietenia, which is much denser than the other hard wood material showed photon linear attenuation more than the other material under study. The linear attenuation coefficient values of the wood samples varied with the physical density of the materials, increased with increasing in density of the material, increased with increase in density for the same gamma ray energy, and this may decreased with increase of gamma ray energy for the same material. Further the mass attenuation coefficients for all the kinds of samples is studied with respect to the corresponding photon energies and is conducted that they follow exponential law.

References

[1] Mortatti J and Nascimento Fillo 1983 *Energia Nuclaer-e-Agricultura Brazil* **185**
[2] De Barros-Ferraz E S and Rezenda M A 1986 *Proceedings of the General Congress of Nuclear energy*(Rio de Janeiro) **2** 443
[3] Mortatti J and Nascimento Fillo 1983 *Energia Nuclaer-e-Agricultura Brazil* **873**
[4] De Miranda *etal* 1977 *Proceedings of the 4th Brazilean Meeting on Nuclear Applications* (Rio de Janeiro) **2** 904
[5] Bucur V 2003 *a Nondestructive characterization and imaging of wood* (Heidelberg: Springer-Verlag)
[6] Bucur V 2003 Ultrasonic imaging of wood structure *Proc 5th World Conf. in Ultrasonics* 299 - 302 (*http : //www.sfa.asso.fr/wcu2003/procs/webside/articles*)
[7] Loos W E 1961 *Forest Products Journal* Madison **11(3)**145
[8] Parrish W B 1961*Forest Science* Washington D.C. **7**136
[9] Loos W E 1965 *Forest Products Journal* Madison **15**102
[10] Woods F W 1965 *Forest Science* Washington D.C. **11**341
[11] Philips E W J 1965 *Methods and equipment for the specific gravity of wood* Australia: IUFRU, Section 41
[12] Hubbell J H 1991 *Jr.Phys. Med. Biol.* **441**
[13] Kaplan M F 1989 *Longrnan Scientific and Technology* England: Longman Group UK Limited, Essex
[14] Saritha B and Nageshwara Rao A S 2014 *Can. Jr. Phys.* **92**1
[15] Saritha B and Nageshwara Rao A S 2014 *Can. Jr. Phys.* **91(3)**221

Energy collection via Piezoelectricity

Ch Naveen Kumar
Department of Physics,Indur Institute Of Engineering And Technology. Telangana, India.
E-mail: naveen.1474@gmail.com

Abstract. In the present days, wireless data transmission techniques are commonly used in electronic devices. For powering them connection needs to be made to the power supply through wires else power may be supplied from batteries. Batteries require charging, replacement and other maintenance efforts. So, some alternative methods need to be developed to keep the batteries full time charged and to avoid the need of any consumable external energy source to charge the batteries. Mechanical energy harvesting utilizes piezoelectric components where deformations produced by different means are directly converted to electrical charge via piezoelectric effect. The proposed work in this research recommends Piezoelectricity as a alternate energy source. The motive is to obtain a pollution-free energy source and to utilize and optimize the energy being wasted. Current work also illustrates the working principle of piezoelectric crystal and various sources of vibration for the crystal.

1. Introduction

Energy harvesting has been a topic of discussion and research for three decades. With the ever increasing and demanding energy needs, unearthing and exploiting more and more energy sources has become a need of the day. Energy harvesting is the process by which energy is derived from external sources and utilized to drive the machines directly, or the energy is captured and stored for future use. Some traditional energy harvesting schemes are solar farms, wind farms, tidal energy utilizing farms, geothermal energy farms and many more. With the advent of technology, utilization of these sources has increased by leaps and bounds [1]. When viewed on a large scale, energy harvesting schemes can be categorized as shown in Table 1.

Table 1. Types of Energy Harvesting Schemes.

Type of Energy Harvesting	Energy Source	Solution	Ultimate Goal
Macro	Renewable sources like solar, wind, tidal etc.	Energy Management solutions	Reduce oil dependency
Micro	Small scale sources like vibration, motion, heat etc	Ultra-low- power solutions	Driving low energy consuming devices

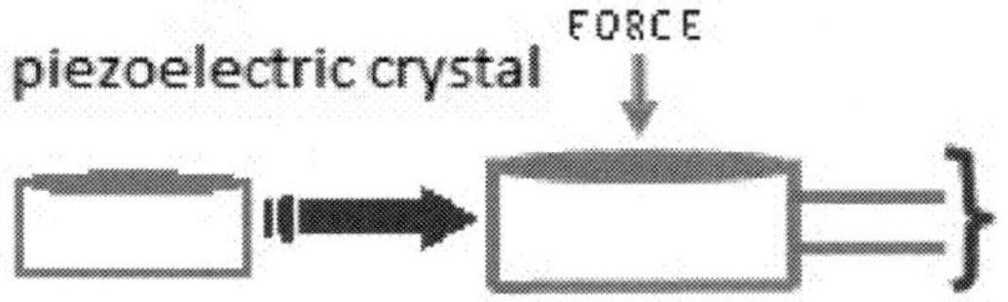

Figure 1. Principle of direct piezoelectric effect.

Piezoelectric energy harvesting is a new and innovative step in the direction of energy harvesting. Not many researches have been carried out till now in this field, hence it is a challenging job to extract energy from Piezoelectricity. Through this research paper, we will describe the basic working of a piezoelectric crystal. Then later in the paper, we have proposed the idea of combining energy from a number of piezoelectric crystals to obtain higher voltages. Certain ways of implanting the crystals at different places have also been cited in the paper. Piezoelectric crystals can be utilized to obtain voltages of very small values and hence can drive low voltage devices. Hence, piezoelectric energy harvesting comes under the category of micro scale energy harvesting scheme.

2. Working principle of piezoelectric energy harvesting

The piezoelectric effect is a special material property that exists in many single crystalline materials. Examples of such crystalline structures are Quartz, Rochelle salt, Topaz, Tourmaline, Cane sugar, Berlinite ($AIPO_4$), Bone, Tendon, Silk, Enamel, Dentin, Barium Titanate ($BaTiO_3$), Lead Titanate ($PbTiO_3$), Potassium Niobate ($KNbO_3$), Lithium Niobate ($LiNbO_3$) etc.[2]. There are two types of piezoelectric effect, direct piezoelectric effect and inverse piezoelectric effect. The direct piezoelectric effect is derived from materials generating electric potential when mechanical stress is applied and the inverse piezoelectric effect implies materials deformation when an electric field is applied. The energy harvesting via Piezoelectricity uses direct piezoelectric effect. The phenomenon will be clear from the diagram shown in figure 1, figure 2 shows the structure of a piezoelectric component being used for energy harvesting.The output voltage obtained from a single piezoelectric crystal is in milli volts range, which is different for different crystals and the wattage is in microwatt range. So in order to achieve higher voltages, the piezoelectric crystals can be arranged in cascading manner, that is, in series. The energy thus obtained is stored in Lithium batteries or capacitors. This is the working principle behind piezoelectric energy harvesting system. Now the extreme engineering lies in optimization of piezoelectric energy, which is done in various ways. A lot of studies are being carried out in order to know which crystal will be the best to obtain maximum output voltage what should be the structure of piezoelectric component, which type of circuit should be used at the output terminals of piezoelectric crystal in order to have maximum wattage.

3. Sources of vibration for crystal previous work

In this section , we have mentioned a number of sources of vibration which are already being used for piezoelectric energy harvesting and a new idea in this direction has been proposed.

3.1. Power generating sidewalk

The piezoelectric crystal arrays are laid underneath pavements, sidewalks and other high traffic areas like highways, speed breakers for maximum voltage generation. The voltage thus generated

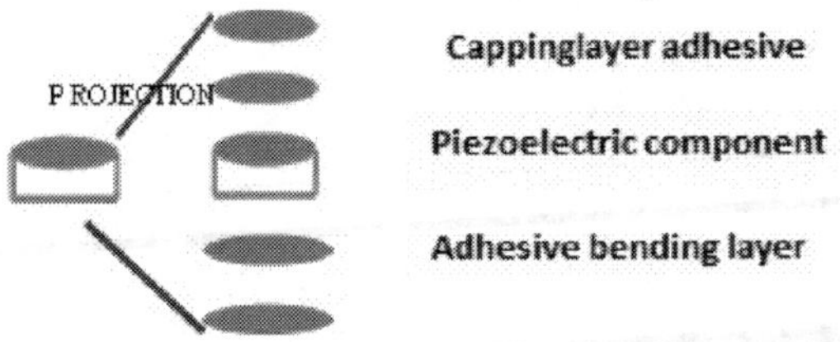

Figure 2. Structure of a piezoelectric component.

from the array can be used to charge the chargeable Lithium batteries, capacitors etc. These batteries can be used as per the requirement [3].

3.2. Power generating boots or shoes
In United States Defense Advance Research Project Agency (DARPA) initiated a innovative project on Energy harvesting which attempts to power battlefield equipment by piezoelectric generators embedded in soldiers' boots [3]. However, these energy harvesting sources put an impact on the body. DARPA's effort to harness 1-2 watts from continuous shoe impact while walking were abandoned due to the discomfort from the additional energy expended by a person wearing the shoes.

3.3. Gyms and workplaces
Researchers are also working on the idea of utilizing the vibrations caused from the machines in the gym. At workplaces, while sitting on the chair, energy can be stored in the batteries by laying piezoelectric crystals in the chair. Also, the studies are being carried out to utilize the vibrations in a vehicle, like at clutches, gears, seats, shock-ups, foot rests.

3.4. Mobile keypad and keyboards
The piezoelectric crystals can be laid down under the keys of a mobile unit and keyboards. With the press of every key, the vibrations created can be used for piezoelectric crystal and hence can be used for charging purpose [4].

3.5. Floor mats, tiles and carpets
A series of crystals can be laid below the floor mats, tiles and carpets which are frequently used at public places.

3.6. People powered dance clubs
In Europe, certain nightclubs have already begun to power their night clubs, strobes and stereos by use of piezoelectric crystals. The crystals are laid underneath the dance floor. When a bulk of people use this dance floor, enormous amount of voltage is generated which can be used to power the equipments of the night club [5].

4. Proposed work
In this section , we propose two methods.

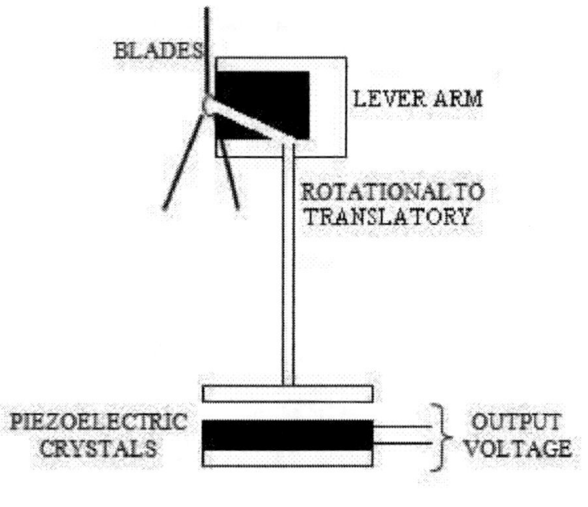

Figure 3. Piezoelectric wind mill.

4.1. Piezoelectric wind mill

In order to energize low power consuming devices, microcells are invariantly used. But these microcells need to be charged once they get discharged. Hence if the devices are placed at remote places like villages, border areas, forests, hilly areas, then continuous charging of the microcells is not possible by conventional charging methods. In such cases, alternative options like solar energy and wind energy can be utilized. But cloudy days and rains restrict the use of solar energy. So, wind energy comes out to be the best alternative [6]. The idea about a piezoelectric wind mill will be clear from figure 3. The piezoelectric wind mill that we have proposed consists of a fan with three blades to effectively capture the wind flow. A lever arm is connected to the windmill fan rotor and a translator is connected with this lever arm to convert rotational motion into translatory motion. A disc is connected at the lower end of translator, such that whenever it moves upwards and downwards, it compresses the piezoelectric crystals. Hence for different speeds of wind also, that is for different frequencies, the Piezoelectric wind mill may function. Hence, it has higher workable bandwidth. The constant compression of piezoelectric crystals causes a huge amount of energy to be generated, which can comfortably drive the remotely placed low power consuming devices [2]. Hence, the concept of piezoelectric wind mill can be used to harness piezoelectric energy very efficiently and effectively.

4.2. Increased bandwidth piezoelectric crystal

In order to increase the workable bandwidth, that is, in order to use piezoelectric crystals over a wide range of vibrations, we are propose a new method. If in place of a single energy source, we make use of more than one, then the efficiency of harvesting system will definitely increase. Hence, we are making use of two energy converting techniques, one is the piezoelectric crystal and other is the electromagnetically induced voltage. figure4 gives the structure of such type of system [7]. The system consists of a flexible strip, over which the piezoelectric crystals are mounted and at one end of the strip, a magnet is mounted. This magnet lies inside a stationary coil. At times, when intensity of vibration is high, voltage is obtained from piezoelectric crystals. Hence, at higher frequencies, piezoelectric crystals give the output. When intensity of vibration is less, the piezoelectric crystals do not give a considerable output. At lower frequencies, the magnet moves inside the stationary coil. This motion causes electromagnetic flux to be generated and hence an output voltage is obtained.

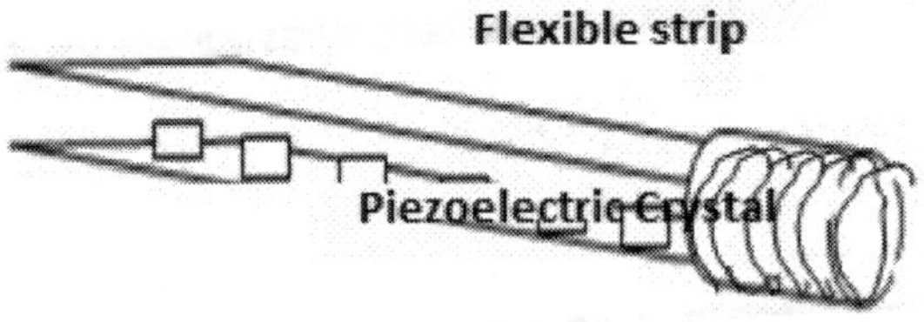

Figure 4. Piezoelectric crystals and electromagnetic energy.

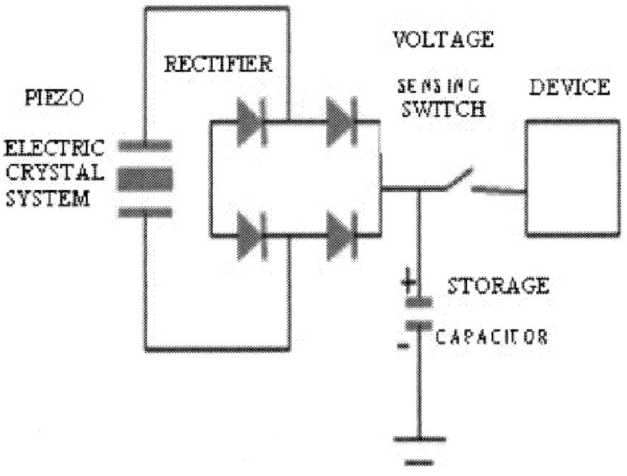

Figure 5. Bridge rectifier type AC to DC converter.

5. Output stage of piezoelectric energy harvesting system

The output of a piezoelectric crystal is alternating signal. In order to use this voltage for low power consuming electronic devices, it has to be first converted into digital signal [2]. This is done with the help of AC to DC converter as shown in figure 5. Figure 5 shows a simple diode rectifier to convert AC to DC. This is followed by a capacitor, which gets charged by the rectifier upto a pre-decided voltage, at which the switch closes and the capacitor discharges through the device. In this way, the energy can be stored in the capacitor, and can be discharged when required. But the energy harvesting capacity of this circuit is not appreciable. Hence, a DC to DC converter is used after bridge rectifier stage, which has been demonstrated in figure 6. The addition of DC-DC converter has shown an improvement in energy harvesting by a factor of 7. A non-linear processing technique "Synchronized Switch Harvesting on Inductor" (SSHI) was also proposed in 2005 for harvesting energy [7]. It consists of a switching device in parallel with the piezoelectric element. The device is composed of a switch and an inductor connected in series. The switch is in open state except when the maximum displacement occurs in the transducer. At that instant, the switch is closed and the capacitance of the piezoelectric element and inductor together constitute an oscillator. The switch is kept closed until the voltage on the piezoelectric element has been reversed. This circuit arrangement of the output circuit is said to have a very high energy harvesting capacity. Figure 7 shows the SSHI technique [6].

6. Implementation

Experimentation has been done on a Piezo-crystal and it is tested with a Light Emitting Diode (LED). The two terminals of the LED are connected with the two terminals of the crystal. Choice

Journal of Physics: Conference Series **662** (2015) 012031 doi:10.1088/1742-6596/662/1/012031

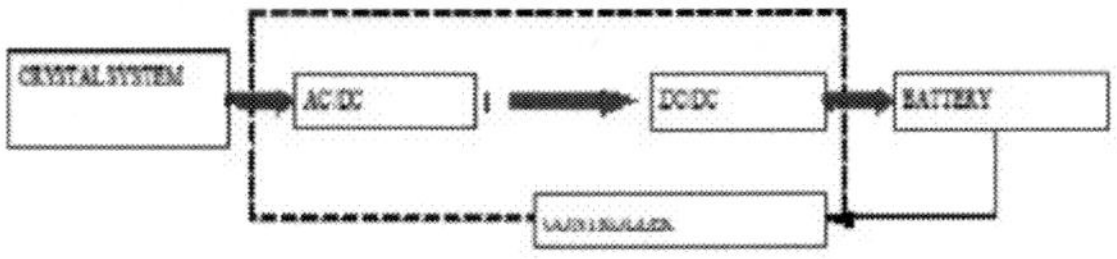

Figure 6. Energy harvesting circuit.

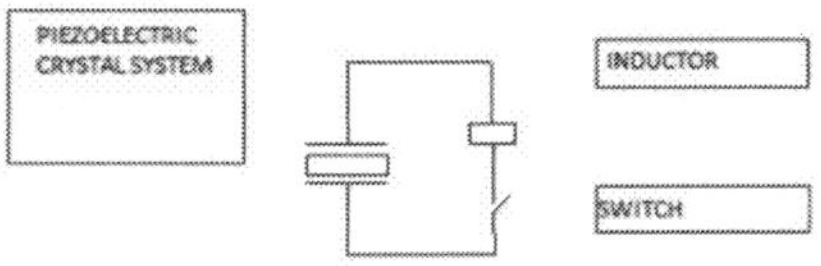

Figure 7. SSHI (Synchronized switch harvesting on inductor) technique.

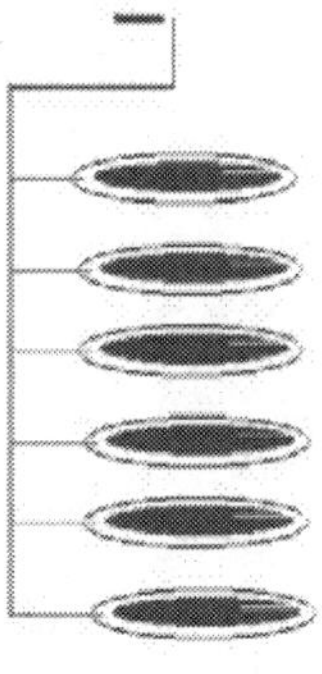

Figure 8. Series and parallel combination of crystals.

of Blue LED is being made for experimentation. Single stroke on the crystal blows blue LED with full intensity. Measured values of output voltage and current from the crystal come out to be 3.5 Volt and 100 milliamps. The only shortcoming of this using a single crystal and a LED was that both the voltage and current obtained exists instantaneously. To increase the range of voltage and current output, an assembly of 6 crystals in series and 6 such series has been put in parallel. When number of voltage sources are put in series, then the net voltage increases, while when a number of voltage sources are put in parallel, then the strength of signal, that is, current increases. This is the concept used behind the assembly. The output of parallel connection is fed to the current amplifier for signal strengthening and the output of series connection is fed to the amplifier for biasing purpose and also to the voltage amplifier. The assembly has been put under a doormat and the output obtained from amplifier has been very encouraging which was around 6 V voltage and 1 ampere current. This magnitude of voltage and current can be certainly used to charge a battery. Figure 8 shows the assembly used in our system.

7. Cost effectiveness

The assembly developed using series and parallel combination of piezo-crystals is very cost effective. A single crystal costs around 23 - 25 Rupees, and hence the cost of whole assembly is very less. It is very encouraging to get a good voltage and current at such a low cost at the same time is utilized.

8. Conclusion

The method used to perform power harvesting is to use PZT materials that can convert the ambient vibration energy surrounding them into electrical energy. This electrical energy can then be used to power other devices or stored for later use. This technology has gained an increasing attention due to the recent advances in wireless and MEMS technology, allowing sensors to be placed in remote locations and operate at very low power [7]. The need for power harvesting devices is caused by the use of batteries as power supplies for these wireless electronics. As the battery has a finite lifespan, recharging needs to be done once discharged. Charging of batteries in order to provide energy to the electronic devices in the applications such as borders or hilly regions is a tedious job to do. Through this paper, we have proposed two new ways of harnessing the piezoelectric energy. Implementation aspects focus on the practical work carried out in this field of piezoelectric energy harvesting. The idea of piezoelectric windmill will solve the problem of continuous microcell discharging in the devices being used at remote places or in rough terrains. The concept of combining two energy sources piezoelectric energy and electromagnetic energy has been proposed in the paper. So these two ideas can greatly help in harnessing the piezoelectric energy thereby improve the results of the system.

References

[1] U K Singh and R H Middleton 2007 Piezoelectric power scavenging of mechanical vibration energy *Australian Mining Technology Conference* 111-118

[2] Takeuchi M Matsuzawa S Tairaku K and Takatsu C 2007 Piezoelectric generator as power supply for RFID-tags and applications *Proc. IEEE Ultrasonics Symposium* New York City USA 2558-61

[3] Ahola J Srkimki V Ahonen T Kosonen A Tiainen R and Lindh T 2008 Design considerations of energy harvesting wireless sensors for condition monitoring of electronic motors *Proc. 5th Int. Conf* Condition Monitoring Machinery Failure Prevention Technologies Edinburgh UK

[4] Roundy S Wright P K and Rabaye J 2003 study of low level vibrations as a power source for wireless sensor nodes *Computer Communications* 1131-44.

[5] Steven R Anton Henry and A Sodano 2007 A review of power harvesting using piezoelectric materials, *Smart Materials and Structures* **16**

[6] Sujesha Sudevalayam and Purushottam Kulkarni 2008 Energy harvesting sensor nodes: survey and implications.

[7] Y C Shu and I C Lien 2006 Analysis of power output for piezoelectric energy harvesting systems *Smart Materials and Structures* **15** 1499-1512

Institute of Physics
Dirac House, Temple Back
Bristol BS1 6BE UK

ISSN: 1757-6588
ISBN 978-1-5108-1815-6

9 781510 818156

CURRAN ASSOCIATES INC.
proceedings.com